Fruit Science

Other Volumes

VOLUME 2

Section 1: Culture and Technology of Major Tropical Fruits

1. The Mango/*V.K. Singh,* **2.** Banana/*K. S. Gill and N. K. Arora* **3.** Guava/ *J.S. Bal,* **4.** Papaya/*S.K. Bhatia and R.B. Kumatkar,* **5.** Pineapple, *P.P. Joy T.A. Rashida Rajuva and R. Anjana*

Section 2: Culture and Technology of Common Tropical Fruits

6. Sapota/*Rajbir Singh Boora,* **7.** Cashewnut/*Adavi Rao Desai*

Section 3: Culture and Technology of Minor Tropical Fruits

8. Custard Apples/*A.P. Gaikwad and S.R. Lohate,* **9.** Jackfruit/*P.M. Haldankar, Y.R. Parulekar, R.A. Patil and N.V. Dalvi,* Colour Plates

VOLUME 3

Section 1: Culture and Technology of Major Sub-Tropical Fruits

1. Citrus/*A.D. Huchche,* **2.** Grapes/*N.K. Arora, Navjot Gupta and M.I.S. Gill*

Section 2: Culture and Technology of Common Sub-Tropical Fruits

3. Pomegranate/*B.N.S. Murthy and Awachare Chandrakant Madhav* **4.** The Ber (Indian Jujube)/*J.S. Bal,* **5.** Litchi/*Alemwati Pongener, S.D. Pandey and Vishal Nath,* **6.** Aonla/*Devi Singh and J.S. Bal*

Section 3: Culture and Technology of Minor Sub-Tropical Fruits

7. Loquat/*H.K. Bons and J.S. Bal,* **8.** Jamun/*Rajbir Singh Boora,* **9.** The Avocado/*M. Tamilselvan,* **10.** Phalsa/*Deepa H. Dwivedi and Munni Gond* **11.** Date Palm/*Anil Kumar, J.S.Bal and Nirmaljit Kaur,* **12.** Fig//*A.P. Gaikwad and S. R. Lohate, Colour Plates*

VOLUME 4

Section 1:Culture and Technology of Major Temperate Fruits

1. Apple/*D.P. Sharma, N. Sharma and Rimpika*

Section 2: Culture and Technology of Common Fruits

2. Pear/*Sukhdev Singh,* **3.** Peach/*Harminder Singh and Indira Devi* **4.** Plum/*J.S. Bal,* **5.** Apricot/*Deepji Bhat, V.K. Wali and Manish Bakshi* **6.** Walnut /*Amit Kumar and K. Kumar*

Section 3: Culture and Technology of Minor Temperate Fruits

7. Almond/*Arti Sharma, V.K. Wali and Simrandeep,* **8.** Kiwi/*J.S. Chandel and Pramod Verma,* **9.** Cherry/*Desh Beer Singh,* **10.** Strawberry/*S.K. Bhatia and R.B. Kumatkar*

Section 4: Culture and Technology of Sub-Temperate Fruits

11. Persimmon/*Girish Sharma, Kiran Thakur, Deepika Negi and Nirmla Chauhan, Colour Plates*

Fruit Science

Culture and Technology

Volume 1: Basic Aspects & Practices

J. S. Bal Ph.D. (Hort.)
Former Professor of Horticulture
Department of Fruit Science
Punjab Agricultural University
Ludhiana-141004, Punjab

NEW INDIA PUBLISHING AGENCY
New Delhi – 110 034

NEW INDIA PUBLISHING AGENCY
101, Vikas Surya Plaza, CU Block, LSC Market
Pitam Pura, New Delhi 110 034, India
Phone: + 91 (11) 27 34 17 17 Fax: + 91 (11) 27 34 16 16
Email: info@nipabooks.com
Web: www.nipabooks.com

Feedback at feedbacks@nipabooks.com

ISBN: 978-93-86546-20-3
978-93-86546-21-0 (Vol. 2)
978-93-86546-22-7 (Vol. 3
978-93-86546-23-4 (Vol. 4)
978-93-86546-24-1 (Set)

Composed and Designed by NIPA

Dedicated to

My wife Dr. Kanwaljit Kaur Bal

for her moral support and continuous encouragement in completion of this task for the cause of students, growers and society

Foreword

India ranks second in area and production in the world occupying 6.4 million hectares of land and annual production of 91.4 million tones of fruits. The country is producing 13.6 per cent of fruits in the world. The average productivity of fruits in India is 12.3 metric tons/ha against 11.4mt/ha productivity in the world. India leads in producing many tropical and sub-tropical fruits viz. mango, banana, papaya, sapota, pomegranate, aonla and lime. Comparatively higher fruit production is due to the adoption of high yielding cultivars and improved techniques in cultural, production and protection operations. The major fruit growing states in India are Maharashtra, Andhra Pradesh, Karnataka, Gujarat, Uttar Pradesh, Madhya Pradesh and Telangana.

In spite of being a second largest producer of fruits after China, horticulture in India is facing some serious challenges to further increase in area and production. Climate change is the major threat and it has direct and indirect impact on fruit growing with special reference to production and protection status of fruit crops. The faulty marketing system is a major hurdle in the way of horticulturists. Middlemen are creating havoc in smooth marketing of fruits and consequently growers get very low profit margin. Retail market is the real culprit which does not let the supply and demand pulses pass from the consumers to and fro producers. As a consequence, producer hardly gets 25 percent of what the consumer pays for the commodity. Also, there is a significant loss of the produce due to the lack of awareness about post-harvest handling, absence of cold chain, poor marketing structure and market gluts in the post harvest periods. The fruits are highly perishable in nature and need careful handling after harvesting. Lack of quality planting material is another problem in further expansion of area under high yielding varieties of fruits. Efficient use of propagation methods, selection of robust and appropriate rootstocks, is the need for further enhancing the area and production of fruits in the country

The book on" Fruit Science Culture and Technology- Volume I is a very useful and well informed work incorporating all the detailed aspects of fundamental horticulture; propagation and rootstocks selection and post-harvest technology. The Scientists/Experts have contributed their valuable knowledge in their respective field in this book .The book will be quite helpful to the scientists working in the field of horticulture as well as students and extension workers in the field. I appreciate the effort of Dr J.S .Bal in editing and compiling of this very useful work for the benefits of all those who are involved in horticultural development in the country.

S.S. Johl
Chancellor
Central University of Punjab

Dated: May 27, 2017
Ludhiana

Preface

India is producing varieties of tropical, sub-tropical and temperate fruits in different regions. Horticulture has emerged as a major area of growth in the agriculture sectors. The cultivation of fruit crops has out performed many other field crops with respect to production and average productivity. In India, the fruit crops now occupies 6.4 million hectares which is about 27 per cent of the total area under agriculture in the country.

Horticulture is also facing certain challenges in the ways of successful fruit growing. Some of them are marketing of fruits, post-harvest losses, climate change, shortage of quality planting material and competition in global markets. To sustain the existing trend in increasing areas and production under fruit crops, efforts have to be made at all levels to make horticulture more lucrative and interesting. Towards this direction, a comprehensive information is compiled in the form of a book on various aspects of fruit culture .The information is shared in four different sections denoted to (i) fruit area and production scenario, (ii) fruit plant fundamentals culture and technology (iii) fruit plant propagation technology and (iv) post-harvest fruit technology. Further various aspects of fruit culture and technology with special reference to fruit area and production status, problems in fruit cultivation and future scope of fruit growing; planning and planting, climatic and soil adaptability, water and nutritional need, physiology of fruit plants, growth and development of fruits; propagation methods, techniques and rootstocks, tissue culture and biotechnology, seed germination and dormancy, use of propagation media, mixture and containers, care of propagated plants and protected cultivation were dealt in detail. In the last section, details about maturity and ripening, grading, packaging, storage and transportation of fruits and treatments for enhancing shelf life of fruits are discussed.

The eminent scientists from different parts of the country have contributed in writing of various aspects of horticulture. Recommendations made in the subject and personnel experience has been shared by the scientists in various chapters. The book will be highly useful to the graduate/post-graduate students in horticulture of Agricultural Universities and Colleges; fruit growers, nurserymen and Horticultural Development Officers. The students preparing for JRF/SRF, ARS examination and various fellowships will be highly benefitted.

I am thankful to my wife Dr Kanwaljit Kaur Bal for her constant encouragement in compilation of this book. I will always remember regular wishing for good health from USA by my son Dr Naveet Singh Bal-wife Dr Jaskiran, daughter Dr Harit Kaur Bal and son-in-law Er Amuldeep Singh from time to time during this task. Morning smiles of my grand kids *Anahat, Armaan and Aarza* made me fresh always. At the end I am highly thankful to the scientists who have contributed in this book whole heartedly.

30 April 2017 **Dr. J.S. Bal**

Contents

Section 3: Fruit Plant Propagation Technology

Section 4: Post-harvest Fruit Technology

List of Contributors

Amit Jasrotia
Division of Fruit Science
Faculty of Agriculture
Sher-e-Kashmir University of Agricultural Sciences and Technology
Jammu-180 009

B.V.C.Mahajan
Punjab Horticultural Postharvest Technology Centre
Punjab Agricultural University
Ludhiana-141 004

Chaitanya R. Belsare
Department of Fruit Science
Dr YS Parmar University of Horticulture and Forestry
Solan-173 230

D.P. Sharma
Department of Fruit Science
Dr YS Parmar University of Horticulture and Forestry
Solan-173 230

Deep Ji Bhat
Division of Fruit Science
Faculty of Agriculture
Sher-e-Kashmir University of Agricultural Sciences and Technology
Jammu-180 009

J.S. Bal
Former Prof. of Horticulture
Department of Fruit Science
Punjab Agricultural University
Ludhiana- 141 004

J.S. Chandel
Department of Fruit Science
Dr Y S Parmar University of Horticulture and Forestry
Solan-173 230

Jagmohan Singh
Division of Food Science & Technology
SKUAST-Jammu Chatha

Jaswinder Singh Brar
Department of Fruit Science
Punjab Agricultural University
Ludhiana- 141 004

K.S. Gill
Department of Fruit Science
Punjab Agricultural University
Ludhiana- 141 004

Krishan Kumar
Department of Fruit Science
Punjab Agricultural University
Ludhiana- 141 004

K.S. Thakur
Department of Fruit Science
Dr Y S Parmar University of Horticulture and Forestry
Solan-173 230

N. Sharma
Department of Fruit Science
Dr Y S Parmar University of Horticulture and Forestry
Solan-173 230

P.P.S. Gill
Department of Fruit Science
Punjab Agricultural University
Ludhiana-141 004

Preet Pratima
Department of Fruit Science
Dr Y S Parmar University of Horticulture and Forestry
Solan-173 230

Parshant Bakshi
Division of Fruit Science
Faculty of Agriculture
Sher-e-Kashmir University of Agricultural Sciences and Technology
Jammu-180 009

P. Mohanpuria
School of Agricultural Biotechnology
Punjab Agricultural University
Ludhiana-141 004

Rimpika
Department of Fruit Science
Dr Y S Parmar University of Horticulture and Forestry
Solan-173 230

Rachna Arora
Department of Fruit Science
Punjab Agricultural University
Ludhiana- 141 004

Satish Sharma
Department of Fruit Science
Dr Y S Parmar University of Horticulture and Forestry
Solan-173 230

S.K. Upadhyay
Department of Horticulture
College of Agriculture
CSKHPKY, Palampur -176 062

S.S.Gill
Department of Forestry and Natural Resources
Punjab Agricultural University
Ludhiana- 141 004

V.K. Wali
Division of Fruit Science
Faculty of Agriculture
Sher-e-Kashmir University of Agricultural Sciences and Technology
Jammu-180 009

SECTION - 1

FRUIT AREA AND PRODUCTION SCENARIO

1

Fruit Area and Production Status

J.S. Bal

INTRODUCTION

The climate of India is favourable for production of a variety of tropical, subtropical and temperate fruits. The area under cultivation, production and productivity of fruit crops has increased significantly during the last fifteen years. Horticulture has emerged as major area of growth in the agricultural sector and has out performed other crops with respect to production and productivity. The increase in fruit production was improved considerably from 43.0 million tonnes in 2001-02 to 91.4 million tonnes in 2015-16 (Table 1). This increase in production has come from an increase in area under different fruit crops and overall increase in productivity. In this way, India has maintained its second rank in global production of fruits next only to China. The fruit area increased from 4 million hectares in 2001-02 to 6.4 million hectares in 2015-16. On the basis of fruit production, it is estimated that per capita fruits availability in India 190 g per day which is still less than the recommended quantity of 230 g per capita per day.

With total production of 91.4 million tonnes, India is now leader in several fruits crops viz. mango, banana, papaya, cashewnut and arecanut. The share of horticultural output is nearly 27 per cent of the total agricultural output. The production of fruit crops registered good increase i.e. 4.9 per cent in 2002-03 and 2.1 per cent in 2015-16. It is clear that a continuous increase in production was recorded every year. The increase in annual area under fruit crop is little variable. It registered higher increase of 18.7 per cent in 2003-04 followed by 7.7 per cent increase in 2004-05 and 5.2 per cent each in 2005-06 and 2007-08. However, a decrease in area was also observed twice during this period, firstly 5.5 per cent decrease in 2002-03 and substantial decrease to the tune of 15.7 per cent in 2014-15. A brief lull in area was observed between the years 2008-

09 to 2013-14. Contrary to this, increase in fruit production was substantial up to 9.7 per cent in year 2004-05, 9.2 per cent in year 2007-08, 8.6 percent during 2013-14 and 8.1 per cent during 2005-06.

India accounts for about 13.6 per cent of the total world production of fruits. Now, the country leads in producing in many tropical and subtropical fruit crops viz. mango, banana, papaya, sapota, pomegranate, aonla and lime. The overall increase in production of these fruit crops is a result of enhanced productivity owing to the adoption of high yielding cultivars and improved modern techniques in cultural, production and protection area.

On account of wide climatic adaptability, a large variety of fruits are grown in India. Mango, citrus, banana, apple and guava occupy largest area in sequence. But, on the basis of fruit production banana, mango, citrus and guava resulted higher production per unit area. In addition to these fruits, grapes, pomegranate, sapota, pineapple, litchi, ber and aonla among the subtropical and tropical fruits and pear,peach, plum, almond, apricot, walnut and strawberry among the temperate fruits are grown in sizeable area. The fruits are grown throughout the country in all the states. The major fruit growing states with respect to area are Maharashtra, Andhra Pradesh, Karnataka, Gujarat, Uttar Pradesh and Telangana. The main fruit producing states on the basis of fruit production are Maharashtra, Andhra Pradesh, Gujarat, Uttar Pradesh, Karnataka, Madhya Pradesh, Tamil Nadu and Telangana.

The increase in area covered by fruit crops during the last one decade i.e. from the year 2006-07 to the year 2015-16 is calculated 13.3 per cent. However the corresponding increase in production of fruits during this period is calculated 34.9 per cent.

RECENT STATUS IN AREA AND PRODUCTION OF FRUIT CROPS

The area under fruit cultivation during 2015-16 is observed 6.4 million hectares which is about 27 per cent of total area under agriculture in India. The area under fruit crop has increased from 5.5 million hectares in 2006-07 to 6.40 million hectares in 2015-16 with a similar corresponding increase in production from 59.6 to 91.4 million tonnes (Table 1.). Maharashtra stood first in terms of fruit production with 12.8 per cent share in total production followed by Andhra Pradesh with 10.5 per cent and Gujarat with 9.6 per cent in 2014-15 (Table 4). Banana is the most cultivated fruit accounting for 32.6 per cent of total production followed by mango at 20.7 per cent, citrus at 13 per cent, papaya at 5.5 per cent, guava at 4.5 per cent, grapes at 3.1 per cent, apple at 2.4 per cent and other fruits 11.9 per cent share in the country.

Some Challenges: Horticulture is facing various emerging challenges with special reference to faulty marketing system, post-harvest losses, global competition, climate change, lack of quality planting material etc. All these challenges need to be taken into concern with a view to sustain existing trend in increasing area and production under fruit crops. Emphasis should be given to implement the existing modern techniques in production and protection technologies. It is felt to raise the productivity of fruit crop by planting new orchard with high density plantations, to produce and supply high grade quality planting material to the growers, supporting protected cultivation and micro irrigation techniques, and rejuvenation of old declining orchard. Post-harvest management of horticultural produce should be ensured to avoid huge losses during transit and cold storage. The fruit growing can stand on strong footing if all such problems should be taken into account either through Govt. by providing necessary subsidy to the growers or direct involvement of cooperative fruit growers associations as a major concern.

FRUIT AREA AND PRODUCTION STATUS IN LAST DECADE

A brief status on area and production of fruit crops during the last one decade from 2006-07 to 2015-16 is described and discussed for better understanding of the horticultural scenario.

A) Status in Area and Production from 2006-07 to 2008-09

Mango was the most important fruit crop covering 38.8, 37.6 and 37.8 per cent of area and accounted for 23.1, 21.3 and 18.6 per cent production of total fruits in the country during the year 2006-07, 2007-08 and 2008-09, respectively (Table 2). The productivity varies from 5.5 to 6.4mt/ha during these years. Major mango producing states were Andhra Pradesh, Uttar Pradesh, Karnataka, Maharashtra, Gujarat and Tamil Nadu.

Citrus ranked second in area and occupied 14.4 per cent of the total area and accounted for 12 percent of the total production of fruits in the year 2006-07. It occupied 14.8 and 15.1 per cent of the total area with 12.2 and 12.6 per cent of the total production of fruits in 2007-08 and 2008-09, respectively. Productivity of citrus varies from 9.0 to 9.3 mt/ha during these years Mandarin, sweet oranges, lime and lemons cover bulk of the area under the fruits and were grown mainly in Maharashtra, Andhra Pradesh, Punjab, Gujarat, Orissa and Uttaranchal (Uttarakhand).

Banana was grown in 6.04 lakh hectares with total production of 20.9 million tonnes annually during 2006-07. It occupied 6.58 and 7.09 lakh hectares area and yielded 23.8 and 26.2 million tonnes of fruits during 2007-08 and 2008-09, respectively. India has first position in the world in banana production. Tamil

Nadu leads other states in area and production of banana. Maharashtra was the second state which has the highest area and production of banana. In productivity, Maharashtra has the highest productivity of 62-63 metric tonnes against India's average of 35-36 metric tonnes per ha during these year. The other major banana growing states were Andhra Pradesh, Karnataka, Kerala and Gujarat.

Apple was the fourth most important fruit crop in India. This fruit was grown in 2.52, 2.64 and 2.74 lakh hectares having annual production of 1.62, 2.0 and 1.99 million tonnes accounted for 4.5 per cent of area each year and 2.7, 3.1 and 2.9 per cent of production of total area and production, respectively under fruit crops during these three years .Apple was largely grown in Jammu and Kashmir, Himachal Pradesh and Uttaranchal.

Guava was the fifth most widely grown fruit crop during these years. The area under guava was 1.76 lakh ha yielded annual production of 18.3 lakh tonnes accounted for 3.2 per cent of area and 3.1 per cent of production in 2006-07. There was marginal increase in area next year but it was recorded 2.04 lakh ha area and 22.70 lakh metric tonnes production during 2008-09. Maharashtra was the leading state in guava production during 2008-09. But in 2006-07, Maharashtra was leading in area and Bihar in production of guava. The other states where guava was grown widely were Bihar, Uttar Pradesh, West Bengal, Punjab and Karnataka.

Pineapple was grown in 86000 hectares with annual production of 13.62 lakh tonnes during 2006-07. Area and production remained almost static during next two years. Its cultivation was done largely in Assam, West Bengal, Karnataka, Bihar, Manipur and Meghalaya. Papaya was grown on 72000 hectares with annual production of 24.8 lakh tonnes in the year 2006-07. This fruit occupied more area to the tune of 98000 hectares with annual production of 36.3 lakh tonnes in the year 2008-09. Production of both pineapple and papaya was the highest among the fruit crops.

Litchi was cultivated in 65000 hectares accounted for 1.2 per cent of area during 2006-07. It was grown on 69000 and 72000 hectares area during 2007-08 and 2008-09, respectively. Bihar, West Bengal, Uttaranchal, Assam, Orissa and Punjab were the important litchi producing states in the country, Sapota was grown 1.49, 1.52 and 1.56 lakh hectares area with estimated production of 12.2, 12.6 and 13.1 lakh tonnes of fruits during the year 2006-07, 2007-08 and 2008-09, respectively. Sapota was mainly cultivated in Maharashtra, Karnataka, Gujarat, Andhra Pradesh and Tamil Nadu states.

Grapes occupied 65000 hectares area and produced 16.9 lakh tonnes annually during the year 2006-07. A marginal increase in area and production in grapes was recorded during next year. In the year 2008-09, grapes occupied 80000 hectares' area and yielded 18.8 lakh tonnes annually. Maharashtra was the leading state and occupied the first position in the production of grapes. Karnataka, Tamil Nadu, Andhra Pradesh and Punjab were the other grapes growing states in the country. Pomegranate was cultivated in 1.17 lakh hectares with annual production 8.4 lakh tonnes during the year 2006-07. A slight decrease in area and production was observed during the year 2008-09. Maharashtra was the main pomegranate producing state in the country. Karnataka, Gujarat and Andhra Pradesh were the other pomegranate producing states.

B) Status in Area and Production from 2009-10 to 2011-12

Total area under fruits in India was estimated 6.3, 6.4 and 6.7 million hectares with total production of 71.5, 74.9 and 76.4 million tonnes during the year 2009-10, 2010-11 and 2011-12, respectively (Table 3). Mango ranked first in area and production followed by the banana and citrus. The area under pineapple, pomegranate and grapes was small but it was credited with high productivity. Sapota has emerged as an important commercial fruit crop. During this period some minor fruits like ber and aonla have emerged as a boon for dry land horticulture.

Mango was grown on about 36 per cent of the area and yielded nearly 21 per cent of total fruit production during these three years. The average productivity varies from 6.5 to 6.8 mt/ha in different years. The largest area under mango was found in Andhra Pradesh and Maharashtra. In production, Andhra Pradesh took the lead during 2009-10 and 2010-11 but Uttar Pradesh recoded higher production during 2011-12. The other mango producing states were Karnataka, Bihar, Gujarat, Tamil and Odisha.

Banana was the second largest fruit on the basis of area and production. The annual production was recorded highest in banana among the fruit crops. The higher production to the tune of 26.5, 29.8 and 28.5 million mt was recorded during the year 2009-10, 2010-11 and 2011-12, respectively. The average annual productivity was noted 35.7 to 37 mt/ha during these years. Banana occupied 11.9 to 13.0 per cent area and 37 to 39.8 per cent production during this period. Tamil Nadu was the leading state and occupied highest area under banana. Karnataka, Gujarat, Maharashtra was the other states having maximum area under banana. During the year 2010-11 and 2011-12, Tamil Nadu resulted highest banana production in the country followed by Maharashtra, Gujarat, Andhra Pradesh and Karnataka.

Citrus fruits occupied second position with regard to area as compared to the other fruits. These fruits were grown on 9.87 lakh hectares with total production of 96.4 lakh tonnes during the year 2009-10. The corresponding area and production was noted 8.46 lakh hectares and 74.6 lakh tonnes fruits during

2009-10 and 9.15 lakh hectares and 79.2 lakh tonnes fruit during 2010-11. Productivity of citrus fruit varies from 8.7 to 9.8 mt/ha during the period. Mandarin was the major crop grown in Maharashtra, Madhya Pradesh, Punjab (Kinnow) and North-Eastern regions. Sweet oranges were largely grown in Andhra Pradesh and Maharashtra.

Apple was the fourth main fruit crop of India and grown at commercial level in Jammu and Kashmir, Himachal Pradesh and Uttarakhand states. It was grown in 2.83 lakh ha with annual production of 17.8 lakh tonnes accounted for 4.5 and 2.5 per cent of area and production, respectively of total fruit crops. The increase in area was registered during the year 2011-12 when it occupied 3.22 lakh ha and yielded 22 lakh tones of apples.

Guava was another major fruit crop which occupied fifth position in area and production. The area under guava was recorded 2.2 lakh hectares yielded annual production of 25.7 lakh tonnes which accounted for 3.5 per cent of area and 3.6 per cent of production. A slight decrease in area and production was noted next year and again in 2011-12 it gained area and production at 2.2. lakh hectares and 25.1 lakh tonnes, respectively. Uttar Pradesh during 2009-10, Madhya Pradesh during 2010-11 and 2011-12 were the leading state in guava production. However, Maharashtra occupied highest area under guava during 2010-11 and 2011-12. The other guava producing states were Bihar, West Bengal, Punjab and Chhattisgarh.

Sapota was the next emerging fruit crop and it occupied 1.59 lakh hectares area with annual production of 13.5 lakh tonnes during the year 2009-10. However, its area and production was noted almost static with marginal increase during the next two years. Maharashtra state occupied highest area under sapota but production was noted maximum in Karnataka. Gujarat, Andhra Pradesh and Tamil Nadu were the other sapota producing states. Papaya and pineapple were the high yielding fruit crops and their productivity was recorded highest among the fruit. The average productivity in papaya was noted 38.1 to 40.9 mt/ha during three years period. The productivity in pineapple was observed 14.7 to 15.9 mt/ha in the same years. Papaya occupied 96000 ha areas during 2009-10 and reached to 1.17 lakh hectares during 2011-12. The corresponding area in pineapple was noted 9200 hectares during 2009-10 and reached to 1.02 lakh ha during 2011-12.

Grape occupied 1.06 lakh ha area during 2009-10 with minor increase the next two years .Substantial increase in production was observed during 2011-12,when it was recorded 22.2 lakh tonnes. Maharashtra occupied the first position in production of grapes. It was followed by Karnataka. The other grape growing states were Andhra Pradesh and Tamil Nadu. Pomegranate crop picked up slowly but its area remained 1.07 to 1.25 lakh hectares during these years. Consequently, its production was also noted between 7.43 to 8.20 lakh tonnes during this period. Maharashtra was the leading state in area and production of quality pomegranate fruits. Karnataka has also sizeable area under pomegranate crop. Litchi cultivation was mainly concentrated in Bihar and West Bengal. The maximum area and production under litchi was noted in Bihar state. The cultivation of litchi was done on 78000 and 80000 hectares area during 2010-11 and 2011-12, respectively. The annual production of fruits was recorded 5.0 and 5.38 lakh tonnes during the years 2010-11 and 2011-12, respectively.

C) Status in Area and Production from 2012-13 to 2014-15

The fruit production showed good increase during these years compared to the previous years. It was recorded 81.3, 88.9 and 89.5 million tonnes from the corresponding area of 6.9, 7.2 and 6.2 million hectares (Table 4) during the year 2012-13, 2013-14 and 2014-15, respectively.

Mango was the first important fruit with respect to area covering 34.7 to 35.8 per cent of area and accounted for 20.7 to 22.1 per cent of total production from 2012-13 to 2014-15. The production of mango was about 40 per cent of the global production. The average productivity ranged from 7.2 to 8.6 mt/ha during these years. The area under mango was observed 2.5 million ha in the first two years but slightly declined and recorded 2.16 million ha during the year 2014-15. Major mango growing states were Maharashtra, Andhra Pradesh, Karnataka, Gujarat, Uttar Pradesh and Telangana during the year 2014-15.

Citrus fruits were grown in 10.42 lakh hectares with total production of ten million tonnes during 2012-13. Area and production in the next year was not increased much. However, a slight decreased in area was registered during 2014-15 and recorded 9.53 lakh hectares. Contrary an increase in total production was observed and it was recorded 11.6 million tonnes which was little higher than the previous two years. Productivity of citrus fruits varies from 9.7 to 12.2 mt/ha during these three years. Citrus fruits were grown mainly in Maharashtra, Telangana, Andhra Pradesh, Madhya Pradesh and Punjab. Sweet orange producing states were Telangana, Andhra Pradesh and Maharashtra where production was noted also higher in same sequence. Maharashtra, Andhra Pradesh and Punjab were the states have largest area under mandarin cultivation. Punjab was the leading state in Kinnow production where average production was the highest to the tune of 11.1 lakh tonnes. Lime and lemon producing states were Andhra Pradesh, Gujarat, Telangana, Karnataka, Odisha and Madhya Pradesh.

Banana ranked third in area and occupied 11.1 to 13.2 per cent of the total area under fruits during the year from 2012-13 to 2014-15. In production, its was placed at first position which accounted for 26.5 to 29.7 per cent during these three years. India has maintained first position in the world in production of banana. Tamil Nadu leads other states in area and production of banana during 2012-13 and 2013-14. However, Karnataka occupied the first position in area which was followed by Tamil Nadu, Kerala, Andhra Pradesh, Maharashtra and Gujarat. On production basis, Gujarat took the lead in banana production over the other states followed by Tamil Nadu, Maharashtra, Andhra Pradesh and Karnataka. Gujarat has highest productivity of 64.5 metric tonnes against India's average of 35.6 metric tonnes per hectare. The major banana growing states were Tamil Nadu, Gujarat, Karnataka, Maharashtra, Andhra Pradesh and Kerala.

Apple was the important temperate fruit and occupied fourth position in area and accounted for 4.3 to 5.1 per cent from the year 2012-13 to 2014-15. It was grown in 3.12, 3.13 and 3.19 lakh hectares with annual production of 19.2, 25.0 and 21.3 lakh metric tonnes of fruits during 2012-13, 2013-14 and 2014-15, respectively. Productivity of apple ranged from 6.1 to 8.0 mt/ha during this period. Apple was largely grown in Jammu and Kashmir, Himachal Pradesh and Uttarakhand. The production of fruit was the highest in Jammu and Kashmir which was 64 per cent of the total production in India.

Guava is another leading fruit crop and ranked fifth in area in India. The area under guava was 2.4, 2.7 and 2.5 lakh hectares which producing 32, 36.7 and 39.9 lakh tonnes during 2012-13, 2013-14 and 2014-15, respectively. Maharashtra had maximum area during 2012-13, Uttar Pradesh during 2013-14 and 2014-15. In Production of guava; Madhya Pradesh was the leading state during three years. The other major guava producing states in 2014-15 were Uttar Pradesh, Bihar, Maharashtra, West Bengal and Punjab. Sapota, pineapple, papaya and pomegranate were the other fruit crops which were next to guava in occupying the area. Sapota emerged another important crop on the basis of area under cultivation. During the recent years (2014-15), Karnataka was the leading state in area and production of sapota. It was followed by Gujarat in both area and production.Tamil Nadu, Maharashtra and Andhra Pradesh were the other states cultivating sapota on commercial scale.

Papaya was grown on 1.32 lakh hectares area during 2012-13 and area under this crop was slightly declined during 2014-15 and occupied 1.15 lakh hectares. The average productivity was recorded 42.7 mt/ha during this year. Major papaya producing states were Gujarat, Andhra Pradesh, Maharashtra, Karnataka and Kerala in recent year (2014-15). Pineapple is cultivated on 1.05, 1.19 and 1.16 lakh hectares area which produced 15.71, 25.85 and 19.84 lakh metric tonnes during 2012-13, 2013-14 and 2014-15, respectively. Assam, Manipur and Meghalaya occupied maximum area but Kerala, West Bengal, Assam and Karnataka produced highest quantity of pineapple during 2014.15. Pomegranate is picking up fast and occupied 1.13 lakh hectares during 2012-13 and 1.81 lakh hectares during 2014-15. Higher production to the tune of 13.46 and 17.89 lakh metric tonnes was recorded during 2013-14 and 2014-15, respectively. Now, major pomegranate producing states are Maharashtra, Karnataka and Gujarat.

Area under grapes cultivation did not shown much variation during these three years. It was recorded 1.23 lakh hectares with annual production of 28.23 lakh metric tonnes during 2014-15. Major grape growing states were Maharashtra and Karnataka. Bihar state leads in area and production in Litchi and occupied 85000 hectares area and produced 5.28 lakh tonnes fruits during 2014-15. The other litchi producing states were Uttarakhand, West Bengal, Jharkhand, Assam, Chhattisgarh and Himachal Pradesh.

During recent year of 2014-15, Maharashtra is the leading fruit growing state in India which accounted for 12.8 per cent of production followed by Andhra Pradesh (10.5%).

D) Status of Area and Production in 2015-16 (3rd Advance estimates)

India is the largest producer of fruits and ranked second in production after China. The annual production is 91.4 million tonnes of fruits from an area of 6.4 million hectares (Table 5). The production has increased to manifold in the last five decades from 6.8 million tonnes to 91.4 million tonnes. Mango, citrus, banana, apple and guava are the major fruit crops of the country.

Mango is the most important fruit crops of India which covers 34.9 per cent of area and accounting for 20.5 per cent of total production. The average productivity is 8.4 mt/ha. Citrus fruits includes mandarin, sweet oranges and lime & lemon. They are grown in 9.35 lakh hectares with total production of 115.2 lakh metric tonnes annually. The productivity of citrus fruits varies from 11.0 mt/ha in mandarin to 16.2 mt/ha in sweet oranges. Banana rank third in area but first in fruit production and occupied 13.21 per cent of the total area and accounted for 31.85 per cent of the total production in India with 34.4 mt/ha productivity. Apple is the fourth important fruit crop which is grown in 3.14 lakh hectares having annual production of 28.7 lakh metric tonnes accounted for 4.9 and 3.14 per cent of area and production, respectively. Guava is the fifth widely grown fruit crop in India. The area under guava was 2.54 lakh ha yielding 40.5 lakh metric tonnes accounted for 3.97 per cent of area and 4.42 per cent of production. The average productivity of guava was recorded 15.9 m/ha.

Pomegranate is placed at sixth place with respect to area under cultivation. It is cultivated in 1.93 lakh hectare with annual production of 21.98 lakh metric tonnes. This fruit occupied 3.01 per cent of area and 2.40 per cent of fruit production with 11.4 mt/ha productivity. Papaya is grown in 1.33 lakh hectares with annual production of 57 lakh metric tonnes. The highest productivity of 42.8 mt/ha was observed in papaya. Pineapple was cultivated in 1.10 lakh hectares accounted for 1.72 per cent of area and 2.15 per cent production. The average productivity is also recorded higher i.e. 17.9 mt/ha. Grapes occupy 1.21 lakh hectares and produced 26 lakh metric tonnes of fruits annually. Maharashtra is the leading state in grape production. Litchi is cultivated in 89000 hectares accounted for 1.39 per cent of area and 0.61 per cent of fruit production.

Ber is a fruit crop which can grow well in dry land climatic conditions. The area under ber was 47000 hectares producing 4.31 lakh metric tonnes fruits annually. Aonla occupied 85000 hectares area and produced 10.9 lakh metric tonnes fruits. Jackfruit occupied area of 1.51 lakh hectares which produced 18.4 lakh metric tonnes fruits annually.

Temperate fruits were grown on limited area under temperate climate. Pear is grown in 51000 hectares, peach on 62000 hectares, plum on 24000 hectares and almond on 13000 hectares only with corresponding production of 4 lakh mt, 1.27 lakh mt, 86000 mt and 8000 mt, respectively. Walnut is grown in 1.16 lakh hectares with annual production of 3 lakh metric tonnes.

Table 1: Area and Production of Fruit crops in India

(2001-02 to 2015-16)	Fruits		Productivity (mt/ha) in Area (%)	Increase (+) Decrease (-) in Prod. (%)	Increase(+) Decrease (-)
	Area (000'ha)	Prod. (000'mt)			
2001-02	4010	43001	10.7	-	-
2002-03	3788	45203	11.9	-5.5	+4.9
2003-04	4661	45942	9.9	+18.7	+1.6
2004-05	5049	50867	10.1	+7.7	+9.7
2005-06	5324	55356	10.4	+5.2	+8.1
2006-07	5554	59563	10.7	+4.1	+7.1
2007-08	5857	65587	11.2	+5.2	+9.2
2008-09	6101	68466	11.2	+4.0	+4.2
2009-10	6329	71516	11.3	+3.6	+4.3
2010-11	6383	74878	11.7	+0.8	+4.5
2011-12	6705	76424	11.4	+4.8	+2.0
2012-13	6982	81285	11.6	+4.0	+6.0
2013-14	7216	88977	12.3	+3.2	+8.6
2014-15	6235	89514	14.4	-13.6	+0.6
2015-16	6405	91443	14.3	+2.7	+2.1

Source: Indian Horticulture Database 2014 and Area Production Statistics NHB.

MAJOR FRUIT PRODUCING COUNTRIES IN THE WORLD

The total area under fruits in the world is 57.5 million hectares with annual production of 654 million metric tonnes (Table 6). Thus, the average fruit productivity in the world is 11.4 mt/ha.

China ranks first in area and production of fruits in the world which occupies 11.8 million hectares areas with annual fruit production of 137 million metric tonnes. India is the second largest country in area and production of fruits. China is producing 20.9 per cent of total world production with average productivity of 11.6 mt/ha. India is producing 13.6 per cent share of fruits in the world. The average productivity of fruits in India is 12.3 mt/ha. Brazil ranked third position by growing fruits on 2.33 million hectares which is producing 38.4 million metric tonnes fruits annually. The average productivity of fruits is the highest (23.3 mt/ha) in USA. However, the production of fruits from USA is only 4.1 per cent of the total world production. The other important fruit producing countries in the world are Indonesia, Philippines, Mexico, Turkey, Spain and Italy. The average productivity of fruits in Indonesia is 22.3 metric tonnes per hectare which is also quite high after USA.

Grape is the most important fruit grown in 6.98 million hectares with total production of 68.4 million mt in the world. It ranks first in occupying the area among other fruits. The average productivity is 9.8 mt/ha. China is the leading grape producing country with a share of 14.00 per cent of total grape production in the world. The total grape production in China is 9.6 million mt. USA is second in rank with the share of 9.7 per cent of total grape production. The other important grape producing countries are Italy, France, Spain and Turkey.

Banana ranks first in total production among the other fruits with annual production of 106.8 million mt annually. However, this fruit crop is grown on 5.0 million hectares where others fruit like grapes and mango occupies more area under growing these fruits. India has first position in banana production in the world. The average per cent of total production in India is quite high i.e. 27.8. China and Philippines are the other countries where production of banana is quite high after India.

Apple is the second most important fruit crop of the world with respect to total fruit production. The annual production of apple is 76.7 million mt growing on 4.8 million hectares area. The average productivity of apple is 15.9 mt/ha. China is a leader in apple production in the world which is producing alone 37.00 million mt annually from the area of 2.06 million hectares. The other apple producing countries in the world are USA, Turkey, Poland and India.

Table 2: Area and Production of Fruit Crops in India (2006-07 to 2008-09)

Fruit Crop	2006-07					2007-08					2008-09				
	Area (000 ha)	Prod. (000 mt)	Per cent		Pdy (mt/ha)	Area (000 ha)	Prod. (000 mt)	Per cent		Pdy (mt/ha)	Area (000 ha)	Prod. (000 mt)	Per cent		Pdy (mt/ha)
			Area	Prod				Area	Prod				Area	Prod	
Banana	604	20998	10.9	35.2	34.8	658	23823	11.2	36.3	36.2	709	26217	11.6	38.3	37.0
Mango	2154	13734	38.8	23.1	6.4	2201	13997	37.6	21.3	6.4	2309	12750	37.8	18.6	5.5
Citrus	798	7145	14.4	12.0	9.0	867	8015	14.8	12.2	9.2	923	8608	15.1	12.6	9.3
Papaya	72	2482	1.3	4.2	34.5	83	2909	1.4	4.4	35.1	98	3629	1.6	5.3	37.1
Guava	176	1830	3.2	3.1	10.4	179	1981	3.1	3.0	11.1	204	2270	3.3	3.3	11.1
Apple	252	1624	4.5	2.7	7.6	264	2001	4.5	3.1	7.6	274	1985	4.5	2.9	7.2
Pineapple	86	1362	1.6	2.3	15.7	80	1245	1.4	1.9	15.5	84	1341	1.4	2.0	16.0
Sapota	149	1216	2.7	2.0	8.2	152	1258	2.6	1.9	8.3	156	1308	2.6	1.9	8.4
Grapes	65	1685	1.2	2.8	25.9	68	1735	1.2	2.7	25.4	80	1878	1.3	2.7	23.6
Pomegranate	117	840	2.1	1.4	7.2	124	884	2.1	1.4	7.2	109	807	1.8	1.2	7.4
Litchi	65	403	1.2	0.7	6.2	69	418	1.2	0.6	6.1	72	423	1.2	0.6	5.9
Others	1016	6244	11.1	10.5	6.1	1112	7321	19.0	11.2	6.6	1083	7249	17.8	10.6	7.1
Total	5554	59563	100.0	100.0	10.7	5857	65587	100.0	100.0	11.2	6101	68465	100.0	100.0	11.2

Source: Indian Horticulture Database 2009 PDF File of Planning Commission

Table 3: Area and Production of Fruit Crops in India (2009-10 to 2011-12)

Fruit Crop	2009-10					2010-11					2011-12				
	Area (000 ha)	Prod. (000 mt)	Per cent		Pdy (mt/ha)	Area (000 ha)	Prod. (000 mt)	Per cent		Pdy (mt/ha)	Area (000 ha)	Prod. (000 mt)	Per cent		Pdy (mt/ha)
			Area	Prod				Area	Prod				Area	Prod	
Banana	770	26470	12.2	37.0	34.4	830	29780	13.0	39.8	35.9	797	28455	11.9	37.2	35.7
Mango	2312	15027	36.5	21.0	6.5	2297	15188	36.0	20.3	6.6	2378	16196	35.5	21.2	6.8
Citrus	987	9638	15.6	13.5	9.8	846	7464	13.3	10.0	8.8	915	7922	13.6	10.4	8.7
Papaya	96	3913	1.5	5.5	40.9	106	4196	1.7	5.6	39.6	117	4457	1.8	5.8	38.1
Guava	220	2572	3.5	3.6	11.7	205	2462	3.2	3.3	12.0	220	2510	3.3	3.3	11.4
Apple	283	1777	4.5	2.5	6.3	289	2891	4.5	3.9	10.0	322	2203	4.8	2.9	6.8
Pineapple	92	1387	1.5	1.9	15.1	89	1415	1.4	1.9	15.9	102	1500	1.5	2.0	14.7
Sapota	159	1347	2.5	1.9	8.5	160	1424	2.5	1.9	8.9	163	1426	2.4	1.9	8.7
Grapes	106	881	1.7	1.2	8.3	111	1235	1.7	1.6	11.1	116	2221	1.7	2.9	19.1
Pomegranate	125	820	1.9	1.2	6.6	107	743	1.7	9.9	6.9	112	772	1.7	1.0	6.9
Litchi	74	483	1.2	0.7	6.5	78	497	1.2	0.7	6.4	80	538	1.2	0.7	6.7
Others	1105	7201	17.4	10.0	6.5	1265	7583	19.8	10.1	6.0	1383	8224	20.6	10.7	5.9
Total	6329	71516	100.0	100.0	11.3	6383	74878	100.0	100.0	11.7	6705	76424	100.0	100.0	11.4

Source: Indian Horticulture Database 2011 Source: Indian Horticulture Database 2013

Table 4: Area and Production of Fruit Crops in India (2012-13 to 2014-15)

Fruit Crop	2012-13					2013-14					2014-15 (Final)				
	Area (000 ha)	Prod. (000 mt)	Per cent		Pdy (mt/ha)	Area (000 ha)	Prod. (000 mt)	Per cent		Pdy (mt/ha)	Area (000 ha)	Prod. (000 mt)	Per cent		Pdy (mt/ha)
			Area	Prod				Area	Prod				Area	Prod	
Banana	776	26509	11.1	32.6	34.2	803	29725	11.1	33.4	37.0	822	29221	13.2	32.6	35.5
Mango	2500	18002	35.8	22.1	7.2	2516	18431	34.9	20.7	7.3	2163	18527	34.7	20.7	8.6
Citrus	1042	10090	14.9	12.4	9.7	1078	11147	14.9	12.5	10.3	953	11655	15.3	13.0	12.2
Papaya	132	5382	1.9	6.6	40.7	133	5639	1.8	6.3	42.3	115	4913	1.8	5.5	42.7
Guava	236	3198	3.4	3.9	13.6	268	3668	3.7	4.1	13.7	246	3994	3.9	4.5	16.2
Apple	312	1915	4.5	2.4	6.1	313	2498	4.3	2.8	8.0	319	2134	5.1	2.4	6.7
Pineapple	105	1571	1.5	2.0	14.9	119	2585	1.7	2.9	21.8	116	1984	1.9	2.2	17.1
Sapota	164	1495	2.3	1.8	9.1	177	1744	2.5	2.0	1.9	106	1339	1.7	1.5	12.6
Grapes	118	2483	1.7	3.1	21.1	110	1737	1.5	2.0	15.8	123	2823	2.0	3.1	23.0
Pomegranate	113	745	1.6	0.9	6.6	131	1346	1.8	1.5	10.3	181	1789	2.9	2.0	9.9
Litchi	83	580	1.2	0.7	7.0	84	585	1.2	0.7	7.0	85	528	1.4	0.6	6.2
Others	1401	9315	18.3	11.2	6.6	1484	9872	20.6	11.1	6.7	1006	10607	16.1	11.9	10.5
Total	6982	81285	100.0	100.0	11.6	7216	88977	100.0	100.0	12.3	6235	89514	100.0	100.0	14.4

Source: Indian Horticulture Database 2014 Source:Area Prod. Stat. -NHBnhb.gov.in>PDFViwer

Table 5: Area and Production of Fruit Crops in India (2015-16)

Fruit Crop	2015-16 (3rd Estimate)				
	Area (000 ha)	Prod. (000 mt)	Per cent		Pdy. (mt/ha)
			Area	Prod.	
Almond	13	8	0.20	0.01	0.61
Aonla	85	1094	1.33	1.20	12.9
Apple	314	2872	4.90	3.14	9.1
Banana	846	29124	13.21	31.85	34.4
Ber	47	431	0.73	0.47	9.2
Citrus	935	11515	14.60	12.60	12.3
i) Mandarin	361	3983	5.63	(4.4)	11.0
ii) Sweet Oranges	217	3520	3.39	(3.8)	16.2
iii) Lime /lemon	258	2978	4.03	(3.3)	11.5
iv) others	99	1034	1.55	(1.13)	10.6
Custard Apple	35	271	0.55	0.30	9.7
Grapes	121	2597	1.89	2.84	21.5
Guava	254	4046	3.97	4.42	15.9
Jackfruit	151	1839	2.36	2.01	12.2
Kiwi	4	11	0.06	0.01	2.8
Litchi	89	556	1.39	0.61	6.2
Mango	2237	18779	34.93	20.53	8.4
Muskmelon	47	936	0.73	1.02	19.9
Papaya	133	5699	2.07	6.23	42.8
Passion Fruit	18	82	0.28	0.09	2.3
Peach	62	127	0.97	0.14	2.1
Pear	51	399	0.80	0.44	7.8
Picanut	1	1	0.00	0.0	1.0
Pineapple	110	1964	1.72	2.15	17.9
Plum	24	86	0.37	0.09	3.6
Pomegranate	193	2198	3.01	2.40	11.4
Sapota	106	1286	1.75	1.41	12.1
Strawberry	1	5	0.00	0.00	0.00
Walnut	116	300	1.81	0.33	2.6
Watermelon	92	2292	1.43	2.51	24.9
Others	320	2925	5.0	3.20	9.1
Total	6405	91443	100.0	100.0	14.3

Source: Area Production Statistics -NHB, nhb.gov.in>P

As far as production is concerned, oranges ranked fourth with annual production of 67.1 million mt in the world from an area of 3.7 million hectares. The highest production of oranges is reported 18 million mt from Brazil. The area under oranges is also maximum in Brazil which occupies 7.3 lakh hectares. However, the average productivity of 32.6 mt/ha is recorded highest in USA. It is closely followed by 24.7 mt/ha in Brazil. The higher production of oranges is also known in China, India, Mexico and Spain.

Mango and guava are the important fruits and occupies 5.6 million hectares area in the world only recorded after grapes. The overall production out of this area is 48.98 million mt annually. India ranks first in both area and production and occupied 2.78 million hectares with annual production of 22.1 million mt. China is second country having more area and production of 4.4 million mt. The average productivity was observed highest to the tune of 48.8 mt/ha in Kenya. The other mango growing countries in the world are Kenya, Thailand, Indonesia, Pakistan and Brazil.

Papaya is grown in 4.4 million hectares with annual production of 12.9 million mt. The average productivity of papaya in the world is the highest among the fruit crops and recorded 29.4 mt/ha. India is producing major share of 43.7 per cent of total production with productivity of 42.3 mt/ha. Brazil is the second country sharing 11.8 per cent of total production but their average productivity is 48.5 mt which is higher than the India. The overall highest productivity of 312.7 mt/ha is reported from Dominican Republic. Therefore, the important papaya producing countries are India, Brazil, Indonesia, Dominican Republic, Nigeria and Mexico.

Pineapple is cultivated on 1.02 million hectares with annual production of 23.6 million tonnes. The highest productivity is reported from Indonesia which is 124.5 mt/ha. Similarly, higher productivity of 59.2 mt/ha is also reported from Costa Rica. Thailand occupies 1.1 lakh hectares area producing 2.65 million mt tonnes annually. The other important pineapple producing countries are Costa Rica, Brazil, Philippines, Indonesia, India and Nigeria.

The overall fruit productivity is compared among the different countries and fruits and shown that it is the highest (29.4 mt/ha) in Papaya. Higher productivity of 23.3 mt/ha and 21.2 mt/ha is also observed in Pineapple and Banana, respectively. The average productivity of oranges and apple is also quite high and reported 18.3 and 15.9 mt/ha.

Table 6: Area and Production under total fruits in the World

Country	Area (ha)	Production (mt)	% of total production	Productivity (mt/ha)
China	11834450	137066750	20.9	11.6
India	7216312	88977134	13.6	12.3
Brazil	2325385	38368678	5.9	16.5
USA	1137779	26548859	4.1	23.3
Indonesia	796530	17744411	2.7	22.3
Philippines	1240370	16370976	2.5	13.2
Mexico	1256730	15917806	2.4	12.7
Turkey	1102662	14974561	2.3	13.6
Spain	1539100	13996447	2.1	9.1
Italy	1125593	13889219	2.1	12.3
Total	57499432	654449438	100.0	11.4

[Source: FAO Website – February 2015 (Data for 2012, 2013 N/A) and for India Data (Data for 2013-14) Department of Agri. & Cooperation}

Table 7: Area and Production under Apple in the World

Country	Area (ha)	Production (mt)	% of total production	Productivity (mt/ha)
China	2060000	37000000	48.2	18.0
USA	132656	4110046	5.4	31.0
Turkey	150847	2899000	3.8	19.2
Poland	194680	2877336	3.8	14.8
India	313044	2497678	3.3	8.0
Italy	54684	1991312	2.6	36.4
Iran	134000	1700000	2.2	12.7
Chile	36500	1625000	2.1	44.5
Russian Federation	183300	1403000	1.8	7.7
France	41051	1382901	1.8	33.7
Others	1533204	19196743	25.0	12.5
Total	4833966	76673016	100.0	15.9

Source: Indian Horticulture Database 2014 [Source: FAO Website – February 2015 (Data for 2012, 2013 N/A) and for India Data (Data for 2013-14) Department of Agri. & Cooperation}

Table 8: Area and Production under Banana in the World

Country	Area (ha)	Production (mt)	% of total production	Productivity (mt/ha)
India	802566	29724548	27.8	37.0
China	400000	10550000	9.8	26.4
Philippines	454179	9225998	8.6	20.3
Ecuador	210894	7012244	6.6	33.3
Brazil	481116	6902184	6.5	14.3
Indonesia	105000	6189052	5.8	58.9
Angola	115749	2991454	2.8	25.8
Guatemala	66000	2700000	2.5	40.9
United Rep. Tanzania	442190	2524740	2.4	5.7
Mexico	72617	2203861	2.1	30.3
Others	1883780	26823720	25.1	14.2
Total	5034091	106847801	100.0	21.2

Source: Indian Horticulture Database 2014 [Source: FAO Website – February 2015 (Data for 2012, 2013 N/A) and for India Data (Data for 2013-14) Department of Agri. & Cooperation}

Table 9: Area and Production under Mango & Guava in the World

Country	Area (ha)	Production (mt)	% of total production	Productivity (mt/ha)
India	2784185	22099225	45.1	7.9
China	460000	4400000	9.0	9.6
Kenya	57021	2781706	5.7	48.8
Thailand	318000	2650000	5.4	8.3
Indonesia	233000	2376339	4.8	10.2
Pakistan	174000	1950000	4.0	11.2
Mexico	195674	1760588	3.6	9.0
Brazil	73310	1175735	2.4	16.0
Bangladesh	124000	945059	1.9	7.6
Nigeria	132000	860000	1.8	6.5
Others	1100295	7990410	16.3	7.3
Total	5651485	48989062	100.0	8.7

Source: Indian Horticulture Database 2014 [Source: FAO Website – February 2015 (Data for 2012, 2013 N/A) and for India Data (Data for 2013-14) Department of Agri. & Cooperation}

Table 10: Area and Production under Oranges in the World

Country	Area (ha)	Production (mt)	% of total production	Productivity (mt/ha)
Brazil	729583	18012560	26.8	24.7
USA	250582	8166480	12.1	32.6
China	475000	6500000	9.7	13.7
India	334939	3886198	5.8	11.6
Mexico	323357	3666790	5.5	11.3
Spain	154000	2933800	4.4	19.1
Egypt	118731	2786397	4.2	23.5
Italy	83514	1770503	2.6	21.2
Turkey	45733	1662000	2.5	36.3
South Africa	45000	1612828	2.4	35.8
Others	1106193	16112401	24.0	14.6
Total	3666632	67109957	100.0	18.3

Source: Indian Horticulture Database 2014[Source: FAO Website – February 2015 (Data for 2012, 2013 N/A) and for India Data (Data for 2013-14) Department of Agri. & Cooperation}

Table 11: Area and Production under Grapes in the World

Country	Area (ha)	Production (mt)	% of total production	Productivity (mt/ha)
China	600000	9600000	14.0	16.0
USA	389349	6661820	9.7	17.1
Italy	696756	5819010	8.5	8.4
France	760805	5338512	7.8	7.0
Spain	943000	5238300	7.7	5.6
Turkey	462296	4275659	6.2	9.2
Chile	204000	3200000	4.7	15.7
Argentina	220000	2800000	4.1	12.7
Iran	215000	2150000	3.1	10.0
South Africa	124000	1839030	2.7	14.8
Australia	148489	1656621	2.4	11.2
Brazil	82603	1514768	2.2	18.3
Egypt	66262	1378815	2.0	20.8
India	118735	2585338	3.8	21.8
Others	1944813	14354594	21.0	7.4
Total	6976108	68412467	100.0	9.8

Source: Indian Horticulture Database 2014 [Source: FAO Website – February 2015 (Data for 2012, 2013 N/A) and for India Data (Data for 2013-14) Department of Agri. & Cooperation}

Table 12: Area and Production under Papaya in the World

Country	Area (ha)	Production (mt)	% of total production	Productivity (mt/ha)
India	133360	5639300	43.7	42.3
Brazil	31310	1517696	11.8	48.5
Indonesia	10450	906312	7.0	86.7
Dominican Republic	2608	815499	6.5	312.7
Nigeria	95000	775000	6.0	8.2
Mexico	14227	712917	5.5	50.1
Congo	14000	230000	1.8	16.4
Taiwan	3800	222000	1.7	58.4
Thailand	13000	215000	1.7	16.5
Guatemala	2300	206500	1.6	89.8
Others	118960	1650252	12.8	13.9
Total	439015	12890476	100.0	29.4

Source: Indian Horticulture Database 2014 [Source: FAO Website – February 2015 (Data for 2012, 2013 N/A) and for India Data (Data for 2013-14) Department of Agri. & Cooperation}

Table 13: Area and Production under Pineapple in the World

Country	Area (ha)	Production (mt)	% of total production	Productivity (mt/ha)
Thailand	105000	2650000	11.2	25.2
Costa Rica	42000	2484729	10.5	59.2
Brazil	60653	2478178	10.5	40.9
Philippines	58442	2397628	10.2	41.0
Indonesia	14300	1780889	7.5	124.5
India	109876	1736739	7.4	15.8
Nigeria	180000	1420000	6.0	7.9
China Mainland	53000	1000000	4.2	18.9
Mexico	17708	759976	3.2	42.9
Columbia	13916	551133	2.3	39.6
Others	360369	6355353	26.9	17.6
Total	1015264	23614625	100.0	23.3

Source: Indian Horticulture Database 2014 [Source: FAO Website – February 2015 (Data for 2012, 2013 N/A) and for India Data (Data for 2013-14) Department of Agri. & Cooperation}

EXPORTS OF FRUITS FROM INDIA

In recent year, the annual production of fruits in India is 91.4 million tonnes from an area of 6.4 million hectares. Mango, walnut, grapes, banana and pomegranate were the main fruits which were exported from India during the year 2012-13. Out of the total export India earned Rs 2503 crores during this year. The major importing countries were UAE, Bangladesh, Malaysia, UK, Netherland, Pakistan, Saudi Arabia, Sri Lanka and Nepal. Apple, banana, oranges, grapes, guava, litchi, mango, papaya, pineapple, sapota and walnut were exported from India during 2013-14. India earned Rs 862.4 million from apple, Rs 1568.3 million from banana, Rs 555.8 million from oranges, Rs 14370.7 million from grapes, Rs 41.5 million from guava, Rs 8.4 million from litchi, Rs 2854.3 million from mango, Rs 351.1 million from papaya, Rs 148.1 million from pineapple, Rs 77.1 million from sapota and Rs 3245.3 million from walnut. Total export value of fruits was Rs 2408.3 crores during 2013-14. The major importing countries were Bangladesh, UAE, Saudi Arabia, Bahrain, Maldives, Kuwait, Oman, Nepal, UK, USA, Malaysia, Netherlands, Qatar, Singapore and Thailand.

Export of Fruits and Value during 2014-15

India exported 20374 tonnes of apples to Bangladesh, China People's Republic, UAE, Saudi Arabia, Singapore, Oman and Bahrain during 2014-15 and earned Rs 52.9 crores. The major share of 56 per cent was exported to alone Bangladesh. Also higher share percentage of 39.4 was exported to China People's Republic. Banana was one of the leading exported fruits from India. Nearly 63274 tonnes of bananas were exported this year of value worth Rs 241.9 crores. UAE was the main country where 36.2 per cent share of export value was marketed. Saudi Arabia was the second largest destination where bananas were exported 18.8 per cent of export value. The oranges export earned Rs 31.94 crores from 172.3 tonnes of produce. The major share of oranges export of 76.9 per cent was made to Bangladesh. However, oranges were also exported in small consignment to many countries in the world.

Grapes were exported in large quantity to several countries and India earned Rs 1086.5 crores this year. The total export quantity of grapes was 1.07 lakh tonnes. The major share of 33.2 and 14.8 per cent of export value was exported to Netherlands and United Kingdom, respectively. Russia and United Arab Emirates were the countries where 10.5 and 10.4 per cent share were sold, respectively. The export of mango fruits also earned good amount of foreign currency. During this year, India earned Rs 302.5 crores from the export of 42998 tonnes of mangos, United Arab Emirates was the leading country from where 71.1 per cent share of export value was earned.

Guava and litchi were exported to the tune of 907.9 and 961.4 tonnes, respectively to various countries. The corresponding export value of guava and litchi was Rs 3.72 and 2.15 crores from the total export. Saudi Arabia and UAE were the main destinations of guava fruits which results 24.9 per cent at 20.8 per cent share of export value. Good amount of guava was also exported to Nepal, Bahrain, Tanzania Republic, Netherlands and Sri Lanka. The major litchi export was done to Bangladesh which was 76.1 per cent of export value.

The export consignment of Papaya and Pineapple fruits was also quite high i.e. 11484 tonnes in Papaya and 3751 tonnes in Pineapple. India earned Rs 38.26 crores from export of Papaya and Rs 16.64 crores from export of Pineapple. UAE, Saudi Arabia and Netherlands were the countries from where 36.6, 22.8 and 12.4 per cent share of export value of Papaya was achieved. In pineapple, Saudi Arabia, Qatar, Nepal and Russia were the main countries where this fruit was exported and 24.4, 21.9, 12.6 and 12.3per cent share of export value was achieved. The quantity of sapota fruits exported to other countries was comparatively less than papaya and pineapple. India earned only Rs 8.18 crores from 1372 tonnes of sapota fruits. The main countries where sapota exported were UAE, Bahrain, UK and Canada. The percent share of 42.9 of export value was sold to UAE. Among the temperate fruits, walnut export was mainly done. Although, the quantity exported was 2666 tonnes but the country earned good amount of foreign exchange to the value of Rs 136.45 crores. This fruits was exported mainly to Vietnam, UK, USA, Egypt, and Netherland. During this year, total export value of fruits was Rs 2746.5 crores.

The export of fresh fruits has been become reality only due to the establishment of cold chain infrastructure and other quality measures. Public and Private sector were involved in this large investment. Apeda helped in setting of perishable cargo and post-harvest handling centres. Growers, processors and exporter initiatives has played great role in establishment of this venture.

Export of Fruits and Value during 2015-16

India exported fruits of worth Rs 3524.50 crores during 2015-16. Mango, walnut, grapes, banana, pomegranate accounted for larger portion of fruits exported from the country. The major destinations for fruits from India to other countries were UAE, Bangladesh, Malaysia, Netherland, Sri Lanka, Nepal, Saudi Arabia, Pakistan and Qatar. Inspite of the fact, India's share in the global market is still nearly 1 per cent only. Now, there is increasing acceptance of horticultural produce from India.

Grape is the important fruit of the India occupying 1.16 lakh hectares area, thus covering 1.70 per cent of the total areas. India is a major exporter of fresh grapes to the world and exported 1.56 lakh mt of grapes for the worth of Rs 1557.32 crores. Grapes were mainly exported to Netherland, UK, Russia, UAE and Saudi Arabia.

India has exported 36329 mt fresh mango for the worth of Rs 317.1crores. The major export of mango was done to UAE, UK, Saudi Arabia, Nepal and Kuwait.

Walnut was mainly cultivated in Jammu and Kashmir, Uttarakhand, Himachal Pradesh and Arunachal Pradesh. This year, India exported walnut for the worth of Rs 117.92 crores with the total quantity of 3291.71 mt. The major importing countries of walnut from India were Germany, UK, Netherlands, Egypt, Arab Republic and USA. The export of other fresh fruits to other countries fetched Rs 1538.2 crores from 308261.23 mt fruits. These fruits were mainly exported to UAE, Nepal, Saudi Arabia, Bangladesh and Iran.

FRUIT MARKETS IN INDIA

The major fruit market in India is Azadpur Fruits Market Delhi. It is Asia's largest wholesale market for fruits. Everyday thousands of trucks bring fresh fruit to this market from the various parts of the country.

Ratnagiri, Nashik, Jalgaon, Allahabad, Kotgarh, Siliguri, Goa, Nagpur, Mahabaleshwar and Dahanu Bordi are the other top fruit markets in India. The best quality mango (Alphonso) are produced in Ratnagiri. The fertile soil of this region produce succulent and incredible sweet mangos. The climate of Nagpur is ideal for citrus. Ambe and Mrig bahar are the two main season for oranges. Jalgaon known a capital of banana, contribute 16% of the total banana production in India. Allahabad is famous for production of finest quality of guava in the world. Nashik is referred as producing good quality of grapes. The seedless variety of grapes from Nashik is very popular. Cold climate of Shimla is excellent for apples and excellent quality of apples are produced in Kotgarh. Mahabaleshwar hill station is good for strawberry cultivation. The main season for its ripening is May-June.Goa is well familiar due to production of excellent quality of cashewnuts. Dahanu Bordi area produce high and good quality of sapota in the country.

Kolkata, Lucknow, Kanpur, Mumbai, Pune, Bangalore, Chennai, Hyderabad, Kanpur, Lucknow, Indore, Jaipur, Pune, Amritsar and Chandigarh are the other important fruit markets in India.

Table 14: Export of Apple from India (2014-15)

Country	Quantity (Tonnes)	Value (Rs in lakh)	% Share of export value
Bangladesh	13807.86	2968.66	56.0
China Ppl. Rep.	6172.21	2088.45	39.4
UAE	289.59	139.18	2.6
Saudi Arabia	42.75	32.43	0.5
Singapore	40.00	31.44	0.5
Oman	22.00	32.28	0.5
Bahrain	0.18	0.18	0.3
Unspecified	0.11	0.11	0.2
Total	20374.70	5292.73	100.0

Source: Horticulture Statistics at a glance 2015- Department of Agriculture & Cooperation – APEDA website, accessed on 10 July, 2015

Table 15. Export of Banana from India (2014-15)

Country	Quantity (Tonnes)	Value (Rs in lakh)	% Share of export value
UAE	19010.88	8755.03	36.2
Saudi Arabia	9314.43	4557.57	18.8
Nepal	8637.73	955.12	3.9
Pakistan	7243.35	1472.21	6.1
Oman	4926.01	2004.43	8.3
Kuwait	4684.43	2241.77	9.3
Iran	3377.60	1671.50	6.9
Qatar	3210.70	1671.64	6.9
Bahrain	1802.09	952.58	3.9
Maldives	664.28	207.02	0.8
Others	402.90	130.90	0.5
Total	63274.40	24619.77	100.0

Source: Horticulture Statistics at a glance 2015- Department of Agriculture & Cooperation – APEDA website, accessed on 10 July, 2015

Table 16: Export of Oranges from India (2014-15)

Country	Quantity (Tonnes)	Value (Rs in lakh)	% Share of export value
Bangladesh	14210.73	2456.01	76.9
Nepal	2845.36	661.37	2.1
Bhutan	82.05	12.78	0.4
Saudi Arabia	42.77	34.47	1.0
UAE	32.05	16.04	0.5
Oman	5.60	1.73	0.0
Qatar	4.82	3.98	0.0
Bahrain	2.70	1.38	0.0
Kuwait	1.80	0.96	0.0
Maldives	1.60	1.89	0.0
Total	17231.45	3194.00	100.0

Source: Horticulture Statistics at a glance 2015- Department of Agriculture & Cooperation – APEDA website, accessed on 10 July, 2015

Table 17: Export of Grapes from India (2014-15)

Country	Quantity (Tonnes)	Value (Rs in lakh)	% Share of export value
Netherlands	32859.42	36167.92	33.2
UK	13475.81	16028.53	14.8
Russia	12625.78	11372.30	10.5
UAE	11860.53	11354.52	10.4
Saudi Arabia	5319.73	4689.93	4.3
Bangladesh	3562.02	1312.34	1.2
Germany	2939.02	3236.63	3.0
Thailand	2260.11	2873.85	2.6
Nepal	2129.54	633.66	0.6
Sri Lanka	2072.47	2064.37	1.9
Others	18153.43	18914.97	17.4
Total	107257.86	108649.02	100.0

Source: Horticulture Statistics at a glance 2015- Department of Agriculture & Cooperation – APEDA website, accessed on 10 July, 2015

Table 18: Export of Guava from India (2014-15)

Country	Quantity (Tonnes)	Value (Rs in lakh)	% Share of export value
Saudi Arabia	290.22	92.73	24.9
UAE	167.45	77.44	20.8
Nepal	87.40	35.03	9.4
Bahrain	48.34	24.08	6.7
Tanzania Repub.	67.80	23.49	6.4
Netherlands	35.00	20.88	5.6
Sri Lanka	51.00	20.36	5.4
Malaysia	35.40	14.94	4.0
Kenya	34.00	12.96	3.5
Oman	26.92	11.40	3.0
Others	64.34	38.56	10.2
Total	907.87	371.87	100.0

Table 19: Export of Litchi from India (2014-15)

Country	Quantity (Tonnes)	Value (Rs in lakh)	% Share of export value
Bangladesh	915.00	163.80	76.1
Nepal	44.60	17.73	8.2
USA	1.30	33.18	15.4
Oman	0.33	0.17	0.1
Kuwait	0.09	0.09	0.1
UK	0.08	0.20	0.1
UAE	0.03	0.01	0.0
Total	961.43	215.18	100.0

Source: Horticulture Statistics at a glance 2015- Department of Agriculture & Cooperation – APEDA website, accessed on 10 July, 2015

Table 20: Export of Mango from India (2014-15)

Country	Quantity (Tonnes)	Value (Rs in lakh)	% Share of export value
UAE	29231.90	21497.85	71.1
Nepal	3574.93	695.42	2.3
Bangladesh	2475.33	473.42	1.6
Saudi Arabia	2171.49	1428.59	4.7
Qatar	998.10	810.81	2.7
Kuwait	787.28	1238.18	4.0
Canada	669.26	429.63	1.4
Bahrain	658.71	505.36	1.7
Oman	605.20	469.27	1.6
Singapore	562.95	588.31	1.9
Others	1263.16	2116.81	7.0
Total	42998.31	30253.65	100.00

Table 21: Export of Papaya from India (2014-15)

Country	Quantity (Tonnes)	Value (Rs in lakh)	% Share of export value
UAE	4378.81	1402.14	36.6
Saudi Arabia	2406.25	870.57	22.8
Nepal	1768.08	163.48	4.3
Bahrain	744.95	246.69	6.4
Netherland	691.79	473.85	12.4
Qatar	520.10	232.41	6.1
Kuwait	429.45	177.53	4.6
Oman	193.97	73.96	1.9
Germany	127.05	85.38	2.2
Bangladesh	102.82	28.40	0.8
Others	121.55	72.00	1.9
Total	11484.82	3826.41	100.0

Source: Horticulture Statistics at a glance 2015- Department of Agriculture & Cooperation – APEDA website, accessed on 10 July, 2015

Table 22. Export of Pineapple from India (2014-15)

Country	Quantity (Tonnes)	Value (Rs in lakh)	% Share of export value
Nepal	1440.71	209.30	12.6
Saudi Arabia	679.17	406.83	24.4
Qatar	563.37	365.20	21.9
Maldives	270.06	162.74	9.8
Russia	248.87	205.16	12.3
Oman	192.97	131.01	7.9
Bahrain	191.05	116.41	7.0
UAE	60.89	27.39	1.6
Pakistan	43.83	5.49	0.3
Others	60.61	35.39	2.12
Total	3751.53	1664.92	100.0

Table 23: Export of Sapota from India (2014-15)

Country	Quantity (Tonnes)	Value (Rs in lakh)	% Share of export value
UAE	535.43	351.02	42.9
Bahrain	357.33	170.07	20.8
Canada	117.89	62.00	7.6
Qatar	75.08	50.21	6.1
UK	72.91	65.05	8.0
Saudi Arabia	50.46	39.06	4.8
Oman	45.18	20.82	2.5
USA	32.80	21.37	2.6
Kuwait	22.64	11.94	1.5
Singapore	16.74	9.48	1.1
Others	45.65	17.07	2.1
Total	1372.11	818.09	100.0

Table 24: Export of Walnut from India (2014-15)

Country	Quantity (Tonnes)	Value (Rs in lakh)	% Share of export value
Vietnam S. R.	479.05	2331.82	17.1
UK	392.50	1941.69	14.2
Netherland	254.66	1419.47	10.4
Egypt, Arab Rep.	241.00	1631.15	12.0
USA	230.01	1735.60	12.7
France	187.00	724.25	5.3
Germany	108.57	570.89	4.2
Nepal	97.65	274.23	2.0
Canada	96.62	728.72	5.3
Australia	81.00	487.67	3.6
Others	497.81	1799.74	13.2
Total	2665.87	13645.23	100.0

Source: Horticulture Statistics at a glance 2015- Department of Agriculture and Cooperation – APEDA website, accessed on 10 July, 2015

PROBLEMS IN FRUIT GROWING

Successful fruit growing involves many aspects which need special attention and care for profitable fruit production. Some of the national problems like alternate bearing and malformation in mango, decline in citrus, wilt in guava, virus diseases in banana and papaya, dieback and powdery mildew in grapes, splitting in litchi fruits, attack of lac insect and powdery mildew in ber, scab in apple, root-rot and sap-wood in rot in pear, crown gall in peach etc. are still needs solution. Production of quality planting material needs to be taken with great efforts. Lack of suitable varieties resistant to biotic and abiotic stress, standardization of suitable rootstock and appropriate horticultural based cropping systems for different agro-climatic zones needs to be developed. However, some of the basic problems being faced by the growers in profitable fruit cultivation are outlined and discussed below.

Marketing of Fruits

The fruits are highly perishable, their marketing presents many problems compared to other agricultural produce. The fruit growers generally gets less return from his produce. Contrary, consumer always pay higher price. Middlemen earn better profit than the producer by exploiting the market in many ways and offer always low price to the growers.

In our country, the marketing of fruits is the major problem being faced by the growers. Marketing system followed for fruit crops is not properly understood and creating main hurdle in the future expansion of horticulture. Some of the commons problems are as follow:

- High post-harvest losses up to 20-30 per cent due to inadequate infrastructure.
- Non-availability of properly controlled cold storage facilities.
- Lack of pre-cooling houses near to the orchards.
- Imperfect competition in fruit trade due to involvement of few known traders in marketing.
- Lack of insufficient marketing information to the growers.

Prevailing Fruit Marketing Pattern: In general, the fruit crop is auctioned by the growers to the contractors. Big contractors sitting in the cities controlled the fruit crop through remote control by involving sub-contractors. The produce is finally sent to terminal markets and both contractors and sub-contractors make huge margin from the sale of produce.

The another channel followed at present is auction of fruit crops to the contractors or few growers market the produce at their own level in the nearby city fruit markets. The middlemen involved in the marketing system earn better profit from the produce.

In normal procedure, growers sell the fruit crop to contractors and sub-contractors They are financed by the commission agents from big fruit markets. It has been observed that the contractor does not pay to the growers more than 40 per cent of the crop value. Major share of 60 per cent is shared by the middlemen in the markets. The produce is purchased by the wholesalers from commission agents.

The bulk purchase in the market is done by these wholesalers They sold the fruits further to the retailers after getting substantial margin. Finally, retailer sell the produce in the open market to the consumer with making enough margin. The whole process of fruit marketing is quite lengthy and cost heavy to the consumer at the end. After following this cumbersome procedure fruit reaches to the consumer after about 2-3 days from actual harvesting. Further, after making retail purchase, consumer either put the produce at ambient temperature or in the refrigerator for daily and regular use. The real quality of the produce is lost slowly and slowly and ultimately some percentage of it lost sheen either at room temperature or at inappropriate cool temperature in the home refrigerator. The food value of the fruit is also affected during this process.

Cooperative system of marketing or concept of Apni Mandi can play pivotal role in putting the fruit industry on sound footing. The growers should bring their produce directly to the market and sell it himself directly to the consumers and can enhance their margin of profit higher even after selling produce at affordable price. The fruit growers and cooperative societies have to struggle hard against the existing marketing system which is exploiting the producers since long. Govt. can play major role in formulating the cooperative system of marketing of fruits.

Marketing Regulations for Fruits

- Proper market regulation with fixed commission monitoring system needs to be introduced- to avoid exploitation of the growers.
- Setting of efficient market information network is important - to get better price of fruits.
- Instituting marketing organisation in fruiting zones –to provide liberal loan facilities to the growers.
- Initiating cooperative fruit marketing units – to sell fruits directly to the consumers.
- Temperature specific cold storage facilities to set up - to suits to individual fruit for enhanced shelf life.
- Strengthening of refrigerated transport facility for efficient marketing of fruits – to reduce post-harvest losses.
- Need base setting of processing units – to dispose of surplus fruits.
- Formulation of cooperative marketing societies – to help marketing of fruits.

- Export of fruits for higher return –to tap foreign markets .
- Introduction of crop insurance scheme – to afford protection from natural calamities.
- Fixation of support price of fruits – to have secured marketing of fruits.

Production of Quality Planting Material

Availability of quality planting material to the growers is most essential for bringing fruit industry as a profitable venture. Major constraint in flourishing horticulture in relation to higher fruit production is non-availability of true-to-type plants of known pedigree. State Agricultural Universities, state horticulture departments and private nurseries are selling fruit plants to growers. However, most of private nurseries are not taking precautions in proper selection of rootstocks and healthy virus free bud wood for propagation. Plants produced under such conditions do not bear to their optimum level and many time become liabilities to the growers after the long run. Plants purchased from unregistered and private nurseries and even foot path sellers proves expensive investment due to their poor pedigree. The outcome from these plants become available after many years only at the time bearing.

Production of citrus and their up keep needs utmost attention to produce plants free from leaf miner and citrus psylla and diseases like citrus canker and citrus scab. Citrus plants should be screened properly against viruses and virus like diseases and insect vectors However, most of the private nurseries do not undertake adequate checks against such maladies. The plants of mango are supplied even without taking care of vegetative malformation. In peach, root knot nematode infested plants are sold ignorantly. It is pertinent that disease free plants known for bearing regularly and uniformly should be replaced with existing poor quality plants in the nursery as a parent plants for taking bud-wood. Popular varieties and outstanding merit plants should be identified for clonal propagation. The superior plant material should be supplied to the growers at right time of planting.

Comfortable Orchard Establishment

The initial establishment cost of new orchard goes very high in present days. It involves cost of digging and filling of pits, high cost of planting material, layout of an orchard, farmyard manure and pesticides. On an average, the original expenditure of planting a new orchard in different fruit plants goes up to 50000 to 75000 per hectare. Such establishment cost is still higher in grapes where erection cost of training structure touches to near Rs 3 lakhs per hectare. Medium, small and marginal famers think twice before establishing a new orchard due to higher cost of material. Most of the farmers shed the idea of planting a new orchard inspite of their inner desire to do so. To solve this problem, liberal loan facility should be provided by the govt. to the farmers on the basis of their land holdings. Repayment of loan can be made flexible for uplifting the fruit industry of the country and to maintain the health state of the people.

Long Waiting Period of Bearing

In many fruit plants, first bearing comes after long period stretching from 5 to 10 years. During this pre-bearing period, no major income is available to the growers Mango, litchi and pear orchards starts giving commercial crop after a decade. Average and small holding farmers unable to wait this incubation period without affordable income. To compensate the economic loss during this pre-bearing period, some filler trees like peach, plum, phalsa, papaya, kinnow and even guava can be interplanted in the permanent fruit trees like mango, litchi and pear. In addition to this, some leguminous crop can be intercropped and basic economic loss can be reduced to good extent.

Transportation of Fruits

Transportation of fruits to the terminal fruit markets and cooling units for extending shelf life, involve high initial cost for the ordinary farmers. Post-harvest losses further aggregated due to available poor and non-specific cold storage facilities as well as due to non-availability of controlled atmospheric storage. Existing system of fruit transportation in most part of the country is not up to the mark and highly injurious to fruit trade. Multipurpose vehicles are operating for transportation of fruits. Such vehicles further enhance post-harvest losses. Paths and roads leading to the orchards are not smooth, thus cause heavy losses to the perishable commodity. Introduction of refrigerated vans and temperature controlled rail transportation can help in extending shelf life of fruits by reducing post-harvest losses.

Storage and Processing of Fruits

It is estimated that 20-30 per cent losses takes place in most of the fruits before appropriate storage. Elimination of field heat after harvesting is very important and cooling house should be established in the fruit orchard pockets. Adequate processing techniques and facilities should be developed to save the post-harvest losses. Most of the fruits are consumed fresh, but useful products can be prepared after processing. Even, many useful products can be prepared from 20-30 per cent waste fruits. Processed products can be marketed in the country or can be exported to earn foreign exchange.

Presently processing facilities are available on limited scale and only numbered processing units are involved in this trade. More processing units on the basis of fruit zones need to be sanctioned to utilise the different graded fruits for useful purpose.

Cooling house and processing units needs to be established both at Govt. and Private sector. Some fruit Growers Association in certain states are doing yoemen service in marketing of fruits for the growers and processing the different products in their processing units. However, efforts should be strengthened to help the fruit growers association at govt level to make the horticulture more profitable in near future. Mandi Board of the different state are playing useful and efficient role in helping the fruit growers. Still this need more organised way to bring the fruit industry on sound footing.

Fruit Buying Capacity

Most of the fruits are now available at very high cost in the markets. Common man and salaried people have low purchasing power to buy the fruits and they do not consume fresh fruits at regular level. Experts suggests to consume atleast 230 g fruits per head per day but at present majority of the people are not able to buy this much quantity due to high prices of fruits in the market. Banana is now available at Rs 70-80 per dozen or Rs 70 per kg, grapes at Rs 100-120 per kg, mandarin and musambi at Rs 60-70 per kg, pomegranate at Rs 100-150 per kg, papaya Rs 40-50 per kg, ber at Rs 70-100 per kg, pear at Rs 80-100 per kg, apples at Rs 100-200 per kg and loquat at Rs 120 per kg. In other words, fruits are not within the range of common man. A family from middle class is unable to spare Rs 2000 to Rs 4000 monthly for the purchase of fruits. Contrary, an average salaried man and person with low income can't even spare Rs 1000 per month for the purchase of fruits. Consumption of fruits is essential for maintaining physical and mental health. The fruits are highly nutritive and their use is best to maintain health and happiness. But, it is not possible for individual to spent even 10 per cent of their income on buying the fruits. Marketing control on price of individual fruit should be streamlined so that all the people, rich and poor could able to buy the fruits and must maintain health for building a strong nation and culture.

Fruit a Perishable Commodity

Fruits are generally highly perishable due to their high water content. Post-harvest technology of fruit crops should be developed more eco-friendly. Appropriate on farm post-harvest handling and storage techniques fit for different fruits and suitable for various agro-climatic zones needs to be developed. Grading and packaging of fruits; transportation and handling systems needs immediate attention for making quality fruits available to the consumers. Development related to making new processed products and proper utilization of processing waste should be reviewed and focussed.

Climatic and Soil Adaptability

India is climatically suitable for planting and producing various tropical, subtropical and temperate fruits. Extensive study on role of temperature, atmospheric humidity, rainfall and wind velocity needs to be strengthened. Some fruits are adapted to high temperature and others to low temperature. The overlapping of fruits to temperature limit can affect the overall development of fruits cultivar. Certain fruit plants are highly sensitive to atmospheric humidity, others to dry climatic and occurrence of rainfall and frost. Proper climatic and soil adaptability of particular fruit crop can leads to high yield and better quality fruits. Research trial on the suitability of fruit crops is under way in many Fruit Research Stations established under ICAR and State Universities schemes. Collaborative and mutual research studies and transformation should be further strengthened for appropriate growth of fruit plants under specific available climatic conditions.

FUTURE SCOPE OF FRUIT GROWING

The important points involved in future scope of fruit growing in India are discussed below:

High Nutritive value of Fruits

The fruits are delicious, refreshing and are the major source of vitamins, minerals and proteins. All these constituents are essential for maintaining health and help in the development of resistance against pathogens. Vitamins are the most important constituents of fruits. The deficiency of any vitamins in the diet for considerable period may lead to disorder condition or ill health. Minerals are found in varying concentrations in fruits. The deficiency of minerals can lead to disturbance of metabolism and results into diseased state. Bones and teeth are mainly made up of calcium, magnesium and phosphorus. Iron is an important constituents of blood. Carbohydrates and fats provide energy to the human body and are essential for keeping good health. Protein builds up the body tissues. Pectin and cellulose contents help in stimulating the intestinal activity. In human health, more than 80 per cent problems arising from under nutrition or malnutrition. These problems can be well managed with indigenous medicines. The fruits can play major role in solving malnutrition in human health.

Fruits are the best coprescription for maintaining better state of health. Mango and papaya are good for patient suffering from night blindness. Aonla, guava and ber are vitamin C rich fruits and are best for those suffering from scurvy disease. Almond, cashew nut, walnut, dry apricot, banana, dates, bael, rainsins and karonda fruits are rich in protein and carbohydrates. The children deficient with protein energy malnutrition (PEM) are prescribed to take all these fruits rich in protein and carbohydrates. Bael is a useful tranquiliser. Bael extract (Sharbat) is very useful in hot summer days as it provide cooling effect on human body. Banana is useful for the patient suffering from diarrhoea. Almond oil extent possess specific medicinal value and proved useful in malnourished and bronchitis-asthma patients in winter. Aonla powder in combination with hard and bahera is useful for constipation.

Keeping into view high nutritive value in fruits, it makes their use in our daily diet imperative to maintain physical and mental health.

Per Capita Consumption

The fruits are best known to contribute to health, happiness and prosperity of the people. The standard of living of the society can be judged from the consumption of fruits per capita and cultivation and production of fruits of that region. The climatic conditions of India are highly suitable for cultivation of tropical, sub-tropical and temperate fruits. But, our per capita per day consumption of fruit is 190g which is less than the prescribed dietary requirement. The fruit consumption in developed countries like USA, Canada, Switzerland, France, Sweden, Netherlands, United Kingdom and Australia is 250-450g per head per day. In comparison to many countries in the world, our per capita consumption is less. So, there is scope to increase the area and production of fruits to maintain our per capita fruits consumption at optimum level and at a affordable price.

High Demand of Fruits

India is the leading fruit producing country and ranked second in the world with a production of 91.4 million tonnes of fruits from an area of 6.4 million hectares. The requirement for fruits is increasing continuously in the cities due to fast growth in Urban sectors. The fruits are now available at very high cost in the markets. In this situation fruits are not within the reach of common people. To meet the daily demand of fruits in the society, there is strong need to be enhance the production of fruits till the country becomes self-sufficient in fruits to meet the recommended per capita consumption. There is need to organise the marketing of fruits with view the fruit prices fall to a level or to be fixed or controlled at which all the people can afford to buy the fruits. This target can only be achieved by rejuvenating the senile orchards and by increasing the area and production under fruits.

Higher Productivity

The productivity of fruit crops per unit area is quite comparable with other agricultural crops. In many fruit plants like papaya, pineapple, banana, grapes and kinnow mandarin average productivity is quite higher. The per hectare annual production is observed 45 tonnes in pineapple, 40 tonnes in papaya and 30 tonnes in banana, 25 tonnes in grapes and 20 tonnes in kinnow which is quite higher than many field crops. The higher productivity of fruit crops also gives higher economic return per unit area. In addition to this, it also offers higher calories per unit area.

Involvement in Specialized Skills

The commercial orchard is now emerged a profitable profession and many progressive farmers are adopting fruit orchard as a profession. Fruit gardening is a labour intensive industry and is highly beneficial to thickly populated country like China and India. In such countries land holdings are less than one hectare. Orcharding involves many specialized skills in production, harvesting, grading and packing, marketing, storage, canning and processing. Thus, it involves many ways and methods of the trade require professional skills. The direct involvement of small and medium scale entrepreneurs is utmost needed which becomes essential for increasing the overall production capacity and profitable fruit marketing.

Multiple Cropping

It is possible to grow more than two crops in the orchard by adjusting their cultural practices. Apart from giving additional income, the multiple cropping in orchards also generates more avenues for employment to skilled and unskilled workers. Some fruit plants come into bearing very late even after 6-7 years after planting in the orchard. The permanent trees like mango, litchi and pear start bearing first crop late. During the pre-bearing period of these crops, the filler trees like papaya, peach, plum, phalsa and kinnow can be interplanted in the vacant space of permanent trees. Leguminous crops and vegetables can also be interplanted in the trees rows in the fruit orchard. Thus, this multiple cropping culture proves useful and handsome income can be obtained during initial years of fruit crops.

Profitable Venture

The growth rate of population in India is high. Therefore, size of holding will remain small. The pressure on land for urbanization is likely to continue in near future. The cultivation of fruit crop in small holding can generate enough income for small families and generate different types of employment for various categories of the society. The fruit crops able to give more production from per unit area and proves more economical from marketing point of view. The cultivation of different fruit crops results higher income and grower can net income of Rs 1.5 to 2.0 lakh per hectare. Definitely higher income is achieved compared to most of the field crops. This indicates that there is future scope for expanding area under fruits in India.

Trade and Industrial Association

Orcharding open many opportunities in trade and industrial development. Many types of associated industries and trade sectors are directly or indirectly affiliated with horticultural produce starting from planting to marketing of fruits. Setting of garden implements and tools, CFB boxes for packaging of fruits, baskets and wooden crates for packaging and transportation, spray pumps; manures and fertilizers; insecticides pesticides and fungicides for plant protection, refrigerated van and cold storage and processing units for enhancing post-harvest life of fruits are the important areas of trade. All these necessary items and articles have continuous and longer association with fruit growing. Therefore, offers variable employment opportunities for many in the coming future.

Suitability for Diversification

This can be stated that orcharding fit well in crop diversification programme at various levels. It is pertinent to explain that wheat rice rotation is no longer remained profitable as water table in agricultural states is going deep on account of continuously sowing of rice crop. The rice crop should be replaced with fruit crop slowly and slowly. In this way, efficient use of water can be availed off. Moreover, fruit crops help to bring high profit and provide nutritional security to the people. On the whole, it is stated that cultivation of fruit crops fit well in diversification process which is the timely need to save water particularly in North Indian states.

Cultural Significance

The fruit crops were adapted to particular climate even on the basis of preference given by the society in a particular locality. Many fruit crops are associated with religious feeling while others are associated with cultural and social significance. Mango, ber, pear, banana, papaya, karonda etc have constant affiliation in certain sects of people and these are planted from time to time to fulfil such needs. Folklores of the people are full of evidence where fruits are correlated according to the occasions, religion, castes and speciific sects.

Mango find special place in the mind and folk song of Doaba region of Punjab. In this way, this crop acclimatised and doing well throughout this belt due to eternal love and affection of the people with the fruit. Ber is favoured by the Sikh Gurus and find its place in most of the Gurdwaras. People from all caste have faith in ber trees as Sikh Gurus stated that its fruit maintain health in summer, give fodder for goats, shade for human being and give strong wood for making carpentry. Pear tree was introduced into India from China during the times of Lord Kanishka (120-170 A.D.) who settled some Chinese hostage in village Harsa Chhina in Amritsar district. After that, this pear tree makes its home town in Amritsar and spread further to other parts of the country.

Different fruits are co-prescribed for curing various ailments in human system. Similarly, fruit trees are correlated from health to happiness at several occasion of life and become integral part of the human system. Constant association of fruit crop with the society prove useful in making that particular fruit popular and bringing more area under cultivation.

LITERATURE CONSULTED

Area and Production of Fruits in India (2015-16). Area Production Statistics – National Horticulture Board –nhb.gov.in>PDF Viwer

Bal, J.S. 2014. Fruit Growing, Third Edition. Kalyani Publishers, New Delhi. Pp:3-13.

Fresh Fruits and Vegetables – APEDA (2015-16). apeda.gov.in>six_ head_productsFFV

Horticulture Statistics at a Glance 2015 – Department of Agriculture and Cooperation. http://www.researchgate.net>publication. Pp: 379-398.

Indian Horticulture Database 2009. PDF File of Planning Commission

Indian Horticulture Database 2011. National Horticulture Board.

Indian Horticulture Database 2013. National Horticulture Board.

Indian Horticulture Database 2014. National Horticulture Board.

SECTION - 2

FRUIT PLANT FUNDAMENTALS CULTURE AND TECHNOLOGY

2

Planning and Planting of an Orchard

Jaswinder Singh Brar and Krishan Kumar

A. PLANNING OF AN ORCHARD

Orchard is a long-term investment and needs proper planning and layout before establishment. Mistakes committed during planning and planting of fruit trees cannot be rectified. While planning and planting a new orchard, it is very important to carefully assess all the factors that will ultimately affect fruit yield and orchard sustainability. The aspects like location and site selection, suitability of soil and subsoil, suitability of fruit crop and varieties, plant spacing, market potential and source of nursery plants must be taken care of seriously. Thus, one should avail of expert advice on the establishment of new orchards and their proper maintenance.

Selection of Suitable Location

For sustainability of success in fruit cultivation, the suitability of site is of utmost important. For selection of suitable site, following points must be kept in mind before establishment of orchards. A well-established fruit growing region for particular fruit crop is always beneficial because one can get benefit of expertise of local growers and also there is more opportunities of organized market system, availability of inputs and other management systems.

- Nearness to the market to the orchard site will be additionally beneficial.
- The selected site should be suitable with respect to climatic factors such as occurrence of frost,hailstorm, drought, flooding etc. may be detrimental for fruit crops.
- There should be availability of assured water supply throughout the year as the fruit crops require assured irrigation at critical stages and irrigation water must be of good quality.

- Site must have proper drainage and in case of heavy rainfall or irrigation water should be drained out of orchard in short period. The water table should be at least three meter from ground level.
- Soil of selected site should be fertile, free from excess salts, deep and suitable with respect to physical and chemical properties.
- As the fruit cultivation is labour intensive and more technical enterprise and requires labour round the year. So, skilled labour must be available in the locality.
- The connectivity with road, railway will be beneficial for local as well as distant marketing.As the fruits are perishable, hence, need quick transport facilities.
- The areas prone to be damaged by wild animals and theft should be avoided for fruit growing.

Selection of Ideal Soil

The suitable soil for particular fruit plant results excellent performance of an orchard over the years. Soil is the primary aspect for proper vegetative growth, fruit yield and quality. Generally, the well-drained soil, free of excess salts with fair humus contents is suitable for most of fruit plants. Most of the fruit plants thrive well in deep, loose, well-aerated soils free of excess salts and having sufficient organic matter content. Although, the fruit crops can grow well in a wide range of soils with varied texture, but, the light sandy to medium loam soils are considered best, because of their high oxygen requirement for roots. Soil properties like soil reaction, soil fertility, drainage, free lime and salt concentration are some important factors that determine the success of fruit cultivation.

Depth of soil: Fruit plants thrive well in deep, loose well-aerated soils devoid of any hard pan layers of calcium carbonate in the root zone. Any soil less than 2-3 meter in depth is not suitable for fruit cultivation. Deep well drained soils with no impervious layers, allow good depth for root development producing trees of standard size, good yield and quality fruits for a long life. On the other hand, shallow soils, produces sub- standard tree with short life span and poor fruit yield and quality. The ideal soil should permit penetration and retention of water at depth of 2 meters.

Water drainage: The fruit plants may be grown on various types of soil provided the sub-soil drainage is good because water drainage is most important aspect for growth and development of fruit plants. Accumulation of water in root zone due to poor drainage results severe injury to roots and incidence of soil borne deceases such as root-rot. So, well drained medium loam to loam soils are most suitable soils for fruit crops provided the soils are fertile and productive. Heavy soils with poor percolation should be avoided.

Soil reaction: Some fruit plants such as citrus and peach are sensitive to soil factors such as high pH, presence of excessive salts in root zone, calcium carbonate concentration, free lime etc. The citrus plants thrives best in soils having slightly acidic or neutral reaction having pH range of 5.5 to 7.5 The most commercial citrus plants fails to grow well in alkaline soils having pH more than 8.5, hence; such soils should be avoided. Therefore, before planting the trees, it would be desirable to have the suitability of the proposed soil tested and appropriate reclamation measures be taken whatever necessary in advance.

Soil Testing for Orchards

The nutritional status and other conditions of the sub soil are also important for the growth of fruit trees. Analysis of soil for orchard plantation should be done to the depth of two meters as follows:

- Collect 500 g soil samples separately for each soil layer i.e. for top 15 cm, 15 to 30 cm, 30 to 60 cm, 60 to 90 cm, 90 to 120 cm, 120 to 150 cm, and 150 to 200 cm.
- If a hard pan or concrete layer is present, note its depth and thickness and collect its sample separately.
- Put different soil samples in separate clean cloth bags, avoiding contamination. Label each bag indicating the depth of the layer from which sample has been taken. Send the sample to the soil testing laboratory or handover the same of the university extension specialist in your district.

Selection of Suitable Fruits and Varieties

S. No	Name of fruit	Varieties	
1	Mango	Dusehri, Langra, Alphonso, Chausa, Bombay Green, Amrapalli, Mallika, Kesar, Fazli etc.	
2	Banana	Dwarf Cavendish, Robusta, Rasthali, Amritpant, Chini Champa, Virupakshi, Red Banana, Poovan, Nendran, Lacatan, Harichal, Gandevi Selection, Grand Naine etc.	
3	Citrus	Mandarin	Kinnow, Nagpur, Khasi, Daisy, PAU Kinnow-1
		Sweet orange	Mosumbi, Blood Red, Jaffa, Valencia, Hamlin, Pineapple etc.
		Grapefruit	Foster, Ruby, Marsh Seedless, Duncan, Thompson, Red Blush, Triumph, Sharanpur etc.
		Lemon	Eureka, Lisbon, Feminello , Monactiello, Bernia, Baramasi
		Lime	Kagzi lime
4	Guava	L-49, Allahabad Safeda, Shweta, Arka Amulya, Arka Kiran, Chittidar, Porugal, Safed Jam, Kohir Safeda, Punjab Pink etc.	
5	Grapes	Beauty seedless, Krishmish Charni, Sharad Seedless, Flame Seedless, Beauty Seedless, Anab-e-Shahi, Dilkush,Cheema Sahebi, Perlette, Thompson Seedless, Tas-e-Ganesh, Manik Chaman, Sonaka Arkavati, Himrod etc.	
6	Apple	Tydeman's Early, Michael, Molies Delicious, Schlomit, Starkrimson, Starking Delicious ,Red Delicious, Richared, Vance Delicious, Top Red, Lal Ambri, King Pippin, Cox;s Pippin, Red Gold etc.	
7	Sapota	Cricket ball, Oval, Vanivalasa, Calcutta, Round, Jonnavalasa-1, Jonnavalasa-2, Baramasi, Kali Patti, Bhuri Patti, Chhatri, Dhola Diwani, Pala, Kirthibarthi, Dwarapudi etc.	
8	Pomegranate	Alandi, Ganesh, Jyothi, Mridula, Jalore Seedless, Ruby, Bhagawa, G-137, Ganesh, Kandhari, Bedana etc.	
9	Papaya	Honey Dew, Sunrise Solo, Washington, Pusa Delicious, Pusa Majesty, Pusa Giant, Pusa dwarf, Pusa Nanha, CO-1, CO-2, CO-3, CO-4, CO-5, CO-6, CO-7, Red lady-786 etc.	
10	Pineapple	Queen, Kew, Giant Kew, Cayenne, Typhone, Abakka, Sugar loaf, Natal, Ripley and Alexandria Mauritius, Red Spanish, Singapore Spanish, Green Selangor, Castilla, P.R.1-67 etc.	
11	Ber	Umran, Gola, Seb, Sanour No. 2, Sanour No. 5, Nazuk, Noki, Sandhura Narnaul, Illaichi, ZG-2, Banarasi Karaka, Pewandi, Meharun, Wallaiti, Dandan etc.	
12	Litchi	Sahi, China, Purbi, Rose Scented, Kasba, Dehra Dun, Muzaffarpur, Saharanpur, Seedless Late, Calcuttia etc.	
13	Pear	Bartlett, Starking Delicious, Max-Red Bartlette, Dr Jules Guyot, Conference, Comice, Hardy, Winter Nellis, Clapp's Favourite, Flemish Beauty, Pathar Nakh, Punjab Beauty, Punjab Soft, Nijjisikki, Punjab Gold etc.	

Contd.

S. No	Name of fruit	Varieties
14	Peach	Alton, Early White Giant, Redhaven, Stark Red Gold, July Elberta, Kanto- 5, Shimizu Hakuto, Sunhaven, J H Hale, Sharbati, Khurmani, Partap, Shan-e-Punjab, Florda Prince, Florda Sun, Early Grande etc.
15	Plum	Kelsey,Red Beaut, Santa Rosa, Meriposa, Rose Scented, Satluj Purple, Kala Amritsari, Kataru Chak, Titron etc.
16	Apricot	New Castle, Early Shipley, Shakarpara, Kaisha, Nugget, Royal, Suffaida, Charmagz, Nari, Suffaida, Moorpark, Turkey, St. Ambroise, Chaubattia Alankar, Chaubattia Madhu, Chaubattia Kesri etc.
17	Cherry	Bigarreau, Schrecken, Black Elton, Black Heart, Cluster Black Bedford Prolific, Black Downton, Black Eagle , Black Tartarian, Early Rivers, Victoria Black, Waterloo, Merton Bigarreau, Van, Bigarreau Napoleon, Emperor Francis, Ving, Lambert Star, Kentish Bigarreau, White Bigarreau, Sue, Victor, Merton Heart etc.
18	Almond	Hybrid-15, Pethick's Wonder, California Paper Shell, Stout Paper Shell, J&K Selection- 55, Sloh etc.
19	Walnut	Drainovasky, Lake English, Wilson, Gobind, Placentia, Eureka, Chakrata Selection, Franquetfe etc.
20	Cashew nut	Vridhachalam-1 to 3,VRI 4,VRI - H-1, Amrutha , Anagha , Anakkayam-1 , Dhana, Dharasree , Kanaka, Madakkathara -1 ,Madakkathara-2, Priyanka , Sulabha, Ullal-1 to 4, Vengurla 1 to 7, NRCC-1, Chintamani-1, Goa-1, Bhubneshwar-1 etc.
21	Jackfruit	GKVK-1, Ceylon Jack, Swarna etc.
22	Fig	Pune, Bellary, Gangam Marseilles, Trojano, Dottato, Brown Turkey, Frage, Lepe etc.
23	Annona	Balanagar, Mammoth, Red Sitaphal, Barbados, British Guinea, Islander, Arka Sahan etc. Atemoya: Hybrid between A. squamosax A.cherimoya.
24	Jamun	Ra Jamun, Paras, PKV - 4,13,14,15, Seedless, Gokak, Konkan Bahadoli etc.
25	Bael	Kagzi, Etawah Kagzi, Deoria large, Mirzapuri etc.
26	Aonla	Banarasi, Chakiya, Hathijhool, NA-4 I(Kanchan), NA-5 (Krishna), NA-6, NA-7, NA-10 etc.
27	Loquat	Golden Yellow, Pale Yellow, California Advance, Tanaka etc.

Land Preparation

After selection of location and site for establishment of orchard, plough the soil deeply followed by harrowing to clear the perennial weeds or shrubs and breaking of clods. Deep ploughing also provide suitable environment to the newly planted fruit plants for healthy development of roots. Laser land levelling with gentle slope in one direction to facilitate irrigation as well as drainage of excess water during rains must be done before layout and planting of plants. In hilly areas, the selected land should be divided into contours depending upon the topography of the land and the levelling is done within the terraces. Terracing protects the land from erosion. Application of organic manures or green manuring should be done in case of sandy or poor soils.

Provision of Wind Breaks and Hedges

To resist high speed and hot winds which may cause severe damage to orchards are very essential to be established on the wind ward sides. The high speed winds may cause breaking of branches, dropping of flowers and fruits and even uprooting of trees if velocity of wind is too high. Moreover, the hot and dry winds may cause severe economic loss as such conditions may cause poor fruit setting due to drying of stigma, disrupting the activities of pollinators etc. During summer months water loss due to higher evaporation and transpirational loss result higher demand for irrigation causing extra economic burden on the orchardists. To avoid such problems, rows of tall trees or wind breaks can be planted close together around the orchard. The wind breaks are quite effective in reduction of wind velocity and diminishing the damage to the fruit trees particularly, in the regions of prevalence of strong winds. Normally, a windbreak provides maximum protection for a distance about five times of its height but beyond this is also effective against high velocity winds. For establishment of most effective wind break double row of tall trees can be planted in alternate fashion. The space between trees planted for wind break should be at least as much as space between the fruit trees. To reduce the water and nutrition competition between wind break trees and fruit trees, a trench of 3 feet deep may be prepared and pruning of exposed roots can be done after 2-3 years. Trees suitable for windbreak should be erect, tall and quick growing, dense, hardy and drought resistant and mechanically strong to offer maximum resistance to wind. The windbreak trees should not harbor the source of pests and diseases of fruit crops e.g. the wind break of rough lemon should not be erected around citrus orchard. The trees which are suitable for growing as wind breaks are eucalyptus, Arjun, shisham, neem, silver oak etc. To prepare the wind break more effective, the medium statured plants such as rough lemon, bougainvillea etc. can be planted in between the rows of tall trees.

B. PLANTING OF AN ORCHARD

Planting operation should be performed with utmost care as any mistake committed during planting cannot be rectified afterward. The each type of fruit plants should be planted following proper method, plant spacing and ideal time. In order to get higher success for establishment of new orchards the following points should be considered. The planting season of different fruit crops vary on the basis of their evergreen or deciduous nature.

Evergreen fruit plants: Evergreen fruit plants such as citrus, mango, guava, litchi, sapota, loquat etc. can be planted during spring (February-March) and rainy season (August-September) under North Indian conditions. The fruit plants such as ber, mango and litchi should preferably be planted post monsoon season i.e. in late September or October when the weather becomes mild and more favourable and there is enough moisture in the soil. Some evergreen fruit plants like kinnow and ber can be transplanted bare rooted during winter season (January-February). However, these plants should be defoliated and the roots must be covered with moist material. Most of the sub-tropical and tropical evergreen fruit plants are suitably planted during the rainy season when the atmospheric humidity is high and sufficient supply of soil moisture is available and the success of establishment of fruit plants is quite high. The planting of evergreen fruit plants can be done during spring season if sufficient irrigation water is available to fulfil initial water requirement of plant for growth and development.

Deciduous fruit plants: The deciduous fruit plants such as apple, apricot, pear, peach, plum, grapes, fig etc. can be planted during winter when they are still dormant. The planting of such fruit plants must be completed before the emergence of new growth otherwise trees may suffer adversely and will not withstand in new environment and sprouts will dry and plant may die in hot weather. Under north Indian conditions, the month of January is highly suitable for planting these fruit plants. However, grapes and pear can be planted up to middle of February as the sprouting in these plants start in late winters.

Layout Plan of Orchard

The layout should aim at providing optimum number of trees per unit area, adequate space for each tree for proper development and ensuring ease in orchard management. The following points should be kept in mind while preparing the plan.

1. Ideal plant spacing to accommodate maximum number of trees per unit area.
2. In case of multi fruit plant orchards, each type of fruit should be planted in separate block.
3. Roads and paths should occupy minimum space, the provision between wind break and first row of trees is beneficial.
4. In case of open irrigation or flood irrigation system, irrigation channels should be laid along the gradients of slope for efficient flow of irrigation water. Slight slope must be provided to drain out excess water in case of excessive rainfall.

5. The fruit crops, which require pollen from another variety for better set fruits must be planned and planted at proper position.
6. Stores and office building in the orchard should be constructed at the centre for proper supervision.
7. A good fence is essential to avoid entry of animals. Live fencing is economic and cheap to other kind of fences. The plants suitable for live fencing should be drought resistant, easy to propagate from seed, quick growing, have dense foliage, should stand severe pruning and should be thorny. Karonda, *Agave, bougainvillea, Prosopis juliflora, Thevetia etc.* can be planted closely for this purpose

Methods of Layout

Layout plan:- The marking the position of fruit plant in the selected and properly prepared field is referred as layout. The objective of layout is to provide adequate space to the plants, accommodation of more number of plants, easy intercultural operations and esthetic aspects. The layout plan of the orchard should be prepared carefully and horticulture expert may be consulted for this purpose. The orchard layout plan includes provision for orchard paths, roads, water channels and farm building in addition to proper position of each fruit plant. A sketch of the proposed orchard should be prepared before the actual planting operation. Fruit plants grown in north-south oriented rows have better light interception than in east-west rows. High density planting requires careful assessment regarding plant vigour, rootstock used, topography, equipment size and availability, labour access etc. The orchards requiring windbreak trees and pollinizers should be planned orderly.

Procedure for Layout

For layout of an orchard, first step is erection of a base line preferably on the straight side of field after keeping proper distance from the boundary of the selected field. The jute or nylon rope can be used to erect base line and position of plants on the first line of proposed plants is marked along this line by fixing wooden stakes at recommended space for particular fruit plant. Another base line at right angle to the first base line, is then marked along with the other edge of the field with the help of a carpenter square or a cross staff. The measuring tape can also be used to draw right angle. For example in case of citrus planting at 6x6 m, one end of the measuring tape can be fixed at three meters distance from the corner along the base line and the tape is then stretched on the perpendicular base line for a distance of four meters. The diagonal distance between these two points will be five meters. Then, the wooden stakes can be put in the ground at the recommended distance along the second line. Three men are required for layout operation; one will fix the peg in the field and other two will mark the rows and plant spacing with rope. After marking the first line and position of plants on each side of the field, the marking of inner plants can be done by marking the cross lines and fixing the pegs on each cross.

System of Planting

The system of layout can be grouped under two broad categories viz. vertical row planting pattern and alternate row planting system. In the vertical planting pattern, the trees planted in a row are exactly perpendicular to those trees planted in their adjacent rows e.g. square system and rectangular system of planting. In alternate planting pattern, the trees in the adjacent rows are not exactly vertical instead the trees in the even rows are midway between those in the odd rows e.g. hexagonal, triangular and quincunx system of planting.

Vertical Row Planting Pattern

- ***Square system***: It is the most commonly used method and easy to layout in the field. In this system, plant to plant and row to row distance is the same. The plants are at the right angle to each other and every unit of four plants form a square. This system facilitates the intercultural operations in two directions after orchard plantation. The plants like citrus, pear, ber, papaya, banana etc can be planted according to this system.

- ***Rectangular system:*** In this system, the trees are planted on each corner of a rectangle. As the distance between any two rows is more than the distance between two trees in a row, there is no equal distribution of space per tree. More number of trees can be planted than square system. The inter culture operations are possible in both directions. High density planting of citrus, guava, peach, pear etc. can be done with this system.

Alternate Row Planting Pattern

- ***Triangular system:*** According to this system, the trees are planted as like that of square system but the even numbered rows are planted midway between those of odd rows instead of perpendicular to them. Triangular system is based on the principle of isolateral triangle. The distance between any two adjacent trees in a row is equal to the perpendicular distance between any two adjacent rows. However, the vertical distance, between immediate two trees in the adjacent rows, is equal to the product of 1.118 x distance between two trees in a row. As compared to square system, each tree occupies more area and hence; accommodates approximately 9 per cent lesser trees per unit area than the square system. The mango cv. Amrapali can be planted as per this system of planting and one can accommodate 1600 plants per hectare.

- ***Hexagonal System:*** *According to* this system, the trees are planted in each comer of an equilateral triangle. In this system, six trees form a hexagon with the seventh tree in the centre known as 'filler' plant. This system is also called as 'septule' as a seventh tree is accommodated in the centre of a hexagon. Although, this system provides equal plant spacing but, it is more difficult to layout. The perpendicular distance between any two adjacent rows is equal to the product of 0.866 x the distance between any two trees. As the perpendicular distance between any two rows is less, this system accommodates 15% more trees than the square system. The limitations of this system are that it is difficult to layout and the cultivation is not so easily done as in the square system. The mango, sapota, litchi orchards can be planted according to this system.
- ***Diagonal or quincunx system:*** This is just like the square system, but one additional plant is planted in the centre of a square. This will accommodate approximately double the number of plants than square system. The central plant or filler plant should be precocious, short lived and fast growing. Fruit trees like papaya, kinnow, phalsa, guava, peach, plum etc. can be planted as fillers in the permanent trees provides an additional income to the grower in the early life of the orchard. This system can be followed when the distance between the permanent trees is more than 25 feet and juvenile phase of main trees is longer. The filler trees should be removed after the main trees start commercial bearing.
- ***Contour system:*** It is generally followed on the sloppy areas or hills where the plants are planted along the contour across the slope. It is suitable for sites of undulating topography and erosion prone areas. The main objective of this system is to minimize land erosion and to conserve soil moisture to make the sloppy area fit for growing fruits crops. The contour line is so designed and graded in such a way that the flow of water in the irrigation channel becomes slow and thus finds time to penetrate into the soil without causing erosion. On the other hand, the terrace system refers to planting in flat strip of land formed across a sloping side of a hill. Terraced fields upswing in steps one above the other and help to bring more area under cultivation in addition to prevent soil erosion. The width of the contour terrace depends on to the nature of the slope. If the slope is sharp, the width of terrace will be narrow and vice-versa. Contour system may not permit uniform distance of planting fruit plants. The plantation crops such as tea can be planted in contours either in single hedge system or in double hedge system. Double hedge contour planting system accommodates nearly 22 % more population than single hedge system. Bench terraces are used where the slope is greater than 10 per cent.

- ***Hedgerow system***: In this system, the distance between the trees in a row is half to one third the distance between rows. This system of planting is followed in European countries such as Italy where dwarf apple trees are plante on double balanced rows due to higher intensity of light.
- ***Adjacent planting***: In this system, two plants are placed closely keeping a double space between the rows and the next set of plants. This system is followed in the arid zone where moisture supply is limited as well as in the less fertile soils where both moisture and nutrients are not easily available to the plants. This system provides large areas between rows for intercropping.

Pits Preparation and Filling

The selected site should be levelled preferably with laser leveller and should be properly laid out. A pit of one meter depth and one meter diameter should be dug manually or with the help of mechanical pit digger. Mark a ring at 50 cm radius from the position of each plant and fix two pegs on each side of the pit along the plant row, so that the exact centre may be located while planting after filling of pits. For filling of the pit, the mixture of top soil and farm yard manure in equal proportion should be used. Fill the pit 2-3 inches above the ground level, so that after watering the loose soil may settle down to the actual ground level. Irrigation should be done before planting of fruit plant for this purpose. Level the settled down pit with soil mixture if needed. Add 15 ml Chlorpyriphos 20 EC mixed in about 2 kg soil against white ants to each pit. The pits should be allowed to wither for few weeks before planting the fruit plants.

Method of Planting

In the filled pit, locate the proper position of plant as per layout plan. For this purpose, planting board can be used. The planting board is usually of 1.15m long, 15 cm wide and 2.5 cm thick with a notch on the one side of the board exactly in the centre and a notch on the either side of the board in a straight line. The planting board is placed in such a way that the pegs fixed on the sides of the pit fits into the notches of board on each end of board. Mark the plant position with the help of central notch and a small pit of about 30 cm deep and 30 cm wide or of the desired dimensions at the position of the plant should be dug.

Transplanting

The fruit plants should be planted at the position of original pegs. After placing the plant with or without earth ball, firmly press the soil around it. Care should be taken that the plant should not be planted too deep or too high. The plants should be planted in such a way that the bud union remains at least six inches above the ground level. Irrigate the soil immediately after planting. A small basin may be made around the tree for the purpose of frequent and light irrigations till the plant get established in the field.

Spacing in Fruit Plants

The distance between any two trees or plants is referred as the planting distance. The principles for deciding the plant spacing is that the fringes of trees may touch each other but the branches should not intermingle. The tree roots may spread over a larger area than tree canopy and there should be proper opportunity for the roots to feed without competition. There are many factors which decide the plant spacing primarily type of fruit tree, tree vigour, cultivar, soil type, method of irrigation, rootstock used, rainfall, pruning and training etc. In general, the recommended plant spacing for fruit plants for normal planting is given in following table.

General Planting

In hexagonal planting system 15% more number of plants accommodated, while,the triangular system accommodates 11% lesser number of plants as compared to square system. The calculation of the number of trees per hectare when planted under square or rectangular system is very easy, and is obtained by the following formula:

Number of plants/acre= Area in square meters/row to row x plant to plant distance in meters.

Table 1: General planting distance in fruit plants.

Name of fruit tree	Planting distance (m)	Number of plants/ha (square system)
1. Mango and Sapota	9.0x9.0	122
Citrus and Pomegranate	6.0x6.0	275
Grape i) Head system	2.0x1.5	3300
ii) Kniffin system	3.0x3.0	1100
iii) Bower system	3.0x6.0	1100
Guava, Peach and Loquat	6.5x6.5	225
Soft pear,Plum and Fig	6.0x6.0	275
Litchi	7.5x7.5	180
Ber, Pear and Aonla	7.5x7.5	180
Date palm and Almond	6.0x6.0	275
Jamun	10.0x10.0	100
Phalsa and Papaya	1.5x1.5	4400
Banana i) Tall varieties	2.7x3.0	1210
ii) Dwarf varieties	1.8x1.8	3050
Pineapple	0.30x0.60x0.90	43500
Custard apple	5.0x5.0	390
Jack fruit	10.0x10.0	100
Acid lime	5.0x5.0	400

High Density Planting

In the scenario of increasing population and urbanization coupled with shrinking natural resources, the strategies of intensive cropping and integrated farming system gained an immense importance. To meet the mounting demand of nutritional security, the area under horticultural crops required to be increased and system of planting must be improved. Presently, the productivity of fruit crops is much below their productive potential. The traditional system of cultivation with large tree canopy and wider spacing has resulted underutilization of ground space, inputs and poor harnessing of solar energy resulting poor output. Crop intensification through high density planting is one such approach to achieve higher productivity per unit of resources used. High density plantation (HDP) is the accommodation of higher number of plants per unit area to increase productivity and cost-benefit ratio with sustained utilization of natural resources. This includes adoption of appropriate plant spacing, canopy management and cultural practices with appropriate methods. High density planting is being used on commercial scales in apple, pear, peaches, plum, guava and banana. In this system, four planting densities are recognized for apples viz., low HDP (< 250 trees/ha), moderate HDP (250-500 tree/ha), high HDP (500 to 1250 trees/ha) and ultra-high HDP (>1250 trees/ha). Recently, super high density planting system has been also established in apple orchards with a plant population of 20,000 trees per ha. In some orchards, still closer, planting of apple trees is followed (say 70,000 trees/ha) which is often referred as 'meadow orchards'.

Early cropping and higher yields per unit area, reduction in labour cost, efficient utilization of natural resources and inputs are main advantages of HDP and high initial investment, more biotic stress and sometime reduction in fruit quality are main disadvantages of high density planting of fruit trees.

The trees of HDP should have maximum number of fruiting branches and minimum number of structural branches. The trees under HDP are generally trained with a central leader surrounded by nearly horizontal fruiting branches. These branches should be so arranged and pruned in such a way that each branch casts a minimum amount of shade on other branches. The height should be one and half its diameter at the base. A key to successful HDP depends upon the control of tree size. High density planting can be achieved by using dwarf rootstocks e.g. in apple use of dwarfing rootstocks and intermediate stocks like MM 106, MM 109, and MM 111 result dwarf size trees suitable for HDP. In pears, Quince A, Adam and Quince-C are commonly used as dwarfing rootstocks. Training and pruning methods also induce dwarfness e.g. apple trees trained with spindle bush, dwarf pyramid, cordon systems are found to limit the growth of trees. Growth retardants like paclobutrazol, daminozide, ethephon, chlormaquat etc. are extensively used to reduce vegetative growth.

The choice of planting system governs the success of high density planting. Generally, rectangular planting system with one, two and three rows planting is followed. In one row planting, the distance between the plants is close, whereas the distance between the rows is wide e.g. 6x3m for HDP in citrus. In two row planting, a wider space is provided after every two rows e.g. 4+2x2m for HDP in peach, whereas, in three row planting, a wider spacing is given after every three rows 4+2x2x2m e.g. in HDP of apple. In meadow orchard system, a bed of 10 to 15 rows is closely planted *e.g.* 40x60 cm and separated by alleys of 2.0 m width between beds. This system is also known as bed system of planting.

High density planting in plum at 6 X 1.5 m spacing
(See colour version on page 419)

High density planting in peach at 6 X 1.5 m spacing
(See colour version on page 419)

Table 2: Plant spacing and accommodation of number of plants per unit area under high density planting system.

Fruit	Variety	Spacing (m)	System of planting	Number of plants/ha
Mango	Amrapalli	2.5 x 2.5	Triangular	1600
Citrus	Kinnow	3.0 x 6.0	Rectangular	550
		3.0 x 3.0	Rectangular	1100
Grapes	-	4.0 x 1.5	Rectangular	660
Banana	Cavendish group	1.2 x 1.2	Square	6900
		1.0 x 1.0 x 2.0	Paired row	
Pineapple	-	25 x 60 x 75cm	Double row	63000
Apple	Spur type on rootstock MM111, MM109	4 x 4	Square	625
	Non-spur type on rootstock MM106, M109	5 x 5	Square	400
	Spur type on rootstock MM106, M7	3 x 3	Square	1100
	Non Spur type on M9	2 x 2	Square	2500
Peach	Shan-e-Punjab	6.0 x 1.5	Rectangular	440
Plum	Sutlej Purple	6.0 x 1.5	Rectangular	440
Pear	Patharnakh	8.0 x 4.0	Rectangular	120
Guava	L-49	3 x 6	Square	550
		6 x 5	Rectangular	330
Papaya	Pusa Nanha	1.25 x 1.25	Square	6400
Acid lime	Kagzi	4.5 x 4.5	Square	480

Selection and Procurement of Fruit Plants

Plant Age

The age of the fruit plant is very important consideration while purchasing the plant. Usually, it is estimated from the age of scion but it does not assume that these should have been budded on old root stock. The ideal fruit plant should be one to two year old. The older nursery plants should not be selected as they do not establish easily, particularly the evergreen fruit plants where the shock of transplanting is greater as compared to the young plants. The deciduous plants such as pear can be planted rather older provided better care and management practices are provided after planting in the field. The nursery plants should preferably be planted as soon as possible after procurement or they may be planted in bags or pots if field is not ready for planting, however, prolonged stay may reduce vigour of plants. Under such conditions, the roots become pot bound, especially in the older plants and thus they have a poor establishment in the field.

Plant Health

Healthy fruit plants, free from disease and insect-pest should be procured from reliable nursery. Plants should be true-to-type and budded or grafted on suitable rootstocks. They must have been propagated from progeny trees of proven merit in regard to yield and quality. In case of evergreen fruit plants, the plants should be lifted with well sized earthen balls. The deciduous fruit plant should be uprooted with plenty of roosts. The plants should be of medium height and have smooth bud union. Tying material should be removed from the bud/ graft union before transplanting.

Source of Availability

The orcharding is a long term investment, any mistake committed regarding source of nursery plant may not be rectified. One can know the fate after the end of juvenile phase of fruit plants. Wrong selection of nursery or fruit plants may cause heavy economic loss. So, always select reliable source for procurement of planting material. The accredited nurseries of public or private sectors should be preferred for purchase of nursery plants.

Transportation of Plants

The young plants particularly evergreen fruit plants are delicate hence; need careful handling while, loading, transportation and unloading. The earth ball of fruit plants should not be broken during these operations. A layer of cushioning material such as paddy straw, dried grass or sand should be spread in the vehicle before loading the evergreen fruit plants. Place the plants carefully in the vehicle and do not pile up the plants haphazardly. For distant transportation sprinkle the water frequently to keep them turgid. Bare rooted plants may be packed in small bundles. The root portion of these bundles can be wrapped in wet gunny bags to avoid dehydration of feeder roots during transit.

Care of Plants Before and After Planting

Before Planting

The planting should be done in the centre of the filled pit with the help of a planting board. During planting, care should be taken that the bud union should remain at least six inches above the ground level. Firmly press the soil around the plant with feet and then apply water. After planting, staking should be done to keep the plant erect. Remove dead, diseased and off shoot if any after planting. Apply half litre chlorpyriphos per acre followed by light irrigation against white ants. Regularly remove the sprouts and off shoots. Keep regular check for insect, pest and disease incidence. Apply recommended doses of fertilizers and follow the spray schedule for insects, pests and diseases.

After Planting

The newly planted fruit plants may tilt in newly filled pit after watering hence; need staking to keep them straight. Remove dead, diseased and unwanted shoots. As the young plants are delicate, so, adequate protection against extreme of weather vagaries is of utmost importance. Protect the fruit plants by erecting thatches, frequent and light irrigations, pruning and training, growing of fast growing plants such as *dhaincha* around the plants etc. Some young plants are subject to considerable injury from sunburn particularly if they have been trained to single stem with no branches for 45 cm or more from the ground. Such trunk can be protected by wrapping them with paper or other material or by painting them with white wash. Prevent the attack of white ants by applying Chlorpyriphos @ half litre per acre followed by light irrigation. Keep examining the young plants and remove stock sprouts, suckers regularly

LITERATURE CONSULTED

Bal,J.S.2014 .Fruit Growing.3rd edition , Kalyani Publishers, New Delhi.

Barooah,S.1998. Modern Fruit Culture . Kalyani Publishers, New Delhi.

Chadha,K.L. and Pareek,O.P. 1996. (Eds.) Advances in Horticulture. Vols. III and V Malhotra Publ. House. Chadha,K.L.2014.Hand Book of Horticulture. I.C.A.R., New Delhi.

Edmond, J.B., Sen,T.L., Andrews,F.S. and Halfacre,R.G.1990.Fundamentals of Horticulture.Tata *McGraw Hill* Publishing *Co., New Delhi*.

Kumar, N. 2014. Introduction to Horticulture. Oxford & IBH Publishing co. Pvt. Ltd.

Naik,B,M.and Thippesh, D.2015. Fundamentals of Horticulture and Production Technology of Fruit Crops by UAHS, Shimoga.

Singh G.2008.High Density and Meadow Orcharding of Guava. Central Institute for Subtropical Horticulture-Rehmankhera, Lucknow.

Singh S., Krishnamurthi,S. and Katyal,S.L.,1963. Fruit Culture In India . I.C.A.R., New Delhi.
Singh, J.2012.Basic Horticulture. Kalyani Publishers, New Delhi.
Singh, S.P.1995. Commercial Fruits. Kalyani Publishers, New Delhi.

3

Suitability of Climate and Soil

Jaswinder Singh Brar

INTRODUCTION

The planning of new orchard is very important aspect of fruit glowing which needs special attention and care to make it more profitable venture. The climate and soil of particular area are the major chief natural factors in the orchard establishment. The proper combination of this governs the success of fruit growing

A. SUITABILTY OF CLIMATE FOR FRUIT PLANTS

Climate refers to an average condition of the atmosphere over a long period, whereas, the term weather refers to the current and temporary atmospheric conditions of an area. Climate is the principal factor for optimum vegetative and reproductive growth of plants. The success or failure of crop in a particular locality depends upon the prevailing climatic conditions. In our environment it is fundamental force which includes various components of climate like temperature, humidity, wind, light, rainfall, hail and frost. These factors should be taken care of for successful fruit production. All cultivated crops including fruit crops have certain natural threshold limits of the climatic parameters for their normal growth. The important climatic components which affect fruit crop production are given below:

Temperature

Temperature is one of the most significant parameter of climate and plays an important role in determining the success or failure of a fruit crop in a particular locality. Most of the physiological activities of plants such as respiration, photosynthesis, transpiration, water and nutrients uptake, reproduction, plant growth and development etc are perform properly at optimum range of temperature. Above or below the normal range, the temperature significantly altered the physiological processes of plants. In case of fruit crops each type of fruit needs a certain range of temperature. The temperate fruit crops such as apple, apricot, almond, cherry, peach, pear, plum, quince, walnut etc. can be grown successfully in the areas of low temperature. In the region of warm winters these fruit trees fail to complete their dormant period or chilling requirement due to insufficient chilling temperature. The tropical fruits like arecanut, avocado. banana, cashew, coconut, jackfruit mango, papaya, pineapple and sapota require a long-growing season and high temperature (65-75° F) without any frost or snowfall. While the fruit crops such as guava, grape, citrus, date palm, phalsa, pomegranate, litchi, loquat require subtropical type climate. The fruit plants can sustain its growth at optimum temperature a very narrow range between minimum and maximum temperature. The temperate fruits like apple, apricot, almond, cherry, peach pear, plum, quince, walnut etc. can grow only on sites having sufficiently cool climate (45-55° F) to induce the trees to shed leaves during winter and after a specific period of rest come into flowering and leafing again. The plant performs well in optimum temperature range. The activities of the plant are affected by very high or very low temperature. The minimum temperature range for plant is 4.5° to 6.5° C (40° - 44° F), optimum range varies between 24° to 27° C (75° - 81° F) and maximum is 29.5° to 45.4° C (85° - 114° F). However, different fruit plants may differ in their tolerance or adaptability to extremes of temperature.

Relative Humidity

The atmospheric humidity plays and imperative role in deciding the amount of moisture required for proper growth and development of fruit plants. Majority of water absorbed by the plants lost through the process of transpiration and the transpiration depends on the relative humidity, temperature, wind, light etc. Hence; it is essential to sustain the health of plant by maintaining the balance between uptake and loss of water. Under north-Indian conditions, high atmospheric humidity during monsoon season is usually associated with high rainfall. In the areas of hot and dry climate, enormous amount of water is lost through transpiration, which results rapid evapo-transpirational loss of water. The plants show symptoms of wilting under such conditions, and need immediate irrigation. However, if the atmosphere is humid, the transpirational loss of water will be very low. High humidity coupled with high temperature promotes vegetative growth and certain fruit crops like banana and pineapple perform very well under such conditions, but, this type of climate may increases pests and disease incidence. High relative humidity at the stage of fruit maturity and ripening can reduce fruit quality and delay harvesting of fruits. High temperature coupled with low relative humidity results fruit cracking particularly in lemon, pomegranate and litchi.

Light

An electro-magnetic radiations coming from the sun to the earth, in a form of kinetic energy is known a light. It is one of the most important aspects affecting plant life. It is an integral part of the photosynthetic reaction as it provides the energy for the carbon dioxide (CO_2) and water (H_2O) combination in the green cells having chlorophyll for the synthesis of carbohydrates with release of oxygen. Intensity and duration of light are important for plant growth and development. Generally, horticultural crops need a lot of light and must be grown in sunny climate. The fruit and plantation crops like young mangosteen, coffee, cocoa and tea need shade during part of their development, while, carambola can be grown under permanent shade. The duration of light between dawn and dusk referred as photoperiod or day length and this exerts substantial effect on reproductive behaviour of plants. Low light intensity decrease the photosynthesis, hence, reduced carbohydrates supply which results poor plant growth and ultimately fruit yield and quality. High light intensity may results plant wilting and adversely affects the physiological activities. Based on the photoperiod, categorization of fruit crops is very limited. Most of fruit plants are day neutral plants while, strawberry, pineapple, can be categorized as short day plants. The sunlight affects the fruit quality and the fruits exposed to sunlight are better compared to fruits bearing in shade. Higher carbohydrate synthesis in the leaves receiving higher sunlight results better fruit quality. However, the fruits exposed to intense solar radiations are prone to 'sunburn; and sunscald'. The temperate fruit plants need proper training and pruning so that solar radiation may penetrate inside the tree canopies for better flower bud differentiation and initiation as compared to tropical and subtropical fruits.

Heat

A non-mechanical energy transfer with respect to difference in temperature between a system and its surrounding environment is known as heat and is measured as temperature by using thermometers. The growth and development of the plants depends primarily on temperature. The average temperature of a place gives an idea about heat units available and the availability of heat units decide the crop for a particular place. The quality and maturity of fruit crops such as 'dates' is governed by the availability of heat units in the area of its cultivation. Temperate fruit crops such as apple, almond, apricot, cherry, pear, peach, plum etc. become dormant in low temperature region and need chilling of various lengths to break dormancy ,while, the chilling are harmful for tropical and even subtropical fruit plants. Under arid conditions extreme heat cause wilting, sunscald, sunburn, cracking of fruits, necrotic spot and even death of plants.

Rainfall

Rainfall is also a very important climatic element for successful fruit cultivation. Well distributed and consistent rainfall is always desirable for an ideal orchard site hence; it is of utmost importance that the amount, distribution and intensity of rainfall in a particular region must be studied before planting fruit plants in particular site. Generally, heavy rainfall is not favourable in the plains, as it leads to water logging which adversely affect the plant growth and promote pest and disease incidence. Rainfall affects plant growth and development in many ways. Rain at the time of flowering is not suitable as it washes away the stigmatic fluid and pollen from the stigma and disrupt the activities of pollinators. On the other hand, in low rainfall areas the cultivation of most of fruits crops is difficult as these require irrigation at critical stages of crop development. Rains at the time of harvesting cause softening of fruits in banana and date palm and induce infection of fruit-fly in guava, pear, peaches etc. The grape cultivation under Punjab conditions is greatly affected due to coincidence of rainfall with harvesting season.

Air or Wind

The velocity as well as season of wind is a decisive force for the success or failure of fruit crop in particular area. The wind causes damage to the fruit trees in several ways. High winds can break the branches and even uproot or tilt the fruit plants. Flower and fruit drop cause severe economic loss to the farmers or contractors. The plants exposed to wind cause a greater evapo-transpirational loss hence; necessitate more irrigation. Hot winds at the time of flowering may cause failure of pollination due to drying of stigma and also reduced the activity of pollinators. High velocity and hot winds cause more damage to young as well as bearing fruit plants. The extent of damage to fruit plants depends on the type of fruit plants. The shallow rooted fruit plants such as guava, litchi, papaya etc are more prone to be uprooting or tilting. Similarly, the plants propagated through cuttings and layering are also more likely to be damaged with high velocity winds. The high velocity winds during fruit bearing phase particularly in ber, mango, citrus etc. cause heavy economic loss. For successful cultivation of fruit crops, establishment of wind breaks by planting tall trees eucalyptus, arjun, seedling mango, mulberry, jamun, shisham on wind side of orchard reduce the velocity of wind. So,this will protect the fruit plants to be damaged by this climatic vagaries. For preparation of more effective wind break, medium statured plants such as rough lemon, bougainvillea etc. can be planted in between the rows of tall trees.

Hail Storm

The hail storm causes severe damage to the fruit crops particularly at the time of fruit maturity as heavy fruit drop results poor economic out put. During selection of site for establishment of orchard, the occurrence of hailstorm in particular region must be considered. It causes damage in fruit crops many ways. The ice pallets shed the flowers and developing fruits in many crops. The developed or mature fruits are badly hit by the ice pallets resulting injuries or bruises on the fruits consequently attack of saprophytic fungal growth and fruit rotting. Hail storm also break the limbs of fruit plants and complete destruction of young plants. Netting of individual plant is may protect the trees from hailstorm damage, however, it is very tedious and costly affair.

FRUIT ZONES OF INDIA

1. Tropical Fruit Zone

The tropical region comprises the areas having hot and humid climate in summer and mild temperature in winter. The fruit crops of this zone are generally evergreen and unable to endure cool temperature but can tolerate warm temperature. Plenty of sunshine and warm- humid climates with mild winter favours the production of tropical fruit crops. These types of fruits cannot withstand frost. Areas under this zone include Maharashtra, Karnataka, Andhra Pradesh, Telangana, Tamil Nadu, Kerala, parts of Madhya Pradesh and West Bengal. Most commercial fruits cultivated in this zone are mango, banana, sapota, pineapple, papaya, pomegranate, grapes,avocado, cashew nut etc.

2. Sub- Tropical Fruit Zone

The sub tropical zone comprises intermediate characters of tropical and temperate region. The temperate occasionally goes below freezing point during winters and the summers are hot and dry. The fruit crops in this zone may be either deciduous or ever green & are usually able to withstand a low temperature but not the frost. However, some fruit crops require chilling for flower bud differentiation. The major fruit crops grown n this zone are citrus, grapes, ber, guava, phalsa, fig, pomegranate, etc. This fruit zone covers the plains of Punjab, Uttar Pradesh, Parts of Bihar, Madhya Pradesh, West Bengal, Maharastra, Rajasthan, Karnataka, Andhra Pradesh, Tamil nadu, Kerala, Orissa. etc. The tropical fruits like mango, banana, papaya etc. can also be grown in this zone, whereas, the temperate fruits with low chilling requiring cultivars such as peach, pear, plum and almond can also be grown in this zone.

3. Temperate Fruit Zone

The temperate fruit grows successfully in the region of low temperature even below freezing point during winters these plants can survive. Temperate plants become dormant during winters and for breaking dormancy or rest period a definite period of chilling is required. The chilling temperature for a specific period in different temperate fruit crops helps the plants for emergence of new vegetative and floral buds. Temperate zone is sub- divided into two regions; higher and lower elevations. Higher elevation area exist between 1500-2500 m above mean sea level and in this region, the fruits like apple, pear, walnut, almond and cherry can be successfully grown. Lower elevation region exist between 1200-1500 m above mean sea level and the fruit crops like peach, plum, apricot and strawberry can be grown successfully. The north-western hilly states of Jammu and Kashmir, Himachal Pradesh and Uttarakhand are well known for cultivation of temperate fruit crops.

4. Arid Zone

The arid zone has high temperature, low relative humidity, very low and erratic rainfall, poor textured soil with low organic matter content. In India, the area of 62% area of Rajasthan, 20 % area of Gujarat parts of the Punjab, Haryana, Karnataka & Madhya Pradesh and Maharashtra are arid. The fruit crops such as ber, date palm, pomegranate, custard apple, lasoda, tamarind etc. can be cultivated in this zone.

5. Semi-Arid Zone

This region exhibits low and erratic rainfall, low humidity and high temperature. In addition to fruits of arid region mango, sapota, guava, jackfruit, avocado etc. can also be grown provided irrigation facilities are available at critical stages.

6. North- Eastern Sub-Tropical Zone

In this zone, the temperature is high and slightly humid during summer. The winters are less cold. The chief regions of north eastern sub-tropical zone are eastern Uttar Pradesh, Bihar, parts of Madhya Pradesh, West Bengal, Jharkhand, parts of Chhattisgarh, Assam, Meghalaya, Nagaland and Manipur. Most of the tropical and sub-tropical fruits can be grown in this region. The major fruits like citrus, mango, guava, fig and litchi are grown successfully. The tropical fruits like pineapple and avocado can also be grown in this zone. Some varieties of pear, peach, plum and apple of temperate zone can also be grown on commercial scale in this zone.

7. North- Western Sub-Tropical Zone

In this zone, the temperature occasionally goes below the freezing point, but not falls below 25°F. The sub tropical fruits grow mostly in the plains where the climate is hot and comparatively dry and the winter is dry and less cold. The types of fruits grown in this zone are citrus, mango, guava, litchi, ber, grape, pear, peach, plum, phalsa, loquat, aonla, datepalm and pomegranate. The chief regions of this zone are parts of Jammu and Kashmir, Himachal Pradesh, hills of Uttar Pradesh, South of Punjab and Haryana.

8. Central Tropical Fruit Zone

During summers, the temperature of this zone is hot and and during rainy season it is humid. The winter is mild and less humid. The fruits like mango, grape, guava, citrus, pomegranate, ber, aonla, papaya, sapota and banana can be grown successfully. This zone includes the southern parts of Madhya Praadesh, Maharahstra, Chhattisgarh, Odisha, parts of Karnataka and Andhra Pradesh, West Bengal, Gujarat etc.

9. South Tropical Fruit Zone

In this zone the climate is hot and humid during summer and mild in winter. The fruits like mango, banana, cashewnut, guava, citrus, grape, sapota, coconut, pineapple and papaya are grown commercially in this region. The major area under this zone includes southern parts of Karnataka, Andhra Pradesh and Tamil Nadu.

10. Coastal Tropical Fruit Zone

The climate in this zone is hot with very high humidity during summer and mild in winter. Mango, banana, cashew nut, guava, pineapple and papaya are major fruit crops of this region. This zone comprises Konkan, Goa, Kerala, Western Ghats, Eastern Ghats in Tamil Nadu, Andhra Pradesh, Daman-Diu and Odisha.

11. Humid Zone

This region is characterized by low temperature and high humidity. The crops are litchi, strawberry, avocado, mangosteen, passion fruit etc.

EFFECT OF ADVERSE CLIMATE CONDITIONS ON FRUIT PLANTS

Effect of High Temperature

Injuries caused

The hot temperature of May-June is unbearable for fruit plants, especially for newly planted fruit plants. Due to high temperature plants fails to coup up the evapo-transipation needs, suffers from wilting, if persists for longer period, drying may also occur. In grown up plants, due to dry weather and high temperature the loss of moisture from tree trunks develops cracks and sometime peels off completely exposing inner sensitive portion. Tolerance to hot weather varies with plant species. Fruit plants like mango, litchi, citrus and papaya are comparatively more prone to damage due to high temperature than other fruit plants. The high temperature during the growing and rest period causes injuries like sun burn, sun scald and fruit cracking.

- **Sun scald:** The hot intense solar radiations fall on the exposed parts of tree facing south-west side, resulting into sun scald injuries. The hot sun rays also scorches the developing fruits, especially those of citrus. The affected portion of fruits first turns yellow, then brown and scorching and later even turns black. Such fruits cease to grow from the affected side, where as the growth on the other side continuous as usual. The fruits thus become malformed and misshapen. The peel of such fruits clings to the segments, which become woody and dry. Such defective cull and deformed fruits do not fetch good price in the market, as their quality is very poor.

- **Sunburn**: The excessive heat often results in the burning of leaves and fruits during hot and dry months. The branches, leaves, fruits and stem portions which are exposed to the sun become much hotter than the atmosphere and are subjected to sunburn injuries during daytime. The bark of the affected trunks and branches may crack or peel off in long strips in extreme cases. Such damage to the bark may result in .the eventual death of trees or they may get crippled. The young and high headed plants are more prone to sunburn injuries.

- **Fruit cracking:** This is a major problem due to disturbed water relations in plants particularly during hot and dry season. The certain fruits such as Baramasi lemon, litchi, ber, pomegranate exhibits cracking in fruits during summer months. High temperature may causes dryness in peel of fruits and with irrigation or rain water, fruits expands, cracking the less elastic peel due to hot and dry weather. Frequent and light irrigation in such fruit plants may be helpful to avoid fruit cracking.

Measures Against High Temperature

The fruit plants, especially young ones require more attention and adequate protection against the hot and scorching weather during summer months. The tolerance of the fruit plants to withstand high temperature varies greatly with species. The fruit plants like those of mango, litchi, lemon, etc. are liable to damage from hot weather as compared with other fruit plants. The following measures should be adopted to save the fruit plants from sun injuries and adverse effects of hot weather.

White washing: This is most important method for saving the trunks of fruit plants from sun injury. The lower bare trunk portion of fruit plants should be white washed during the months of April. White washing may have to be repeated in June-July if it gets faded or is washed out by rains. In the white wash material, little amount of copper sulphate should be added as it helps in warding off the fungal disease. The effective white-wash mixture can be prepared by dissolving 25 kg slaked lime 500 g Copper sulphate, 500 g Gum suresh in 100 litres of water.

Providing shelters: Young fruit plants can be conveniently protected by erecting thaches or *kullies* made of *Sarkanda*, Rice straw or date-palm leaves. Quick growing temporary shade plants like *jantar*, arhar planted around the basins provide shade and thus minimise sun-burn injuries to the young plants.

Wrapping of bare trunks: Wrapping of lower portion of trunks of fruit trees with old gunny bags or with other farm waste material like *parali* (Rice straw) or *Khori* (sugarcane thrash) is also effective against heat injury to tree trunks of fruit plants especially in case of young plants.

Low headed trees: The plants trained to low headed framework escape the ill effects of hot sun better than high headed trees. It is therefore, desirable that the fruit trees should be trained to develop low headed trees.

Wind breaks: By growing wind brakes on the western side of the orchards, damage by hot winds can be avoided. The windbreak should be planted before planting the fruit plants. Eucalyptus, mango seedling, jamun seedling, arjun, shisham, mulberry etc can be grown as a suitable wind break.

Frequent irrigations: The irrigation at a regular interval is quite effective against the hot weather and sun injury. The frequency of irrigation should be increased during the months of May-June, when hot and dry winds generally blow particularly in arid and hot region.

Effect of Low Temperature

Low temperature or frost is common feature in northern plains and hilly areas of India. The bright sunshine during day, quick decline in temperature during calm evening and below 1.1°C temperature during the night is indicator of frost occurrence. The frost or low temperature adversely affects fruit plants particularly young plants. However, the extent of damage depends upon tree age, soil moisture content, site of orchard, growth stage, time, severity and duration of frost occurrence. Older trees having mature wood are less prone to be damaged by than young plants having immature growth. Moist soil or irrigated soil raises the soil temperature and provides protection against mild frosts therefore; trees of such growth and age growing under dry soils conditions are severely damaged by frost than to those growing in wet land. Low temperature or frost not only damage the vegetative growth but also affect the fruit yield and quality in different fruit plants. Hence; the protection of fruit plants particularly young plants and frost sensitive plants require protection during winter months in this region.

Injuries caused

Under Northern plains, December- January are the coldest months and frost occurrence is most common. Sometimes frost may also occur in the month of February. Both early and late frosts are more dangerous and impose heavy damage to fruit plants. The evergreen fruit plants such as papaya, banana, mango, litchi, aonla, guava, etc. are highly susceptible to cold injury. The newly planted evergreen fruit plants are more prone to frost injury. The vegetative growth resulting from late application of irrigation and nitrogenous fertilizers to fruit trees is most likely to be damaged by low temperature or frost. The general injuries on fruit crops due to low temperature or frost are given below:

Black heart: The wood of tree trunks or branches is darkened while the cambium and the bark remain alive. However, the growth of the trees continues and recovers from the injury. The nursery plants are more susceptible to black heart injury.

Collar injury: The fruit plants are damaged by frost at the ground level or at collar region. Cracks appear on the bark of stem and in extreme case plants particularly young one are killed. The freezing of sap in plant tissue results death of plant parts.

Crotch injury: Crotch injury occurs when tissues get killed in the crotches or fork of primary or secondary branches.

Bark splitting: In case of severe frost, bark splitting even in older trees occurs. The entire trunk or branch may be injured under prolonged cold or frost conditions.

Drying of branchs and twigs: The newly emerged young shoots or twigs of frost prone fruit plants such as papaya, aonla, litchi, mango etc. may be killed or dried under frost conditions. The sub-tropical and tropical fruit trees are more sensitive to be damaged by frost if occurs in early spring. In such cases, normal growth cannot be resumed for a long time.

Measures Against Low Temperature

Permanent measures

- **Plant selection:** Selection of fruit crops and varieties which are tolerant to frost and will mature within the available freeze-free period should be done in frost prone areas. Deciduous fruit trees like pears, peaches, almond, plums,grapes, etc. do not suffer from frost damage. In citrus, select more resistant varieties. For example, lemons are least tolerant to frost damage, followed by limes, grapefruit, tangelos and oranges, which are most tolerant. Similarly, the papaya and banana are also very cold sensitive plants. The young plants of aonla, mango and guava are also prone to frost damage hence; require adequate frost protection measures .
- **Wind breaks:** Wind breaks around the orchards particularly in western and northern side will protect the plants from cold currents as well as advective frost. The wind breaks should be established by planting tall trees such as eucalyptus, mango, jamun, arjun etc. around the orchards. For complete protection the rough lemon, bougainvillea or other small trees can be planted in between these tall trees.
- **Healthy trees:** Healthy trees can withstand frost damage much more effectively than weak and sick plants. The fruit trees may be kept healthy by adopting proper management practices and plant protection measures.
- **Selection of site and soil:** The site and soil more prone to the frost damage should be avoided for sensitive fruit plants. More open areas to advection and convective frost may result greater damage to fruit plants. The sensitive plants grown on sandy soils are comparative more susceptible to frost damage.

Temporary measures

- **Light and frequent irrigations:** The soil moisture will influence the risk of frost damage to both above and below ground parts of the fruit plants. To protect the fruit plants, light and frequent irrigation to the fruit plants during winter months is of utmost importance. Irrigation water raises the soil temperature as well as improves the microclimatic conditions within the orchards which protect the fruit plants from frost damage. With application of light and frequent irrigation to the evergreen fruit plants particularly young plants and the plants having fruit load during this season will be highly benefitted.

- **Mulches:** This method reduces heat loss from the surface. The material such as straw mulch or plastic can be used as mulching material to reduce the heat loss from the surface. Clear plastic may transmit some long-wave radiation whereas dark, opaque covers do not. All the covers will be effective in reducing heat loss by convection. If possible, the plastic covers should be removed during the day as air humidity would be higher under the cover and this would increase the danger of certain plant diseases.
- **Sprinkling water:** Sprinkling of water through sprinklers can be effective in preventing freeze damage through the release of heat during cold period. Sprinkling of the fruit crop should begin with the onset of freezing conditions and a film of water continuously maintained during the frosty period. The backyard plantation can be protected by sprinkling irrigation water during this period.
- **Trunk wrapping and painting:** The bark of fruit plants splits when there is more difference between day and night temperatures. Painting the trunks during these days will reduce this problem. White paint, insulation and other wraps are known to improve hardiness against frost damage in fruit plants. Wrapping tree trunks with paddy straw, sugarcane trash and dry leaves and stems of maize, gunny bags etc resist heat transfer and will protect young trees from frost. The main drawback to trunk wraps is increased disease problems, so the bud unions should be at least 15 cm above the ground. Applying fungicide sprays prior to wrapping helps to reduce disease incidence.
- **Smoking:** Fruit plants can also be protected by smoking in the orchards. Collect the dry leaves, paddy straw, dry grass etc. during winter season and burn to smoke on the day of possibility of frost occurrence.
- **By preparing thatches:** Thatches of *sarkanda* or leaves of date palm can be prepared around the newly planted fruit plants. Keep the sun side open so that plants can receive solar radiations during winter season. Remove the thatches in early spring season.
- **Pruning and training of plants:** Pruning of trees at low head withstands frost than high headed trees. Hence; proper training and pruning of fruit trees can save the plants from low temperature or frost.

B. SUITABILITY OF SOILS FOR FRUIT PLANTS

The orchard soil governs the success or failure of fruit plants thus; its physical and chemical properties are of prime importance. The capacity of the soil to provide all the essential elements and having congenial physical and chemical properties is prerequisite for success of fruit plants. The fruit plants require suitable soil for excellent performance over the years. Requirement of suitable soil is the primary aspect for plant growth, higher fruit yield and quality. Most of fruit trees thrive well in deep, loose, well-aerated soils free of excess salts and having sufficient organic matter content. Although, the fruits can grow well in a wide range of soils with varied texture, but, the light sandy to medium loam soils are considered best. Before planting an orchard the soil should be examined for suitability in respect of physical parameters such as depth, drainage, compacted layers, water table, texture etc. and chemical properties such as soil fertility, pH, electrical conductivity, $CaCO_3$ content, soil reaction, lime concretion etc.

Physical Properties of Soil

Soil Texture: The soil texture is of an utmost importance for success of fruit plants and management practices required for cultivation. The extremely light textured and extremely heavy soils are not favourable for most of fruit crops. The medium textured soils are more suitable. The majority of fruit plants thrive well under sandy loam to loamy soils. The soil must have enough water holding capacity, in light soils the water will drain down quickly while, in heavy soils it will not percolate to the root zone and will be subjected to run off. Some fruit plants such as lasoda, kair, ber, aonla, datepalm etc. can be grown in light textured soil. Some fruits like papaya, citrus, strawberry require well drained soils as standing water even for one day may be detrimental.

Depth of Soil: Most of fruit plants thrive well in deep, loose well-aerated soils devoid of any hand pan layers of calcium carbonate in the root zone. Any soil less than 2-3 meter in depth is not suitable for cultivation of most of fruit plants particularly citrus. Deep well drained soils with no impervious layers, allow good depth for root development producing trees of standard size, good yield and quality fruits for a long life. On the other hand, shallow soils, produces sub-standard tree with short life span and poor fruit yield and quality. The ideal soil should permit penetration and retention of water at depth of 2 meters.

Water Drainage: The fruit plants may be grown on various types of soil provided the sub-soil drainage is good because water drainage is most important aspect for growth and development of fruit plants. As the roots requires oxygen for proper growth and development of plants, the accumulation of water in root zone due to poor drainage results injury to roots, poor uptake of essential nutrients and incidence of soil borne deceases. So, well drained medium loam to loam soils are most suitable soils for fruit plants provided the soils are fertile and productive. Heavy soils with poor percolation should be avoided.

Water Table: The water table should be at least three meters from ground level and it should be constant for good growth and development of fruit plants crop. Fluctuation water table causes deposition of salts in the root zone of plants which result poor nutrient and water uptake. Some fruit plants such as papaya is quite sensitive to water stagnation even 1-2 day water stagnation may result killing of plants. Similarly citrus is also sensitive to water logging. Plants subjected to continue flooding found to be short statured than in well drained soils. Poor drainage in heavy clays can be more detrimental for plant growth.

Soil Moisture: Water stress generally, reduces the yield and quality of fruits but it induces flowering in many fruit plants. Deficit soil moisture results reduction in vegetative growth particularly when water stress period is prolonged. However, in many fruit plants such as citrus, pomegranate, guava such conditions results flower induction and initiation. After the initiation of flower panicles, the fruit set can be achieved with light irrigation, however at peak flowering heavy irrigation should be avoided. As the fruit set is too high and most of the fruitlets abscise prematurely immediately after setting. However, water stress after fruit set may cause excessive early fruit drop, resulting heavy economic loss to the orchardists. Moisture stress during summer months particularly under north Indian conditions, results fruit drop in citrus, mango pear etc. Excessive soil moisture during rainy and post rainy season results pre harvest fruit drop particularly in citrus. The pests and disease incidence also increases if soils remain moist for prolonged period. The fruit size and quality reduced drastically in fastly growing fruits such as peach, pear, plum, hence; irrigation at 4-5 days interval is required under north-west Indian conditions. Fruit cracking is another disorder associated with deficit soil moisture and high temperature and low relative humidity. Hence; the optimum soil moisture content during critical stages of vegetative and reproductive growth of different fruit crops is of utmost importance.

Chemical Properties of Soil

Soil Reaction: The performance of fruit plant depends upon the reaction of the soil solution. The soils can be acidic, neutral or alkaline in nature. The pH of the soil at particular site governs the selection of suitable fruit plant. Soil pH affects availability of essential nutrients and in general the fruit plants thrive well at pH value of 5.5-7.5. Most of the fruit plants are sensitive to soil factors such as high soil pH, although, the fruit plants respond differently to soil pH.

The citrus and peach plants fails to grow well in alkaline soils having pH more than 8.5, hence; such soils should be avoided. Acid soils with pH between 5.5 and 6.5 are generally considered as ideal for litchi but it can tolerate soils with a pH even less than 5.5 easily. Certain fruit plants such as blueberry can be grown at fairly high acidic soils having pH 4.5 to 5.5. The fruit plants such as date palm, lasoda, ber, aonla can be grown in soil with higher (upto 9.0) pH. The pH of soil may be changed with irrigation water, addition of manures and fertilizers and certain soil amendments. Lime induced chlorosis and presence of excessive free lime render phosphorus, iron, manganese and zinc deficiency. In citrus growing areas, where the soil pH is beyond 8.5, either regular application of Gypsum or heavy use of organic manure or both can be used as soil amendment on regular basis. Therefore, before planting the trees, it would be desirable to have the suitability of the proposed soil tested and appropriate reclamation measures be taken whatever necessary in advance. The fruit plants are categorized into three classes viz. high salt tolerant, medium salt tolerant and low salt tolerant on the basis of salt tolerance:

High-Salt Tolerant	Medium Salt Tolerant	Low Salt Tolerant
Guava	Fig	Peach
Ber	Olive	Mango
Aonla	Phalsa	Apple
Date-palm	Jamun	Grapefruit
Bael	Cashewnut	Mandarin
Coconut	Pomegranate	Lemon
-	Grapes	Apple
-	Almond	Pear
-	-	Plum
-	-	Apricot
-	-	Strawberry
-	-	Avocado

Soil Organic Matter

Soil organic matter consists of decomposed plant and animal residues in the upper surface of soil. The fallen plant parts, weeds, crop residues become a part of soil humus. Crop residues such as weeds, grasses, fallen tree leaves, twigs in combination with soil microorganisms like bacteria, fungi and actinomycetes and worms are part of nutrient enrichment process in the soil. Although most cultivated soils contain only 0.5-1.0 per cent organic carbon mostly in the top 6-12 inches cm of soil. The slight enrichment of soils with organic matter can amend its chemical and biological properties. Soil organic matter is constantly undergoing changes and it should be replenished continuously to sustain the soil health and crop productivity

• Soil Testing for Orchard Plantation

Orchard establishment is long term investment hence; the suitability of soil for particular fruit plant is of prime importance. Before the selection of fruit plant for establishment of orchards soil must be tested. Soil analysis provides information on nutrient availability, soil pH, electrical conductivity (EC), organic carbon (OC), lime concretion, hard pan, presence of salts and problem of salinity, alkalinity, sodicity etc. and to suggest suitable reclamation measures. It also provides the physical and chemical status of sub soil up to two meter depth. For analysis of soil, the soil sample should be collected as per following procedure.

Soil analysis

1. In case of analysis of soil proposed for orchard plantation collect 500 g soil samples separately from each layer i.e. for top 15 cm, 15-30 cm, 30-60, 60-90 cm, 90-120 cm, 120-150 cm and 150 –200 cm with the help of auger or digging a pit.
2. Put different soil samples in separate clean cloth bags, avoiding contamination. Label each bag indicating the depth of the layer from which sample has been taken. Send the sample to the soil testing laboratory or handover the same of the university extension specialist in your district.

Following points should be kept in mind while analyzing the soil.

a. The samples should not be taken from recently fertilized soil, areas near wells, water channel and compost piles.

b. If a hard pan or concrete layer is present, note its depth and thickness and collect its sample separately.

c. For better results take 3-4 samples from proposed field for orchard establishment and mix it to prepare composite sample.

d. If topography of site and soil texture is different, collect the samples separately.

e. Soil analysis is used primarily as an indicator of the chemical and physical parameters of soil of the site selected for establishment of orchard. Selection of fruit plants for orchard establishment should be done on the basis of soil analysis report. Advice of expert may be taken in this regard.

• Soil Suitability Limits for Growing Fruit Crops

The success of an orchard is governed by suitable site and soil. As the roots of the fruit trees generally penetrate deep into the soil so, the characteristics of sub-soil also play an important role in the growth and development of plants.

The soil acts as storehouse of water, nutrients as well as oxygen for root growth. The capacity of the soil to support plant growth and development is governed by soil fertility and physical condition. Therefore, the soil must be suitable and a suitable soil has the capacity to nourish and sustain plant growth by providing mineral elements in an available form to plants through their interaction with soil moisture, soil air, soil microbes and humus content.

Suitability limits of soils for fruit trees under North-Western region of India

S. No.	Characters	Citrusand peach group	Other fruit plants
1.	Conductivity (mmhos/cm)	< 0.5	<1.0
2.	$CaCo_3$ (%)	< 5.0	<10
3.	Lime concentration (%)	< 10.0	<20
4.	pH	< 8.5	Up to 8.7 in upper 2 feet soil profile and up to 9.0 beyond this depth.

Suitable Soil: The suitable soil must have following characteristics:

- For success of fruit plants the soil should be deep, well-drained, sandy-loam to loamy and rich in nutrients.
- The soil should be free of hard pan or concrete layer up to two meter depth.
- The water table should be three meter from ground surface and it should not be fluctuating.
- Avoid water-logged, marshy, saline, alkaline or acidic soils for fruit cultivation.
- The ideal pH for most of fruit plants ranges between 6.5 to 7.5, however, certain plants such as blueberry need acidic (4.5-5.5 pH) soils.
- Certain fruit plants such as datepalm, ber, aonla and guava can grow successfully even in soils having pH 8.7 and even more.
- All the major elements such as N, P, K, S, Ca and Mg should be in optimum range.
- Normally, when pH is higher than 8.7, application of gypsum is recommended for successful cultivation of fruit trees. If the soils are problematic like poor aeration or drainage, sodicity, alkalinity, acidity and salinity, they require improvement or reclamation before taking up crop

production or the venture would fail. The selection of fruit crop should be based on the soil analysis report and fruit crops can be selected as per following categories:

Salinity tolerant crops: Aonla, Ber, Date palm, Fig, Kair, Khirni, Sapota, Wood apple etc.

Sodicity tolerant crops: Aonla, Ber, Bael, Date palm, Fig, Guava Karonda, Phalsa, Pomegranate, Tamarind, Wood apple etc.

Drought tolerant crops: Ber, Aonla, Custard apple, Fig, Phalsa, Kair, Lasoda, Karonda, etc.

LITERATURE CONSULTED

Bal, J.S., 2014. Fruit Growing third Edn. Kalyani Publishers, New Delhi.

Barooh, S.,1998. Modern Fruit Culture. Kalyani Publishers, New Delhi.

Bose, T.K. and Mitra, S.K. 1996 Fruits: Tropical and Subtropical. Naya Prokash, Kolkata, Chadha, K.L. 2014. Hand Book of Horticulture. I.C.A.R., New Delhi.

Edmond, J.B., Sen T.L., Andrews, F.S. and Halfacre, R.G..1990. Tata McGraw Hill Publishing Co., New Delhi.

Hemla Naik, B. and Thippesh, D. 2015. Fundamentals of Horticulture and Production Technology of Fruit Crops. UAHS, Shimoga.

Singh, A.,1995. Fruit Physiology and Production. Kalyani Publishers, New Delhi.

Singh, J. 2012. Basic Horticulture. Kalyani Publishers, New Delhi.

Singh, S., Krishnamurti, S. and Katyal, S.L.1963. Fruit Culture in India. I.C.A.R., New Delhi.

Singh, S.P. 1995.Commercial Fruits. Kalyani Publishers, New Delhi.

4

Soil Cultural Practices Management

S.K.Upadhyay

INTRODUCTION

Soil is one of the most important factors associated with the success or failure of fruit production. Orchard soil management refers to those practices in the soil that aim at getting maximum productivity/production of quality fruits within the limits of soil, water and fertility management over the years. The decision regarding the practices of orchard management should always be taken well in advance before planting of an orchard. The practices for orchard management should be executed in a timely manner during the lifetime of the orchard. An efficient orchard management system always ensures higher return to the grower. However, no single management practice can be recommended for all orchards. The management systems varies from orchard to orchard, locality to locality, soil to soil, from one set of climatic conditions to another set of climatic conditions etc. These systems also have their own merits and demerits. Thus the choice of best suitable system depends on many factors such as climate, topography, rainfall, tree spacing, planting system/ orchard design, root depth of tree, economic condition of orchardists, fertility status of soil and type of main fruit crop etc.However, it should be combined with pruning, nutrient and water management, provision for pollination, fruit thinning and disease and pest management in order to obtain desirable tree growth and vigour with the aim to promote precocity with high productivity. The objectives of an ideal soil management system should be:

i) To create favourable moisture conditions in the soil

ii) To maintain a high level of soil fertility

iii) To keep the soil in porous and friable conditions

iv) To prevent runoff and soil erosion

v) To maintain organic matter/humus content in the soil

vi) To reduce root injury during severe winter/summer

vii) To cause efficient weed control

DIFFERENT CULTURAL PRACTICES

The orchard soil management affects the growth, yield and quality of fruits indirectly through its effect on the availability of nutrients and water to the plants. The important soil management practices usually adopted are given as under:

I. SOIL CULTURE

a. Tillage: Physical manipulation of soil with tools and implements for obtaining conditions ideal for better seed germination, seedling establishment and growth of plants is called tillage. The objectives of tillage are as under:

- To facilitate adequate soil aeration for gaseous exchange in the seed and root zone
- Adequate seed-soil contact to permit water flow to seed and seedling roots
- A non-crusted soil to permit seedling emergence
- A low density soil that permits root elongation and proliferation
- To create an environment that provides adequate light to the seedling (weed free environment).
- To create pest and pathogen free environment
- To mix the applied manures and fertilizers with the soil
- To remove the hardpan, if any to increase the soil depth for water absorption

b. Clean Cultivation: This is the most extensively adopted system of soil management all over India. It is commonly followed in young orchards. In this system, the inter space between the trees is kept clean by tillage and removal of weeds. The cultivation of orchard should not be deep but only shallow harrowing may be given. Deep cultivation is likely to harm the roots. Clean cultivation does not provide satisfactory results especially during rainy season. Clean cultivation for long periods is detrimental to trees as well as soil. The advantages and disadvantages of clean cultivation are:

Advantages

i. Improves soil aeration and physical conditions.

ii. Controls weeds and thus reduces competition for light, nutrients, and moisture.

iii. Eliminates or avoids alternate hosts for pests and diseases.

iv. Improves the infiltration of water through breaking the hard soil surface crust.

v. Enhances biological activity in the soil through better aeration.

Disadvantages

i. Depletion of organic matter.

ii. Soil erosion by water and wind.

iii. Loss of nutrients due to excessive leaching.

c. Mulching: Mulching refers to spreading of dried leaves hay, straw, or any other organic and inorganic materials around the tree basin to cover the soil surface. A wide variety of materials can be used as mulches, which are broadly classified as organic and inorganic mulches. All the organic residues viz. straw, hay, crop residue, leaves, saw dust and organic wastes viz. farm yard manure, poultry manure, spent mushroom compost etc. and inorganic mulches viz. black polythene can be used as mulches. In some arid fruit growing areas, pebbles can be used in the event of non-availability of organic material. These mulches act as a surface barrier against evaporation, conserve soil moisture, check weed growth and enrich organic matter to the soil. However, mulches should be removed at the onset of monsoon.

Advantages

i. Conserves soil moisture by reducing evaporative losses from soil surface and controls weeds.

ii. Organic mulches add organic matter and nutrients to the soil upon decomposition thus save fertilizer requirements.

iii. Improves the soil structure, nutrient availability and micro flora.

iv. Controls soil erosion.

v. Improves water infiltration and controls weed growth due to etiolation.

vi. Reduces soil temperature fluctuation.

vii. Mulch acts as a cushion for dropping fruits and thus avoiding physical injury.

viii. Growth, production, colour and quality of fruits are improved.

Disadvantages

i. Cost of mulch is high.

ii. Dry grass, hay or pine needle mulches are a fire hazard.

iii. Mulch may harbour insects, diseases and rodents.

iv. Growth may be extended in some seasons through fall with reduction in fruit quality.

Organic mulches also improve root growth, increase the infiltration of water, and also improve the water-holding capacity of the soil. Organic mulch provides an ideal environment for earthworms and other beneficial soil organisms. Organic mulches are more useful for warmer climates i.e. valley areas and low and mid-hills areas of Himalayan region, as these keeps the soil temperature cooler by 2-3° C in summer.

Inorganic mulches are usually preferred in fruit orchards located in the high hill areas having cooler climate. It also helps in raising the soil temperature in summer, required for greater nutrient absorption and roots metabolic activities. However, inorganic mulches lack soil improving properties of organic mulches.

One of the benefits from winter mulching is the reduction in the freezing and thawing of the soil in the late winter and early spring. These repeating cycles of freezing in night and then thawing in the warmth of the sun causes many small or shallow rooted trees (e.g. young trees) to be heaved out of the soil. This leaves their root systems exposed and results in injury or death.Mulching helps prevent the rapid fluctuation in soil temperature.

How and when to apply mulch

Time of application of mulch depends on the purpose for which it is to be applied. In strawberry, mulching is done to avoid the contact of berries with soil and it should be applied at the time of appearance of flowers. About 15 tonnes of straw per hectare is sufficient for mulching in strawberries. Mulching by providing an insulating barrier between the soil and the air, moderate the soil temperature. This means a mulched (organic) soil in summer will be cooler than the adjacent non-mulched soil; while in the winter the temperature of mulched soil may not drop as deeply as otherwise it would have been. Around 10 cm thick dry grass, hay, straw or pine needles are spread over the tree basin areas in spring immediately after completing the tillage operation and applying second dose of nitrogenous fertilizers in the basins. However, about 10-15cm area away from the tree trunk should be kept open, in order to protect the trunk from the damage of any harbouring insect-pests, rats etc. Among the inorganic mulches, use of black polythene sheet of 200 micron thickness around the tree basins or in strips is common, keeping sufficient vacant space for working between the rows. Mulches can be applied late in the winter after the temperature falls below freezing point but before the coldest temperature arrives. Applying mulches before the coldest temperature may attracts rodents looking for a warm over-wintering site. Delayed applications of mulch should prevent this problem as, hopefully, the creatures would already have found some other place to nest.

d. Sod Culture: This practice is followed in the orchard located on sloppy land, particularly when the gradient of the slope is greater than 10 per cent.In this system, a suitable grass or sod cover is grown in the interspaces between the trees without tillage or mulching. Sometimes soil is loosened with rake during spring which improves aeration and water intake capacity of the soil. Growing of clovers such as red clover (*Trifolium pretense)*, white clover (*Trifolium repens*) with grasses is useful for improving soil fertility.The key to good sod management is periodical mowing of grasses to reduce competition for water and nutrients with the trees. Orchard grass (*Dactylis glomerata*) is suitable for areas facing north, tall fescues (*Festuca arundinacea*) a drought tolerant perennial grass for areas facing south, and Timothy (*Phleum pratense*) for cold temperate region.This system prevents soil erosion to the greatest extent; however it should not be followed in orchards of young or dwarf trees because such trees are shallow rooted and compete for nutrients and water with grasses.The roots of large trees on the other hand penetrate the soil to a greater depth and are less likely to suffer from such competition.Tree basins are kept free of weeds by tillage or application of herbicides. The advantages of sod culture are:

i. Controls soil erosion.
ii. Maintains optimum soil moisture to ensure water and nutrients supply to trees.
iii. Maintains organic matter.
iv. Provides better aeration to the roots.
v. Enhances microbiological activity in the soil through better aeration
vi. Avoid bruising of falling fruits.

e. Sod Mulch: In this method, grasses are allowed to grow in the entire area of the orchard.Tree basins are mulched with straw, instead of tilling.The grasses are mowed down regularly and are spread to maintain the mulch in basins.This method is best suited in orchards on steep slopes to prevent soil erosion. The advantages of sod mulch are:

i. It has all the advantages of sod-tilled method.

ii. Most suitable for orchards on slopes having greater gradient.

iii. Highly effective for controlling soil erosion.

iv. Conserves moisture and controls weeds in tree basins.

v. Improves fruit colouration and enhances fruit maturity.

Sods suitable for different areas

Legumes		Non-legumes	
Common name	Botanical name	Common name	Botanical name
Kudzu vine	*Pueraria* sp.	Timothy grass	*Phleum pratense*
Lucerne (alfalfa)	*Medicago sativa*	Orchard grass	*Dactylis glomerata*
White clover	*Trifolium repens*	Rye grass	*Lolium* sp.
Red clover	*Trifolium pratense*	Toowamba canary grass	*Phalaris aquatica*
Subterranean clover	*Trifolium subterraneum*	Tall fescue grass	*Festuca arundinacea*
		Kentucky blue grass	*Poa pratensis*

II. INTER CULTURE

a. Cover Crops: The crops, which are raised between the tree rows during rainy season for protecting the soil from erosion, are referred as cover crops. These may or may not be turned into the soil. Leguminous crops like green gram, black gram, cowpeas, cluster bean, soya been during kharif season should be used. Pea, fenugreek, broad bean during rabi season can be preferred, as these add N to the soil by fixing atmospheric nitrogen in their nodules. The advantages of cover crops are:

i. Increase water retaining capacity and biological complexes of soil.

ii. Increase organic matter in soil.

iii. Improve soil condition.

iv. Improve soil fertility.

v. Check soil erosion and thus also nutrient losses from soil.

b. ***Green* Manuring *Crops:*** Crops grown for the purpose of restoring or increasing the organic matter content in the soil are called green manure crops. Use of Green manure crops in cropping system is called 'Green Manuring' where the crop is grown *in situ* or brought from outside and incorporated when it is purposely grown. Green leaf manuring consists of gathering green biomass from nearby location and adding to the soil. The objective of green manuring is to add N to the companion or succeeding crop and add or sustain organic matter in the soil.

Classification of green manures

Legumes		Non-legumes	
Green manure	Green leaf manure	Green manure	Green leaf manure
Daincha	Gliricidia	Sunflower	Calotropis
Sunhemp	Cassia	Buck wheat	Adathoda
Kolinji	*Pongamia glabra*		*Thespesia*

The desirable characteristics of legume green manure crop are

- It should be for multipurpose use, short duration, fast growing, high nutrient accumulation ability
- It should be tolerant to shade,flood, drought and adverse temperatures.
- It should have wide ecological adaptability and efficient in use of water
- It should have early onset of biological nitrogen fixation and high N accumulation rates as well as timely release of nutrients and high N sink in underground plant parts.
- It should have photoperiod insensitivity, high seed production and high seed viability
- It should be easy to incorporate and able to cross-inoculate or responsive to inoculation
- It should be pest and disease resistant

Leguminous green manures: Sesbania, Dhaincha, Sunhemp, Wild Indigo, Pillipesara, Cowpea, Cluster bean (Guar), Green gram (Mung bean), Berseem and Madras Indigo

Commonly used shrubs and trees as green leaf manures

Shrubs	Trees	Green Leaf Manures
Cassia auriculata	*Thespesia populnea*	*Leucaena leucocephala*
Derris indica	*Neem*	*Calotropis gigantea*
Ipomoea cornea	*Glyricidia*	*Delonix regia*
Jatropha	*Cassia tora*	*Cassia Occidental*
Tephrosia candida	*Vitex negundo*	*Hibiscus viscosa*

Burial of green manure crop at specific time provides maximum nitrogen and organic matter. This specific stage is when plant is immature and has started flowering, as the basic aim of green manuring is to provide maximum succulent green matter at burying. During early period of crop growth N content, protein, water soluble constituents are maximum, whereas fibre, hemicelluloses, cellulose, lignin and C:N ratio are less. Therefore, tissues of immature plants usually decompose more rapidly as compared to matured plants. Delay of even 15- 20 days reduce nitrogen content and increase C:N ratio, fibre, hemicelluloses, cellulose, lignin making it difficult for soil microorganisms to act and decompose

Before a crop is sown the green manure is dug back into the soil. Green manures should not be ploughed in as this buries the plants and the nutrients too deep. They should be turned in just under the soil surface. Here, it decomposes and the nutrients held inside green manure plants are released. Immature crop can decompose at any depth, but mature crop should be buried at less depth. If the weather is dry green manure crop should be buried at more depth compared to moist season. If moisture in soil is less water should be supplied externally. Green manure crop should be buried at higher depth in sandy soil and in heavy soils at less depth for proper decomposition.

The time interval between burial of green manure crop and the sowing of the next crop should be such that it allow complete decomposition of the turned in green manure crop before planting of the next crop. Time interval depends on the following factors such as weather conditions and nature of buried green material. Crop should be sown after 35- 45 days of burial of green manure crop as green manure crop takes about 4-6 weeks for complete decomposition.

c. Inter Crops: The inter crops generally refer to the crops grown in an orchard for increasing the initial income from the land/orchard. These intercrops may be vegetables, pulses and strawberries. Sometimes short duration fruit crops are raised as inter crops, which are also referred as 'fillers'.

During early years of orchard, when it is young and non-bearing, intercrops are recommended for cultivation between the rows of main fruit trees as an additional profit making enterprise. Under Indian conditions, it is essential to grow intercrops in the orchard for additional income since most of the orchardists are poor. The advantages of inter-cropping are:

i. Generates supplementary income during the initial juvenile years of orchard plantings.

ii. Increases organic matter, soil nutrient status by fixing atmospheric N to soil by leguminous crops.

iii. Reduces weed growth.

iv. Improves soil structure.

Suitable intercrops for different fruit crops

Fruit crop	Suitable inter-crops
Mango	Phalsa, papaya, guava, banana, peach, strawberry, pineapple
Banana	Green gram, cowpea, cauliflower, cabbage, yam, elephant foot, onion, Black gram, turmeric, brinjal, colocasia, dioscorea, chillies, lady's finger
Citrus	Beans, carrot, tomatoes, berseem, onion, potato, chillies, pulses, cucurbits,lady's finger, gram, peas, tomato, cabbage
Litchi	Turmeric, ginger, pointed gourd, sweet potato, tomato, radish, cabbage,turnip, brinjal, cucurbits, green gram, black gram, cowpea
Guava	Cauliflower, peach, Frenchbean, cowpea, clusterbean, blackgram
Papaya	Cabbage, cauliflower, chillies, radish, tomato
Grape	Vegetables relevant to area
Pomegranate	Berseem, Lucerne, cowpea, green pea, cucurbit, cabbage,cauliflower, bean, peas, tomato, carrot, onion, potato, brinjal
Sapota :	Banana, papaya, pineapple, broad bean, tomato, brinjal, cabbage, cauliflower, spider lily.
Ber	Green gram, moth, cluster bean, cowpea, cumin, chillies
Date palm	*Citrus medica*, guava, sapota

Principles of Intercropping

- Intercrops should occupy a secondary place in the orchard, primary consideration being given to the main fruit trees.
- Short duration and less exhaustive crops should be selected as intercrops
- The rows of intercrops should be away from the rows of main fruit trees but this distance can be decreased as the age/vigour of main crop increases.
- Water requirements of the intercrops should as far as possible coincides with the requirement of fruit trees.
- Select those intercrops that do not exhaust the nutrients and moisture from the soil, so essential for the growth of fruit trees.

- Perennial or exhaustive crops should be discouraged as an intercrop in the orchard. This may have devitalizing effect on the growing trees. For example, sugarcane, pigeon pea, maize, jowar should invariably be excluded from an intercropping programme in the orchard.

Selection of Intercrops: The selection of intercrops depends on the kind of fruit plant, location, type of soil, amount and distribution of rainfall, irrigation facilities, availability of labour, transport facilities and availability of packing material etc.Main crops should always be given proper care in order to avoid root restriction, damage and infection, undue exhaustion of the soil, perpetuation of virus, fungal and nematode infection. Intercrops should therefore, receive secondary importance and fulfill following criteria.

- Should not be tall growing and spreading type.
- Should not be exhaustive.
- Should not function as alternate host for common pests and diseases.
- Water requirement schedule should match or phenology of crop should match so that operation could be synchronized.
- Should match for climatic requirements with main plantings.
- Separate provision for nutrients should be made for inter-crop to avoid competition.
- Intercrops should be preferably legumes or shallow rooted vegetables.

Multistory Cropping or Crop Modelling

To increase the cropping intensity, crop model or multistory crop combinations consisting of crops of varying canopy orientations and rooting habits have also been developed. One such crop combination is Coconut+ Black pepper+ Cocoa+ Pineapple; the explanation of combination is as under:

Coconut: Grow at a height of 10-30m forms the top floor of this crop building

Black pepper: Grown and trailed on coconut trunks upto 6-8m height with little lateral spread and forms the second floor

Cocoa: With a canopy height of 1.5-2.5m forms third floor, as it is relatively less stature crop

Pineapple: Forms the fourth or ground floor

The number of plants accommodated per hectare is Coconut: 175 plants; Black pepper: 600 plants; Cocoa: 600 plants and pineapple: 3500 plants.

In such high intensity cropping programme a higher efficiency in the utilization of solar energy incident in a given area is obtained as it is intercropped at vertical intervals by the canopies of these crops. Rooting pattern of these crops are also mutually exclusive horizontally as well as vertically. Water use efficiency could also be enhanced considerably under such programme, because the total irrigation requirement of any crop at a potential production level is divided by the evaporative demand of the climate pertaining to the area. Therefore, irrespective to the number of crops grown in the area, the water requirement of the area remains almost same. Similarly, expanses on weed control and inter-cultivation, which is a major component, cost of cultivation also remains the same.

The requirements of inputs, such as fertilizers and other agro-chemical and human labour etc. per unit area also correspondingly high in such cropping programme. Such cropping systems enable to raise the productivity from the land considerably. The magnitude of this increase can be seen from the fact that the yield of crop models raised in the interspaces is an additional item of production. This increase in productivity is often more than proportional to the expenses and consequently the net return per unit amount of investment, is considerably enhanced.

Under South Indian conditions, the total labour requirement for the maintenance of a hectare of adult bearing coconut plantation is estimated to be 15 man days/ year. Whereas, in multistory crop model combination, the total labour requirement per year is 360. On the contrary, crop diversification is an insurance against such a risk and help to the farmer and his family.

Table 1: Time required for occupying allotted space and suggested period of intercropping in various fruit species.

Fruit crop	Spacing	Time to complete coverage of space	Period of intercropping
Mango	10x10 m	7 years	5 years
Citrus	6x6 m	5 years	4 years
Banana	1.8x1.8 m or 2x2 m	6 months	5 months
Guava	6x6 m	5 years	3 years
Apple & Pear	6x6 m	6 years	4 years
Stone fruits	6x6 m	4 years	2 years
Litchi	10x10 m	10 years	7 years
Sapota	10x10 m	10 years	7 years
Grape	3x3 m	3 years	1 year
Papaya	2x2 m or 2.5x2.5 m	5 months	5 months
Ber	6x6 m	4 years	2 years
Date palm	8x8 m	8 years	5 years
Pineapple	0.90x0.60x0.25 m	9 months	6 months
Pomegranate	6x6 m	5 years	3 years

Crop modelling is intended to maximize land space and light use efficiency to generate supplemental income particularly during the initial years of unproductive orchard life. To protect the interspace from losses through weeds, erosion, impact of radiation, temperature, wind, water and enriching it by nitrogen fixing leguminous crops or leaves growing intercrops is better in the interspace. The period of intercropping vary depending on crop species and time required to occupy allotted space. However, if the main plantation is not given priority care, serious losses may occur as a result of root restriction damage and infection under exhaustion of soil and perpetuation of viral, fungal and nematode infections.

Advantages of crop modelling practices

1. It has higher employment potential as being more labour oriented. Fruits proved a higher energy values/unit of cultivated area than cereals.
2. It is a conservation farming practice which can help in the control of erosion of catchment areas.
3. Shifting cultivation practiced in North-East Himalayan region is more of the major causes of land degradation. A 3-tier farming system incorporating the concept of jackfruit+ Pineapple +Banana production was found economically viable and socially acceptable alternative to farming in this area.
4. Intensification of monoculture of crops in rainfed areas accelerates runoff and erosion and other degradation processes causing permanent damage to the ecosystem.
5. Ber, guava, custard apple, bael and aonla as fruit trees are competent alongwith other crops like cowpeas, green gram and black gram. They are highly promising for stabilizing the income of farmers in semi-arid drylands.

III. MANURES AND FERTILIZERS

For optimum growth and production of horticultural crops, application of fertilizers is most important. Horticultural crops require nutrients for its growth and development which are absorbed through soil. In order to get maximum benefit from manures and fertilizers, they should not only be applied in proper time and in right manner but there are many other aspects, which should also be given due consideration. Different soils react differently with fertilizer application. Similarly, the N, P, K requirements of different crops are different and even for a single crop, the nutrient requirements are not the same at different stages of growth.

There are 17 elements which are essential for plants namely, carbon (C), hydrogen (H), oxygen (O), nitrogen (N), phosphorus (P), potassium (K), calcium (Ca), magnesium (Mg), sulphur (S), Iron (Fe), manganese (Mn), zinc (Zn), copper (Cu), molybdenum (Mo), chlorine (Cl), boron (B) and Nickel (Ni) . Manures and fertilizers are major nutrient supplying sources to plants. A deficiency of any of these nutrient elements can limit plant growth and development and, ultimately, yield. Most soils contain sufficient amounts of the micronutrients needed to support plant growth. However, soils may be lacking in some of the macronutrients, particularly nitrogen, phosphorus and potassium. Therefore, it becomes necessary to ensure the presence of all the essential elements supplied by the soil in the right quantities and the right chemical forms for plant use. This is done by supplying organic matter and by the judicious use of fertilizers and manures in order to supplement the nutrients required by the plants from soil to increase crop yield *vis-a-vis* to maintain/ improve the soil fertility. Continuous cropping and several other factors necessitate the use of manures and fertilizers. The dose, method and time of application depend on crop, soil, fertilizer/manure and climatic factors of the region.

Integrated Nutrient Management

It is the use of chemical fertilizers, organic manures, slow-releasing fertilizers, nitrification inhibitors, vesicular arbuscular mycorrhiza and nutrient efficient rootstocks either singly or in combination, all at a time in a definite sequence during the growth developmental stages of plant for economic and efficient use of nutrients without having any adverse effects on soil health and environment.

Principles of Fertilizer Application

The basic principle of fertilizer application is to make the nutrients readily available to the plants as per their requirement without much wastage and harmful effects on soil. Usually larger quantities of fertilizers are added to clayey soils at longer intervals than to sandy soils because clayey soils are richer in humus than sandy soils and both clay and humus have a high capacity to retain nutrient ions by a phenomenon called Base Exchange. These adsorbed nutrient ions are not lost by leaching, can be gradually taken up by the plant roots. If a heavy dose of water soluble fertilizer is applied to sandy soils, most of it will be leached down by high rainfall in the humid regions.

Quantities of Fertilizers to be Applied

Different crops require different quantities of nutrients. Fertilizer dose is calculated on the basis of requirement of plants for its various physiological activities. While applying fertilizers, nutritional status of soil is taken into account.

After deducting contribution of soil, rest amount is replenished externally by the application of fertilizers.

Time of application of fertilizers

Generally, the manures are applied in the field before the onset of monsoon. With the appearance of rain, the manures are decomposed well and their nutrients become readily available to plants. The fertilizers are applied at the time of active growth of plants so that the nutrients may be absorbed by the roots. Generally, the plants remain active during February- March and July and new growth emerges on the plant. During these times, fertilizers must be applied. In case of bearing plants, the manures and fertilizers are applied to cope with nutritional requirement. Under such circumstances, the manures and fertilizers must be applied one month before the commencement of flowering in plants. Phosphatic fertilizers being less soluble should be applied about 20 days before the commencement of new growth. Nitrogenous fertilizers being highly soluble and hence prone to leaching losses should be applied in split doses. Half of the recommended dose of nitrogen should be given before commencement of flowering and rest half dose is given after fruit set. This holds true especially in case of fruits. In case of vegetables, the recommended doses of fertilizers are applied at the time of sowing, transplanting and also during growing. Nitrogenous fertilizers are given in splits to support production.

Precautions in Fertilizer Use

- It is better to get the soil tested from soil-testing laboratory. The amount of fertilizer(s) should be calculated based on soil test for balanced use of nutrients.
- Secondary nutrients like sulphur should be used either alone or through sulphur bearing fertilizers. In acid soils, calcium and magnesium should be maintained at the optimum level.
- Micronutrient should be applied whenever necessary. In acidic soils boron and molybdenum, and in alkaline soils, iron, zinc and manganese should be made available. Phosphate rich calcareous soils may show zinc deficiency problems.
- Fertilizers should be selected on the basis of soil characteristic. Avoid acid fertilizers in acid soils and basic fertilizers in alkaline soils.
- Improve soil structure through the addition of organic manure and gypsum. Black and alluvial soils should be deep ploughed.

- Use of high yielding varieties, irrigation at an appropriate time and amounts, removal of weeds, spacing and plant population etc. should be given due consideration.

Fertigation: It is the application of fertilizer or chemicals through the irrigation system. It is a controlled system to supply soluble plant nutrients at the root zone of the irrigated crops. Fertigation is done through tank, ventury or pump systems. The most practical method of applying of fertilizers through the irrigation system is by creating a 10% bypass flow of the main line flow, through an artificial fertilizer mixing tank.

Advantages of fertigation

a. An opportunity for placement of fertilizer at the vicinity of root zone of the crop along with irrigation water which increases in water and fertilizer use efficiency.

b. Deeper penetration into the soil.

c. Avoids volatilization from soil surface.

d. Easy coordination with specific crop demand.

e. Decrease in labour and energy cost in fertilizer application by making use of water distribution system.

f. Improves the availability of nutrients and their uptake by roots.

g. Trace elements can also be applied along with major nutrients.

h. Saving of fertilizers.

Limitations of fertigation

a. Possibility of clogging emitters if the pH of irrigation water and fertilizer sources is not managed carefully.

b. May result in possible contamination of the drinking water supply if devices are not used to prevent back flow of nutrients into the well or other water sources.

c. Some of the chemicals are quite corrosive to metal and also cause skin burning is safety devices are not provided to protect the workers against unplanned discharge or spilling of chemicals.

IV. IRRIGATION OF FRUIT CROPS

Horticultural crops require significant amounts of water due to their perishable nature. In fruit trees, water stress not only affects the current season's crop, but future crops as well due to their perennial nature. Since water is such a critical component of the growth and development of horticultural crops, it is recommended that none of these crops be established without assured irrigation facilities. Water requirement of a plant depends on its growth habit and life period. Plants differ greatly in their water requirements.

Aim of irrigation to fruit plants

Irrigation may be defined as artificial supply of water to support plant growth and production in the absence of adequate supply of water through rainfall. Water is the main constituent of plants. It performs the following important functions in plants:

- Water is essential for the germination of seeds and growth of plants.
- During the process of photosynthesis, plants synthesize carbohydrates from carbon dioxide and water. Therefore, water is one of the essential components for the plant.
- Water acts as a solvent for fertilizers and other minerals, which are taken up by the plant roots in the form of solution. Thus, water serves as the medium in which plants absorb soluble nutrients from the soil.
- Water serves as medium for transport of chemicals to and from cells.
- Water pressure in plant cells provides the firmness to the plants.

Irrigation is very important in horticultural crops as sufficient moisture must be maintained in the soil for obtaining the optimum yield of quality fruits. The aim of irrigating the crops should be to wet the entire root zone without allowing any wastage of water beyond the root zone. The irrigation system has to be properly devised so that the water requirements of the crops are met at the minimum expenditure without any wastage of water. Various types of irrigation techniques differ in how the water obtained from the source is distributed within the field. In general, the goal is to supply the entire field uniformly with water, so that each plant has the amount of water it needs, neither too much nor too little.

Many factors determine the suitability of irrigation system for a particular crop. Several methods are employed for the irrigation of horticultural crops depending on the type of fruit crops grown, age of the tree, the soil topography and the availability of irrigation water. Thus the system of irrigation must be decided in relation to the varying field conditions. Choose the correct system for a particular crop and situation.

When to irrigate?

- The time when a plant needs irrigation can only be judged by a keen observing eye.
- The plants need water when their new leaves begin to show a wilting appearance. A little before the trees show the sign of wilting.
- The shedding of broad leaves in orchard shows distress symptoms.

How much to irrigate?

- If water supply is limited, only a light irrigation can be given at a time with higher frequency of irrigation.
- If water is available in plenty, the irrigation may be heavy with longer intervals between successive irrigations.
- However, inadequate irrigation reduces the growth and fruiting of the trees while, over irrigation serves no useful purpose and it may even prove to be harmful.
- It may create water logging, the nutrients may get leached and fruits may become watery and develop poor quality.
- Plants which have suffered from drought should not be given liberal doses of irrigations all at once. That may result in the splitting of fruits and even the splitting of bark of the branches and trunk.

Irrigation Scheduling

When to irrigate and how much to irrigate, form the schedule of irrigation. In order to avoid water stress in plants and to obtain good yields of quality fruits, proper irrigation schedule should be followed. Scheduling of irrigation in plants is governed by soil, climatic and plant factors. Some practical approaches used to determine schedule of irrigation are soil moisture depletion, cumulative pan evaporation and sensational etc.

Assured supply of water is required at certain stages of plant growth, which are referred to as critical period for crops. If water is not available at critical stages, yield is greatly reduced. Critical stages vary from crop to crop such as fruit setting and development stage for fruit crops.

Irrigation is generally applied to horticultural crops by flooding on the field surface (Surface irrigation), applying beneath the soil surface (sub surface irrigation), spraying under pressure (sprinkler irrigation) or by applying in drops in the crop root zone (Drip irrigation). Several water application methods are practiced to suit different soil types, the topography of the land, crops to be irrigated and costs.

V. WEED CONTROL IN ORCHARDS

Horticultural crops are high value crops and the challenge of weed management is one of the important factors to successfully grow them. Weeds are unwanted and undesirable plants which interfere with the utilization of land and water resources and thus adversely affect human welfare. They can also be referred as plants out of place. Weeds compete with crops for light, nutrients and water, resulting in stressed plants, poor fruit quality and yield. Stressed crops are also more susceptible to disease and insect problems, while excessive weed growth itself creates higher humidity in the foliage enhancing disease spread and inviting unwanted insects.

In fruit crops, weed pressure must be addressed throughout the life of the planting. When compared to annual crops, perennial culture is a greater challenge, as fall, spring, summer and winter weeds need to be managed efficiently. Understanding seasonal weed thresholds and integrating cultural and chemical management becomes even more important in year-round culture. However, because of different cultural requirements, each crop should be considered individually.

Common Weeds in Horticultural Crops

Echinochloa colonum, Cyperus esculentus, Cyperus rotundus, Digitaria sanguinalis, Eleusine indica, Cynodon dactylon, Amaranthus viridis, Phyllanthus niruri, Ageratum conyzoides, Setaria glauca, Phalaris minor, Avena fatua, Poa annua, Vicia sativa, Medicago denticulata, Ageratum conyzoides, Lantana camara, Imperata cylindrical, Rosa moschata, Berberis, Rubus spp.

Weed management includes prevention, eradication and control by regulated use, restricting invasion, suppression of growth, prevention of seed production and complete destruction. Thus weed control is one of the aspects of weed management. The various methods of weed control employed in horticultural crops are described as under:

(a) Mechanical/Physical method

In this method, weed control is done using tillage, hoeing, hand weeding, digging, cheering, mowing, burning, flooding, mulching etc.

(i) *Tillage*: It removes the weed from the soil. It causes injury to root and pruning to shoot of weed. The weeds in tilled field lose their regenerative capacity. Weeds are also get buried at the time of tillage. The effectivity of tillage in controlling weed depends upon a number of factors. A field infested with deep rooted perennial grasses like thatchgrass*(Imperata* *cylindrica),* motha*(Cyperus rotundus),* quackgrass*(Agropyron repens)* having sufficient food reserve in their underground rhizomes and tubers, needs more cultivation with deeper ploughing than a newly infested field with annuals.

(ii) *Hoeing: It* is widely used weeding tool for centuries. Hoe is quite effective in controlling weed in row crop. It is very useful for annual and biennial weeds. In case of perennials, it destroys the aerial growth. The underground growth is not much affected.

(iii) *Hand weeding: It* is done by pulling out weeds from the field. Pulling out is done with the help of *khurpi.* Hand weeding is effective against annuals and biennials. The perennial weeds along with underground portion are not pulled out completely and hence they regenerate in the field. In fields which cannot be sprayed with weed controlling chemicals, can be made weed free using hand weeding.

(iv) *Digging: It* is practised especially for the removal of shrubby and stubby woody perennials. *Prosopis juliflora* (Kikar), *Calotropis gigantia* (Aak or Madar), *Zizyphus nummularia* and *Z. rotundifolia* (Jharber) etc. which grow in horticultural crops in arid regions are removed by digging pit. This is ofcourse, costly and time consuming but very useful technique for the removal of very hard perennial woody weeds.

(v) *Cheering: It* is done using cheer hoe similar to a spade with long handle. It cuts and scrap the above-ground weed growth. As the underground part of the weed is left undamaged, cheering is not useful in perennial crops which are regenerated by roots, tubers, rhizome etc. Cheering is widely used in India in plantation crops, particularly in tea.

(vi) *Mowing:* It is practised to keep the growth of weed under check especially in lawn. The process of mowing is achieved by mower machine and hand blade having long cutting edge of about 1 metre. This method is successful only in case of short weeds growing close to the ground.

(vii) *Burning: It* is very powerful technique of weed control in uncultivated field. It destroys aerial portion of the weed directly through flame of the fire and underground portion through the heat effect.

(viii) *Mulching*: It is getting popularity in controlling weed in a variety of horticultural crops. It excludes light from photosynthetic portion and. thus, inhibits the top growth of weeds. Mulching is very effective against annual crops. The mulches such as straw, saw dust, paddy husk, paper, plastic etc. are common in use. The mulches should be about 10 to 15 cm thick for better efficacy.

Materials such as black polyethylene have been used for weed control in a range of crops. Plastic mulches have been developed that filter out photo synthetically active radiation, but let through infrared light to warm the soil. These infrared transmitting mulches have been shown to be effective to control weeds

(b) Chemical methods

In this method, chemicals are used for weed control. The chemicals employed for weed control are commonly referred to as weedicides or herbicides. Chemical weed control functions on the basis of the fact that certain chemicals are capable of killing weeds without any significant injury to the crops.

A large number of herbicides are now available commercially to tackle various weed problems. A herbicide may be selective or non-selective. A selective herbicide kills a particular type of weed or plant e.g. 2, 4-D (selectively toxic to broadleaf weeds). A non-selective herbicide (for example, Paraquat, Glyphosate) kills the whole vegetation on which it is applied. Nonselective Herbicides (also known as Broad spectrum) are formulated to control both broadleaf and grass weeds. According to the type of action, herbicides may be classified as contact or translocated. The contact herbicide kills the plant parts where it comes in contact (for example, Simazine), whereas the translocated (Systemic) herbicide goes into the plant system and affects the whole plant (for example, Glyphosate). Use contact herbicides on annual weeds while systemic herbicides are more effective on perennial weeds than contact herbicides

The chemical weed control offers following advantages:

- In crop rows, where harrowing or cultivation is impossible, herbicides can be easily applied to control weeds.
- Herbicides give quick response in terms of checking the growth of weed and by that time crop gets an escape and grow.
- Hand weeding injures the root system of crop also and thus damages the crop.
- Herbicides reduce the need of pre planting tillage. In case of minimum tillage, herbicides are very much successful.
- The use of herbicides reduces the requirement of manual labour.

Control measures of weeds infesting fruit, vegetable and flower crops are given in table below:

Crop	Recommended herbicide	Dose (kg/ha)
Mango	Diuron or	2.25
	Oxyfluorfen	4.00
	Glyphosate	6.00
	Paraquat	12.00
Banana	Diuron or	2.25
	Oxyfluorfen	4.00
	Paraquat or	12.00
	Glyphosate	6.00
Grapes	Diuron	3.00
	Glyphosate	4.00
Citrus	Glyphosate	4.00
	Paraquat	3.00
Apple	Diuron	5.00
	Paraquat	3.00
Pear	Diuron	4.00
	Glyphosate	3.00

(c) Biological methods

Reduction in infestation of weeds by direct or indirect actions of biological entities such as plants, parasites, predators and pathogens is covered under these methods.

(i) *Plants as bio-agents of weed* control: The differential growth habits, adaptiveness and competitiveness of the plant over weed, forms the basis of weed control by growing a particular crop. Crops which grow vigorously have an advantage over slow growing ones which are sensitive to weed. These competitive crops, also known as smothering crops, are very effective in weed control. Growing of corn and ginger either alone or in combination, has smothering effect on the growth of weed. They grow more rapidly and fill inter-row space with their canopy faster than the weed. This gives smothering effect on the weed growth.

(ii) *Parasites, predators and pathogens*: In this method, the natural enemies of plants like insects and disease causing organisms are used. Insects kill the weed by exhausting the plant food reserves and through destruction of photosynthetic parts. Of the wide range of enemies, insects have been proved effective. Parasites like viruses, higher plants and fungi that develop only on live plants are also potent agents of weed control.

Lantana camera weed has been controlled successfully by the larvae of *Crocidosema lantana*—the totricid moth, which bores into flower-stem;

Agromyza lantanae, the seed fly which eats flower and fruit and *Thecla echion* which destroys flower. *Cuscuta* spp. (dodder) which grows on old fruit plants is controlled successfully by insect *Melanagromyza cuscutae* and *Smicronyx cuscutae.*Nutgrass or Motha *(Cyperus rotundus),* grows in almost all horticultural crops is controlled by insect *Bactra vermosana.*The larvae of insect *Diacrisia obliqua* feed voraciously on *Parthenium hysterophorus* and destroy the whole plant.

Pathogenic organisms damage the host plants through enzymatic degradation of cell constituents, production of toxins, disturbance of hormone system, obstruction in translocation of food and minerals and malfunctioning of physiological processes. As a result plants die.

The fungi *Alternaria macrospora* and *Puccinia heterospora* control anode *(Anoda cristata),* a weed of malvaceae family. The fungi cause significant reduction in plant height, seed pod per plant, seed per pod and seed per plant. Fungus *Cercospora roclmanii* is effective against water hyacinth *(Eichhornia crassipes)* a problematic aquatic weed for waternut or singhara and lotus.

LITERATURE CONSULTED

Bal, J. S.2014. Fruit Growing.Kalyani Publishers, New Delhi.

Chadha, K. L. 2001. Hand Book of Horticulture. ICAR , New Delhi.

Chadha, T. R.1998. A Text Book of Temperate Fruits. ICAR, New Delhi.

Chattopadhyay, T, K.1999.A Text Book of Pomology. Vol. I-IV. Kalyani Publishers,New Delhi.

http://agritech.tnau.ac.in/org_farm/orgfarm_weed%20mgt.html

http://www.carolinafarmstewards.org/

http://www.puricare.co.za

http://www.miami-dade.ifas.ufl.edu

www.yourarticlelibrary.com/soil/objectives-of-orchard-soil-management.../24644/ ecoursesonline.iasri.res.in/course/view.php?id=151

http://ecoursesonline.iasri.res.in/mod/page/view.php?id=95625

Jayakumar,R. and Jagannathan, R. 2007. Weed Science Principles.Kalyani Publishers.New Delhi.

Jindal, K.K. and Gautam,D.R. 2001. Enhancement of temperate fruit production.Compendium of summer school.

Khokhar, U.U. 2001. Recent trends in intercropping and weed management of temperate fruit orchards. In: Productivity of temperate fruits. Eds K. K Jindal and D. R. Thakur.Pp: 259-280.

Kumar, N.1997. Introduction to Horticulture. Rajalakshmi Publications, 28/5 – 693, Vepamoodu Junction, Nagercoil.Pp: 15.47- 15.50.

Kunte, Y.N, Kawthalkar,M.P.and Yawalkar,K.S.2005.Principles of Horticulture and Fruit Growing.Agri-Horticultural Publishing House, Nagpur.

Mitra, S.K., Rathore, D. S. and Bose,T.K. 1991. Temperate Fruits. Horticulture and Allied Publishers, Calcutta, India.

Singh C.M, Angiras, N.N and Kumar, S.1996 Weed Management. M D Publications Pvt. Ltd, New Delhi.

Singh, J. 2012. Basic Horticulture.Kalyani publishers, New Delhi.

Westwood, M.N. 2009.Temperate Zone Pomology. Timber Press.

5

Fruit Plant Canopy Management

N. Sharma, Chaitanya R. Belsare and D.P. Sharma

INTRODUCTION

Canopy management is one of the most important management practices in an orchard. It is the 'art' of fruit growing involving manipulation of tree canopies to optimize the production of quality fruits. In fact it forms the basis of precociousness, productivity and longevity of the fruit trees. Canopy management is an important aspect of fruit culture due to its effect on yield, quality, vigour, and the prevention of diseases.

In orchard, light is critical to the growth and development of trees and their fruits. The green leaves harvest the sunlight to produce carbohydrates and sugars which are transported to the sites where they are needed – buds, flowers and fruits. Better light penetration into the tree canopy improves tree growth, yield and fruit quality. Training and pruning, affects the tree shape and facilitate exposure of leaves to sunlight. An ideal training strategy centres around the arrangement of plant parts, especially, to develop a better plant architecture that optimizes the utilization of sunlight and promotes productivity. Besides, development of strong scaffold system helps the trees to produce heavy crop loads of quality fruits without any major limb breakage.

In some fruits e.g. peach, grapes, pruning is practiced every year to ensure regular crop. Various fruit production problems, such as colour development, uneven fruit ripening, sunburn, and frost damage, can be addressed by skilful canopy management. During excessive fruiting year, blossom or fruit thinning ensure increased return bloom and fruiting in the following year. In grapes, leaf removal after berry set is practiced to open up the interior of the canopy to light and air to help promote fruit ripening, control diseases and increase spray coverage. It is particularly important in cool-climate areas. It alters the micro-climate within each vine by allowing more air circulation and light penetration.

In addition to pruning and leaf trim, the canopy is often trained on trellis systems to guide its growth and assist in access for ongoing management and harvest.

Besides, tree spacing and planting systems are important components of canopy management. The density and orientation of planting also impact light penetration in an orchard. In many fruit crops, increase in production with enhanced fruit quality is obtained by managing canopies of short statured trees. Small trees capture and convert sunlight into fruit production in a better way than larger ones. Tree size control with rootstock, training and pruning and chemical means induce precocity, increase productivity and fruit quality and reduce labour costs in various culture operations like pruning, spraying and harvesting and ensure higher return to the growers.

Generally, in close planting, quicker shading becomes a problem. An east-west row orientation results in more shading as compared to the western and southern orientation of trees. In many fruit crops, increase in production with enhanced fruit quality is obtained by managing canopies of short statured trees. Small trees capture and convert sunlight into fruit production in a better way than larger ones. The main controlling factors are amount of incoming radiation, and percentage of radiation intercepted by tree canopies for the production of fruit biomass.

Principles of Canopy Management

The basic principles of canopy management are:

1. Maximum utilization of environment (light) and land for fruit production.
2. Avoidance of built-up microclimate congenial for diseases and pest infestation.
3. Convenience in carrying out the cultural practices.
4. Maximizing productivity with quality fruit production.
5. Promote regular cropping.

What is the need of canopy management?

Question may come in our mind, what will happen if a fruit plant is just left over to nature to take care of it? Will it not bear a crop without proper training and pruning? The answer is it will fruit even without our intervention. However, the purpose of nature may not always be same as ours. In nature, the function of a fruit is to provide food for humans, animals and birds and to develop, protect, nourish, and ultimately disperse the seeds away from the parent plant through birds and animals for perpetuation. Nature does not care how many good sized, high quality fruits are born by trees. However, for commercial fruit growing, it is important to regulate the growth and development of trees in a scientific manner so as to obtain higher yield of better quality fruits for higher return. It is achieved by initially giving proper shape to the tree and later optimizing a balance between vegetative growth and fruit production and improving light penetration into the interior of tree canopy. Properly trained and pruned tree also aid in easy cultural operations.

Training and Pruning of Fruit Plants

Training and pruning are often considered as one and the same and are often most neglected practices. Pruning in broader sense means cutting of plant parts and include both real pruning as well as pruning for training, however these two terms differ from each other.

Training

Pruning is judicious removal of such plant parts as limbs, branches, twigs, shoots, or roots to increase its usefulness. It is a skill acquired through knowledge, practice and observation. Pruning of young trees is often referred as ***training***. It is the technique that controls the shape, size and direction of plant growth or the pruning conducted during the early years of plant with the purpose of providing it desired shape and frame work so that it can function more efficiently and effectively. In the nonbearing years, growth of the young tree should be directed into branches that will be a permanent part of the mature tree and all the superfluous branches that should be removed later should be avoided. This technique involves orientation of plant in space through pruning (e.g. apple, peach) or techniques like tying, fastening, staking, supporting over a trellis or pergola in a certain fashion (e.g. grapes, kiwifruit). Training includes summer training and summer pruning as well as dormant pruning. Training is essential for proper tree development of young tree.

The framework should facilitate the development and maintenance of optimum tree shape. Training programme should also produce trees that will be easy to manage in the later year. A properly trained tree requires only moderate annual pruning when it reaches bearing age. Neglected young trees often require removal of large branches later, which however, invites infectious disease organisms. Moreover, neglected trees also suffer more damage from fruit load and storm breakage than properly trained trees. Therefore, it is better to direct tree growth with training than to correct it with pruning.

Objectives of Training

i. To develop a well spaced system of scaffold limbs with strong crotches that shall have ability to bear a heavy crop load of fruits.

ii. To promote early production.

iii. To direct the growth of the tree so that various cultural operations, such as spraying and harvesting are performed at the lowest cost.

iv. Properly trained trees allow maximum penetration of sunlight into the tree.

v. To protect the tree from sunburn and wind damage.

vi. To secure a balanced distribution of fruit-bearing parts on the main limbs of the plant.

Pruning Techniques: There are two basic types of pruning cuts—i) Heading back and ii) Thinning out, with each serving a specific function.

Thinning Cut – This type of cut involves the removal of entire branches or shoots back to their point of origin. Thinning cuts do not invigorate the tree in comparison to some of the other pruning cuts.

Heading Cut – It involve removal only the terminal portion of a shoot. This type of cut induces lateral branching and help in stiffening of the branch. When lateral branches are headed into one-year-old wood, the area near the cut is invigorated. The headed branch is much stronger and rigid, resulting in lateral secondary branching. Young trees and branches where heading cuts are made will be referred to as headed.

Bench Cut - removes vigorous, upright shoots back to side branches that are relatively flat and outward growing. Bench cuts are used to open up the centre of the tree and spread the branches outward. This is a major cut and should only be used when necessary.

Renewal Pruning: It is a technique best used for long-neglected fruit trees to renew and rejuvenate these. It best used for the strong laterals on the outer part of the tree where there is room for such growth. However, this is a lengthy process and involves removal of not more than one third of the branches each year.

When making pruning cuts, it is important to use techniques that will allow the cut surface to heal quickly. Rapid healing minimizes the incidence of disease and insect infection. Pruning cuts should be flush with the adjacent branch without leaving stubs. Also, when large horizontal cuts are made, they should be slightly angled so that water does not set on the cut surface, allowing the growth of rot and disease organisms.

PRINCIPLES OF TRAINING AND PRUNING

The important principles of training and pruning systems are as follows:

i. Modification of Apical Dominance

We can alter the natural characteristic shape and size of a plant and its response by pruning. Some tree grow tall due to unintrupted growth of their growing point (the terminal bud) without allowing the growth of laterals, a phenomenon referred as apical dominance. However, the intensity of apical dominance varies from one plant species to another and it influences not only the number of shoot-forming lateral buds and the lengths of lateral shoots formed but also the angle at which the shoots emerge from the main limb. Growth of lateral buds is directed by the terminal bud, which produces a hormone called *auxin*. Auxin moves downward in the shoot from the shoot apex and inhibits the growth and development of lateral buds. The natural hormones or auxins produced in the growing leaves are inhibitory to lower growth. As they are translocate down the stem, they inhibit or reduce branching and growth from lateral buds. When we cut back the apical growing point, the supply of auxin is temporarily stop and lateral buds break and their growth is encouraged. The lateral growth of branches will resume at a faster rate and the natural growth pattern of tree gets altered. Therefore, we can change vertically growing tree into bush form by giving more heading cuts, such trees are easily manageable.

ii. Balance of Root and Top

The ratio between root top, or vice versa, has considerable influence on the growth, flowering and fruiting of plants. Pruning reduces the size of the above-ground portion of the plant in relation to the root system. The root area remained the same maintaining the rate of absorption of moisture and nutrients. As a result, the undisturbed root system nourishes a smaller number of shoots and buds, as a flush of growth (re-growth) occurs. Generally, the more severe the pruning (greater size or number of limbs removed), the greater the resulting re-growth. In essence, the plant is re-growing in an attempt to restore a balance between the top and the root system.

iii. Dwarfing Process

Pruning is a dwarfing process and can be used to maintain any desired tree size. Removal of a branch removes not only stored carbohydrates but reduces the potential leaf surface as well. Therefore it reduces root growth by supplying lesser carbohydrates. Smaller root system can supply lesser quantity of water and nutrients to the top and thus reduce growth.

TRAINING SYSTEMS

Pruning starts immediately after planting the tree in the field. At the time of planting, the whip should be headed back to about 60-90 centemeter from the soil surface. It will ensure better survival of the tree by balancing the size of the above ground and the damaged root system during digging and promote lateral bud break. Height of the heading cut depends upon the type of training system to be followed. Heading back the plant too low may induce excessive vegetative growth from all growing point, and the branches will be too low in position and cannot become a part of tree structure. In contrast, heading back at higher than 90 centimeter height will result in short and weak laterals at the top and poor growth further down the trunk. These laterals will be difficult to be used as a framework structure. Subsequent pruning procedure depends upon the type of training system to be developed. Proper training from the beginning will help promote a strong, productive tree which requires a minimum of pruning later in life. There are different types of training methods employed in fruit crops, which depend upon the natural growth habit of tree, climate and resources available. Three most common structures adopted for the standard trees are open centre, central leader and modified leader.

1. Open centre system (vase open): In this system the main stem is allowed to grow to a certain height and the leader is cut to encourage lateral scaffold from near the ground giving a vase shaped plant. This is common in peaches, apricots and ber .

The open centre tree (Plate 1) has 3 to 5 major limbs, called scaffolds, coming out from the upper end of trunk. However, selected branches should not originate too

Plate 1: Open Centre (Nectarine) (See colour version on page 420)

closely to each other, because these might split in the event of heavy crop load or snow. The branches are trained to equal dominance by pruning each year. This training system allows for adequate light penetration into the tree, which minimizes the shading problem prevalent in higher vigour trees such as peach.

At planting, peach trees are generally headed approximately 75 to 80 centimeters above the soil surface. During the spring, new branches will come from the buds that are 15 to 20 centimetres below the heading cut. Remove branches that are too low. After the new vegetative growth is approximately 10 to 20 centimetres long, it is time to select the shoots that will become the major scaffolds. The lowest scaffold should be 60 to 75 centimeters above the soil surface to avoid interfering with cultural work under the tree, such as harvesting and weed control. It is best to select 3 to 4 scaffolds that are uniformly spaced around the tree, with wide branch angles, and not directly across from another scaffold. However, if there are less than 3 scaffolds the tree should be cut back to a whip and the side branches removed.

Advantages

i. Affords maximum light penetration, which gives highly coloured and better quality fruits

ii. A low spreading tree develop, which facilitate pruning, thinning, spraying, and harvesting.

Disadvantages

i. Since scaffold branches are attached at the same location to the trunk, heavy crop load leads to breakage of limb

ii. Fruits are more prone to hailstorm damage

iii. The trunk is more susceptible to winter injury

2. Central leader system: In this system the main trunk of plant is allowed to grow uninterrupted and well spaced side branches are allowed to grow around it. The dominant main trunk is referred as the *leader* and the side branches as *subordinates*. This system is also known as closed center system and followed in fruit crops having upright growth habit. Because apical dominance is favoured, this result in charismas tree shape with spreading branches. It is used common in use in apple, pear, pecans, litchi and sapota .

In this system, the branching angles are very important because the base scaffolds support heavy crops once bearing begins. Narrow crotch angles are very weak, and may break easily from the trunk due to heavy crop load. Branches having 45 to 60° angle from the trunk will be stronger and support heavier crops than those with narrower angles.

Method: At the time of planting in winter, the tree is headed back to 0.7-0.9 m from the ground. Summer pruning starts during the spring, when the new vegetative growth has reached 8-10 cm in length. It is the optimum time to select leader and remove undesirable side shoots. The upright shoot near the top of the tree is selected as leader and all other competing shoots for approximately 10 cm below it are removed. Subsequently, the branches lower than the desired height i.e. 60 centimeter are also removed. The remaining side shoots (laterals) should be spread out using wooden sticks, toothpicks or spring clothespins to form an angle of 60 to 70 degrees between the leader and the side shoot. This angle is referred to as the branch or crotch angle. Narrow angled branches are too vigorous and have a weak point of attachment to the leader and are likely to break due to heavy crop load. Furthermore, practice of spreading slow down the growth of lateral branches to desirable level and promote the development of secondary or side shoots on these and promote early production.

Summer pruning should be done once a month through July to remove unwanted growth. Tree greatly benefits from summer pruning during tree training. It encourages the growth of the selected leader and side shoots favourably located on the trunk. Summer pruning reduces the amount of wood to be removed during dormant pruning by discouraging the growth of less desirable shoots. If growers neglect this aspect during the first year, it will result in improperly trained tree and will require drastic pruning during winter to correct tree structure.

During the next winter, 2 to 3 well spaced branches with wide crotch angles, called a *scaffold limbs*, are selected among those allowed to grow during the previous summer. The selected scaffolds should be uniformly spaced around the trunk, not directly across from or above one another. The lowest branch on the leader should be about 60 centimeter from the soil surface to allow movement under the tree, and preferably it should be facing south. Because, when the nearest branch to the ground is towards the south, it will provide shade the main trunk and prevent sun burn injury.

All the selected laterals are headed hack to $1/4^{th}$ of their length to encourage branching and for stiffening. Dormant pruning should be done in late winter. The same pruning procedure is followed in subsequent years to select the remaining side branching. Secondary branches adequately spaced on the limbs are also selected from the second dormant season onward and these are also headed as the primary scaffolds. Any upright shoot on the limbs should be removed. Any broken or diseased limbs should also be removed. Remove weak, twiggy growth. Keep the leader dominant by shortening competing branches.

The process of training will be completed in 3-4 years, after which the central axis weaken and tree eventually attain a pyramid shape, which is wide at the base and narrow at the top.

Advantage

1. Crotches are strong

Disadvantages

i. Tree becomes too large to manage

ii. Poor light penetration

iii. Lower productivity and fruit quality

3. Modified leader system: The most commonly used training system is modified leader. It combines the best qualities of the *central leader* and *open center systems.* Initially, the pruning procedure is the same as followed in developing the leader system. However, a leader develops on the tree until it reaches the height of 2 to 3 m. It is dominant until the tree is formed and is then restricted by cutting back the leader to a weak growing lateral. Lateral are selected to ascend in a relatively spiral fashion up to the central trunk, and these are cut back until the proper number (generally 5-7) and distribution of branches have been obtained. The lower lateral is 60 cm or more above the ground. When the central leader is removed, the tree develops a top that is rounded and open. The limbs are low and well spaced, and fruiting wood is well distributed, so this is not drastically different from the *central leader system.* This system is adopted in apple, pear, cherry, guava, plum (cv. Santa Rosa), apricot, walnut, persimmon etc.

Advantages

1. It has the advantage of central leader in having good crotches
2. Afford better light penetration than leader system, therefore ensure better fruit quality
3. As the tree height is kept lower than the leader system, it facilitate various operations like pruning, spraying and harvesting.

In grapes, common training followed includes Bower (Arbour or Pergola), Vertical Trellis, Kniffen.

Training Systems for Dwarf Trees

High density plantation require tree size control, which is principally accomplished with size controlling rootstocks, genetically dwarf scion cultivars, training and pruning technique. In HDP systems, training and pruning techniques require special attention and must be modified to contain tree size and promote early production. The main aim should be to have more of production units and less of structural units. Branches are often fruitful on the entire length. Sometimes, fruit buds are also directly developed on the main trunk. Trees grown on dwarf, size-controlling rootstocks require training and pruning primarily during the summer to control tree size and encourage fruiting early in the life of the tree with minimal dormant pruning. In Europe, mostly well branched feathered trees are planted to begin fruit production a year earlier. Branch spreading is must for growth control and early production. Trees of various shapes are developed with the main aim of intercepting maximum light. Limb number and their arrangement play an important role in providing a particular tree shape. With uniform distribution of light in the entire canopy, fruits of uniform size, colour and quality can be obtained. There are various systems of training proposed for HDP systems; some are explained as here under.

1. Cordon system: The term 'cordon' simply refers to a single stem with short side shoots (the fruiting spurs). Trees can be trained singly vertically or at an angle of to 45 degrees (oblique cordon). When the tree is trained horizontally, it is referred as "step over". Other forms are multiple vertical 'U' or double 'U' cordons. It is suitable for spur bearing apples and pears, grape vines incapable of standing on their stem, passion fruit, red current, and gooseberries. An oblique cordon system helps in dwarfing the tree/vine and is more productive. Vigorous apple cultivars on vertical cordon are not suitable.

Cordons need a permanent support system. The trees are trained against a wall, fence or on wires between free-standing posts. In grapes, a single arm is allowed to develop from the trunk which will be trained along a stretched wire towards one side, or the main truck itself may be bent on the wire at a suitable height. Sometimes, trunk may tide to the wooden pole. This form is less popular and used for maintaining catalogue of varieties in the orchard.

Advantages

i. Plants are easier to manage.

ii. This system ensures better quality fruit production.

Disadvantage

i. Cordons are lower-yielding.

2. Dwarf pyramid: It is a low headed compact central leader tree, with a central stem about 2.5 m tall. The lowest branch is at about 30-40 centimeter from the ground and the successive branches radiating at interval along the central stem, gradually diminishing in length from bottom to top, thus giving a pyramid effect. Fruiting spurs are developed on the short branches. Trees are raised on dwarfing rootstocks. Apple trees are planted just 1.75 m apart for M 9 and 6m apart for M 26. Pears are also planted 2 m apart on Quince C. If more than one row is required, the rows should be 2 m apart. In hedgerows, trees should be spaced one metre apart in rows, whereas, rows are kept two to three metres to facilitate mechanical operations. Summer pruning is an essential part of success with dwarf pyramids with minimal winter pruning to restrict the growth. The weight of the fruit will pull the branches down and further limit growth. Once the crop has pulled the branches down, they will stay bent. This is the shape of many fruit trees.

Method: At the time of planting, one-year-old (maiden) tree is cut back to about 50 centimetres. Cut just above a bud, and then rub out the bud that is second from the top; otherwise it may produce a second leading shoot. Select three or four of the buds in different directions below the topmost bud to form the tree's first scaffolds. Remove the unwanted buds or shoots. The upward growing topmost bud will form the tree's central leader, and notch above the two lower buds. Developing shoots from lateral buds are selected to form strong branches. In the second winter cut back the central leader to about 45 centimetres from the first year's cut. As in the previous winter, rub out the bud below the top one and another tier of branches are developed from the three or four buds below it. As in the previous year, notch the two lower buds using a sharp knife. Repeat this training each winter until the main stem has reached a height of about two metres. After this, cut back branch leaders in summer only, except during the year experiencing excessive growth due to wet summer. In such situation, winter pruning may be practiced to remove such growth in order to maintain the pyramidal shape of the tree.

Advantages

i. Early fruiting, higher yield and better fruit quality.

ii. Because of smaller size, trees are easier to manage.

3. Spindle Bush: A spindle bush tree is a variant of the pyramid form having a single central stem (a leader) with the branches spreading outwards from this. This system was designed for dense orchards of by Otto Schmitz-Hubsch and Heinrichs in Germany in 1936 and suited to densities up to 2,000 trees/ha at 2 to 3m in height and 3 to 4m x 1 to 2m apart. It has been most popular training system for dwarf apple and pear trees. It can also be used in plum and cherries.

One of the common features of this system is that the lateral branches are tied down to a horizontal position. Positioning of shoots and branches towards more horizontal induce earlier fruiting. Once new shoots have been tied or weighted down for a year or so, they will stay in place on their own. Another principle is that fruiting is the most effective way of reducing growth and, once a shoot has started to fruit, it will continue to do so.

Method: In Europe, spindle systems are usually planted using well feathered two-year old nursery trees. Similar feathering can be induced in the one year old Whip (unfeathered plant) by heading it back to 90cm after planting. At planting of feathered tree, a number of strongest laterals, usually four are retained to form part of the permanent scaffolds in the bottom third of the tree. The lowest one should be no less than 60cm (2ft) from the ground. These are all cut back by half and remaining feathers are cut back to their point of origin. Competing laterals that develop at the end of the unpruned central leader have to be removed in a very early stage. The central leader is shortened by a third during winter. Any strong growing narrow angled lateral should be removed as and when noticed. As the leader grows more scaffolds are selected and spaced equally. The shoots to be selected as scaffolds should be wide angled and these should be tied down horizontally to promote early fruiting. However, Leader dominance is important and if it is lost will result in a reduced tree canopy. On the contrary, if it becomes too strong lateral growth and development of lateral will be reduced. In such case, leader vigour is reduced by thinning it out to a weak lateral and then the weak lateral is tied up to the supporting pole into the position of the leader and left intact.

When the tree has reached a height of 2-2.5m, its growth is restricted in winter by heading back upward growth to a weak growing lateral. In this system, fruiting starts earlier. Initially the crop is taken on the shoots (laterals) growing out from the trunk and later on side shoots that develop on these. The branches are shortened after taking crop for four or five years, or else these are these may cut out completely to make room for younger and more fruitful ones. Some of the lower branches are allowed to stay for longer; this is simply because there's nothing below them to be shaded.

Advantages

i. These are dwarf trees and easy to manage.

ii. It is one of the most productive systems, in which fruiting start earlier.

iii. The fruit is borne on the outside of the tree, where it gets better exposure to light, and hence potentially better quality and flavour.

4. Vertical axis: Developed by J.M. Lespinasse in France, this has become a widely used system in Europe and United States. The system relies on high density, dwarfing rootstocks, minimal pruning, limb renewal, and an effective support system. A vertical central axis (Plate 2) is developed with relatively 'weak' fruiting branches arising around the leader. In this a narrow canopy along the axis with 3-5 feet is maintained in a single row is maintained to ensure better distribution of light and better fruit quality. Malling 9 is an excellent rootstock for this system, however G.41, G.11, G.16 are equally good. Well featured trees are planted at a spacing of 4 to 5m × 1 to 2m, accommodating 1,000 to 2,500 trees/ha. Tree require minimum pruning at planting except removal of vigorous laterals. All laterals should be headed back by 1/3rd and the leader headed 45 centimeters above the top scaffold. Maintaining apical dominance is important in the vertical axis system particularly during early stages of development to ensure weak fruiting branches – therefore no heading of the leader occurs. Once fruiting start, branches are systematically renewed from the axis leaving a "bevel cut" to prevent them from becoming permanent scaffolds and shoot replacement. Various variants of support system can be used. Although, single high wire with a conduit stake at each tree is the preferred, but sometimes 2 to 3 wire trellis is required. Height can reach up to 3m.

Plate 2: Vertical Axis training in apple (See colour version on page 420)

Advantages

i. Start early production.

ii. High quality fruit.

iii. Lower management costs than free standing systems.

Disadvantages

i. Tall system that requires working off the ground.

ii. Vigour control is difficult. It is difficult to maintaining sufficient vigor in the bottoms and controlling excess vigour in the tops in vigorous apple cultivars.

5. Slender spindle: The slender, spindle-type tree (Plate 3) is a modified miniature central leader tree with many more limbs per tree. It is suited for densities of 2,000 to 5,000 trees/ha, with tree spacing of 3 to 3.5m x 1 to 1.5m. In this, every individual tree must have a permanent support system to the height of the entire tree to promote rapid tree growth and support the tree with heavy crop loads. Post is usually 3 meter long wooden (5-7 cm dia.) or of conduit or galvanized pipe (2.5 cm dia.) which is placed approximately 6 inches away from each tree and driven 2 feet (60 cm) into the ground. Alternatively, trellis systems can also be used with a smaller posts at each tree (i.e. bamboo stake or 1 cm conduit) attached to a top wire approximately 2.75 meter above the ground with large anchored posts every 15-30 meter or more to stabilize the top wire. The trunk or central leader is maintained continuously tied loosely with post to a maximum height of 3 meter.

The system is followed using feathered apple plants on M 9 rootstocks in Europe. Geneva rootstock series rootstocks such as G 41, G.11, and G.16 have also been found appropriate rootstocks elsewhere. More vigorous rootstocks can be used for spur type (dwarf) apples. The branching should begin at approximately 60 to 75 centimeters above the soil surface. During the formative years all lateral branches are left on the tree to optimize early production, branches are thinned out to the two permanent whorls as shading becomes a problem in the 3rd to 5th year or later as required. Above the second whorl of branches, all branches are temporary and are renewed every 2 to 4 years. It is important to maintain leader growth as well as encourage lateral branching to contain tree to an allotted space and ensure optimal productivity. It involves more severe pruning than the vertical axis, as the

Plate 3: Tall Spindle training in apple
(See colour version on page 420)

maximum branch spread of this tree outward from the leader is allowed within 1-1.75 meter and and tree height is usually restricted up to 3m. The slender spindle is a more conical shaped tree with a distinct, supported, vertical central leader branch. At the base of the tree there is a permanent frame of a varying number of branches, above which exist more or less horizontal fruiting branches shorter in length than the frame branches. Leader control is important. Strong leaders and lateral growth in the top of the trees should be avoided. Generally the leader is headed back to a weaker lateral when it becomes too dominant.

Advantages

i. It maximizes profitability through early yield, improved fruit quality, reduced spraying, pruning, and training costs.

ii. It is easy to rapidly turn over apple varieties from those less profitable to those more profitable.

6. Super spindle: It was first developed the late eighties in the Bodensee region of Germany for super high densities apple orchards with the objectives of obtaining higher and early production, less reliance upon chemical inputs, reduced labour cost, easy picking and increased profits and the possibility to introduce quickly new valuable varieties into the market. In pear, the super spindle system has been utilized for super high-density orchards on Quince C rootstocks. It is suited for densities of 4000 trees/ha with a spacing of less than 3 m in a row and less than 0.8 m apart. However, it can be used for densities up to 12,000 trees/ha. This system has closely spaced compact trees with short fruiting wood or spurs evenly spaced along the central leader. Super spindles require a multi wire support system. Usually whips with a number of short feathers along the leader are used. Even cheaper trees such as budded rootstocks (i.e. sleeping eye trees) can be used.

Advantages

i. Greatly higher yield than the traditional slender spindle system.

ii. Higher return in the long run due to tree management cost.

iii. New varieties can be introduced as quickly as possible to meet market demands.

Disadvantages

i. Higher investment at planting time.

ii. Fruit colour is poorer than slender spindle.

ii. Fruit size smaller than Slender Spindle trees.

7. Palmette: The palmette training system was introduced in mid 1950s by Baldassari in Ferrara, Italy. Later, it became successful because it improved orchard productivity by inducing earlier bearing and allowed mechanical pruning and harvesting. Palmette refers to the shape of a palm leaf, an open fan or an open hand with spread fingers. It is formed by developing lateral scaffold branches from the trunk beginning at 30 cm from the surface of the ground. These laterals are in pairs and are equally balanced in opposite directions. Other scaffolds are in turn vigorously developed but are kept about 30 cm apart.

A number of variants of palmette training (Horizontal, Oblique, Candelabra, Fan Palmette) have been developed, each with a central leader with scaffolds in the plane of the row only. Tiers of scaffolds are chosen each season and tied to wires to reduce vigour and promote spurring. These systems have been popular because the bending of branches on trellises controls growth and provides a balance of fruiting and vegetative growth.The palmette is best suited to medium-high planting densities (700–1,100 trees/ha in peach, and 1,500–2,000 in standard apple cultivars), with tree heights ranging from 2.5 m (apples on weak rootstocks) to over 4 m (peach in fertile soils). It is widely used in deep, fertile soils for those species (e.g. peach, plum, apricot, pear on seedling) whose vigour control is more difficult due to lack of suitable rootstocks.

The modern palmette greatly differs from the original concept, because of cost considerations. It is developed as freely as possible (minimizing pruning, and thus improving early yield) in the first 3–4 years, given the constraint that a hedgerow must be formed. It is more labour-intensive to establish than other systems but it is less expensive to maintain once the tree is formed.

8. Tatura Trellis: This is an advanced system for peaches, developed for managing more than 1500 trees per hectare without the use of dwarfing rootstock. Here emphasis is laid upon developing more of fruiting branches or surfaces and less of structural surfaces. At the time of planting, one year old plant is headed back to 20 cm above the ground level. In next growing season two limbs or branches are selected in opposite directions and these branches are trained across the inter row space at an angle of 60 degree from the horizontal, forming V-shaped canopy. The canopy is fastened on wires supported by poles or angle iron inserted in soil at both ends of the row in X cross. Generally three wires are fastened through holes of the each of the upper arm of the X cross. The secondary branches are developed along each primary branches forming fruiting canopy. In this, maximum amount of sun radiation falls on the tree canopy, and none fall on the alley ways. This ensure higher yield per unit area.

Another modification is V shaped canopy in which trees are planted in inclined position alternately in opposite direction in a row. It is also referred as V hedge.

Advantage: Efficient system for high density planting established on seedling rootstocks.

9. Hedgerow Plantation: This system is adapted to mechanical cultivation. A hedgerow planting involves establishing plantations very closely (1- 1.3 m) in a row. The distance between the rows is kept wide (4m) to accommodate machinery and management. Adapted to compact trees, staked free standing on trellised growing north to south. Trees are grown as small bush that develop into a continuous or hedgerow habit. Hedgerow plantation can also be developed without support using semi dwarf or semi-standard rootstocks. Height of the tree is maintained to 3 meter.

Pruning of Bearing Trees

The severity of pruning varies with tree age and vigour and with kind of fruit. Old trees in low vigour can be pruned more heavily than the younger and more vigorous ones of the same kind. Pruning techniques to be required also depend upon the kind of fruit because of different bearing habits. The fruits which bear on one year old shoot e.g. peaches require heavier pruning than those bear on older spurs e.g. cherries.

Principle of Pruning: To admit more sunlight, remove unproductive branches which are producing few or no fruits and also to keep the plant in its proper vigour, vitality and to obtain optimum yields of good quality fruits.

Objectives: In general, bearing trees are pruned to meet the following objectives:

i. To regulate a balance between vegetative growth and production.

ii. To improve size, colour and quality of fruits.

iii. To control insect-pest and diseases.

iv. To increase longevity of the tree

Steps in pruning a bearing tree

- Start from the top of tree and work downward, cutting strong upward growing limbs to horizontal weak laterals.
- Remove or shorten any large limbs that are crowding other limbs or growing parallel to them.
- Remove water sprouts (vertical shoots) growing on limbs, except where these are required to fill the vacant space.

- Remove all diseased/ insect pests' affected, broken and dead wood. Leave well spaced fruiting wood along the limbs weak growing shoots on the limbs after removing thin and weak growing shoots and twigs.
- Where required prune back limbs sufficiently for stiffening.

All the trees require corrective pruning. However, some fruits require heavier pruning to adjust crop load, improve fruit quality and ensure regularity in bearing. Type of pruning cuts to be given depends upon the kind of fruit, because of their different bearing habit. Some fruits require more thinning cuts while others, the heading back. Apple bears terminally on 1- year old shoot and spurs and rarely laterally on 1- year old shoot, therefore require thinning cuts. Heading cuts will remove fruit buds and stimulate vegetative growth. In spur bearing cultivars, heading cuts will induce elongated shoot growth instead of spur development. In apples, 40% of one year old shoots should be thinned out, and heading cuts are given only to remove upright growing branch from the point where a shoot is growing outward. This will help to spread the tree and improve light penetration in the canopy. In peach and nectarine, flower buds develop laterally on one year old shoot. These require a combination of both thinning out leaving only 40-60 thick and healthy shoots per tree and heading back of the retained shoots to 10-12 nodes. This practice will not only improve fruit size and quality, but will also promote new shoot growth that will become a bearing surface in the following year.

Flower buds on almond, apricot, cherries and plum develop laterally on 1- year old shoot and spurs. In plums and apricots, 30% removal of 1-2 year old shoots is thinned out and rests are headed back by 1/3rd to ½. Pear, cherries, almonds, walnuts, pecannuts require comparatively less pruning, that is related to annual shoot extension growth, and restricted to thinning cuts. Grapevines bear on new growth that emerges from the last season's mixed buds. Regular pruning is practiced to regulate the number of buds per cane and bearing canes per tree to regulate yield, size of bunches and quality of berries; however intensity of pruning depend upon cultivar and method of training. Evergreen trees such as mangoes, citrus, guava, etc. require little pruning, wherein pruning is mostly restricted to removal of dead and diseased wood, suckers and crowded branches.

Time of Pruning: In temperate fruits, usually pruning is conducted during dormant winter months. It is a convenient time as tree structure without leaves can be easily seen and pruning cuts to correct the shape can be made more easily. Careful detailed pruning in winter help to achieve our objective to develop strong scaffolds on the tree. It can be done any time between December to February, however, optimum time is January. In areas where winter injury to pruning wound is experienced, dormant pruning should be done as late as possible.

Summer pruning is an essential practice during training of young trees. In bearing trees, it can also be practiced to improve light penetration into the interior of tree canopy, fruit colour and/ or fruit firmness. In high density plantation, fruit trees are summer pruned and winter pruning is minimal, as the later invigorate the tree.

LITERATURE CONSULTED

Barritt, B. H. 2000. Selecting an Orchard System for Apples. *The Compact Fruit Tree* ,Vol.33 (3): 89-92.

Buler, Z. and Mika, A. 2009. The influence of canopy architecture on light interception and distribution in "Sampion" apple trees. *Journal of Fruit and Ornamental Plant Research*, Vol. 17 (2): 45-52.

Chadha, T R. 2001. *Text Book of Temperate Fruits*. Directorate of Information and Publications of Agriculture, ICAR, New Deldi. 217p.

Childers, N. F., Morris, J. R. and Sibbett, G. S. 1995. *Modern Fruit Science: Orchard and Small Fruit Culture.* 10th Ed. Gainesville, Fla. : Horticultural Publications. 632p.

Lespinasse, J.M. and Delort, J.F. 1986. Apple tree management in vertical axis: appraisal after ten years of experiments. *Acta Horticulturae*, 160:139-155.

Musacchi, S. 2008a. Bibaum: a new training system for pear orchards. *Acta Horticulturae*, 800: 763-768.

Özkan, Y. and Kücüker, E. 2009. Pruning and Training Techniques in Dwarfing Apple trees. Tabad 2 (1): 1-9.

Richard, P. Marini. 2014. *Training and Pruning Apple Trees*. Virginia Cooperative Extension. Agricultural Research and Extension Center, Virginia Tech. www.ext.vt.edu 9p.

Robinson, T.L., Hoying, S.A. and Smith, W. H. 1997. Training strategies for high density orchards of the future. New England Fruit Meetings, 103: 27-36.

Robinson, T.L., Hoying, S.A. and Reginato, G.L. 2006. The Tall Spindle apple planting system. New York Fruit Quarterly, New York, 14(2): 21-28.

Robinson, T.L. 2011. Advances in apple culture worldwide. Rev. Bras. Frutic., Jaboticabal - SP. Volume Especial, E.037-047.

Tustin, D.S. 2000. The Slender Pyramid tree management system–in pursuit of higher standards of apple fruit quality. *Acta Horticulturae,* 513: 311-319.

Westwood, M. N. 2009. *Temperate-Zone Pomology: Physiology and Culture*, Timber Press; 3rd edition, 536 p.

6

Nutritional Needs of Fruit Plants

P.P.S. Gill

INTRODUCTION

The chemical compounds needed for growth and metabolism in plants are termed as nutrients. Mineral nutrition of fruit plants is a vital soil management practice in an orchard. The nutrients are converted to cellular constituents or used for energy purposes through metabolic process. Broadly the essential plant nutrients may be divided into macro-nutrients and micro-nutrients. Macro nutrients are needed in plants in relatively higher amounts than micronutrients.

Macro nutrients: C,H, O, N, P, K, S, Ca, Mg.

Micro nutrients: Fe, Mn, Cu, Zn, Mo, B, Cl, Ni.

Altogether 17 elements are considered essential for normal growth and production of a plant.While elements like cobalt, sodium and silicon have been established as essential elements for some plant species and not all. C-H-O matrix accounts for over 90% of the dry matter and is usually 20 to 50% of the fresh weight of the whole plant. Chloride is a most unusual nutrient as it is often present in plants in high concentrations yet its essential requirement is very much lower. Chlorine like Na and K plays an important role in osmoticum. The nutritional needs of fruit trees can be assessed through leaf and soil analysis, tree growth and cropping, and past experience. The plant is in state of nutrient deficiency when it contains below optimal level of an essential element, it is said to be in a state of deficiency. Deficiency results in physiological disorders in leaf and fruit which can be identified by visual symptoms but acute deficiency of any essential element can lead to death of the plant. The loss of nutrients in soil can be replenished by addition of fertilizers and manures to the soils. Fertilizer is any material of natural or synthetic origin that is applied to soils or to plant tissues (usually leaves) to supply one or more plant nutrients essential for the growth of plants.Fertility of the soil can improved steadily by using farmyard manure (FYM), oil cakes and fertilizers. Organic manure provides all the nutrients that are required by plants, but in limited quantities and also improves the physical, chemical and biological properties of the soil. Commercial fertilizers and animal manure are used to increase crop yields and to replace soil nutrients removed with harvested crops.

TYPES OF MANURES

1. Organic Manures

- Organic manures are valuable by-products of agricultural farming and allied vocations. Organic fertilizers are known to be the safest and long lasting soil improvisers and conditioners. These manures consist of natural and bio-degradable constituent and elements that can easily be worked upon by soil microorganisms and thus taken up by crops and plants in a safe manner. They do not exert harmful and polluting effects on the soil and plants. Organic manure is highly rich in nitrogen.
- These are very important for horticultural plants, annuals, perennials and tree fruits. The addition of organic matter provides material from which microorganisms produce vital compounds such as gums and resins, and make minerals available. Organic manure is easy to find, easy to store and does not require any special treatment as it is not reactive or inflammable.

Some of the sources of organic manures are:

Farmyard Manure (FYM): For successful cultivation of fruits, a heavy quantity of organic matter should be added particularly in soils having low organic matter. Plant materials after passing through animals may be returned to the soil as FYM which decomposes and in turn supplies the soil with minerals important for maintaining its fertility. Farmyard manure includes mainly dung of dairy cattle and animal litter. FYM is good for all soils and crops. It must be applied 4-6 weeks before the sowing of crops depending on the weather conditions. Evenly spreading over the field and mixing with the soil without delay give better results. The use of FYM is decreasing because of its unavailability in sufficient quantities; high cost and the nutrient available per unit value are far less than commercial fertilizers. Moreover, with the advent of mechanical agricultural operations, the number of cattle has decreased thus, reducing the production and availability of FYM.

Compost: Compost is prepared from cattle-shed wastes. Composting has been adopted since past many years. It is the process of reducing vegetable and animal refuse (rural or urban) to quickly utilizable condition for improving and maintaining soil fertility. Good organic manure similar in appearance and fertilizing value to cattle manure can be produced from waste materials of various kinds, such as cereal straws, cotton stalks, groundnut husk, farm weeds and grasses, leaves, leaf-mould, house refuse and other similar substances. These materials are rich in cellulose and other readily decomposable carbohydrates. The direct application of such under-composed, low-nitrogen organic matter as manure brings temporary deficiency of mineral nutrients (specially nitrates and ammonium compounds) in the soil by stimulating the growth of microorganism which compete with crop plants for available nitrogen, phosphorus and other elements. Hence, it is necessary to compost or partially decompose them, the process which lowers the C/N ratio from about 10 or 12 to 1.

Poultry Manure: Poultry manure has high nitrogen content and has long been recognized as one of the most desirable manures. It supplies essential plant nutrients and adds organic matter, which help to improve the soil moisture and nutrient retention. Manure is high in phosphorus and in soils testing high in phosphorus concentration the manure should be applied on the basis level of phosphorus rather than nitrogen. The actual nutrient composition may vary depending on the manure- to litter-material ratio, litter (manure) handling, and the type of bird, feed, and litter material.

Oil Cakes: Oil cakes can be grouped into two classes, i) edible oilcakes-suitable for feeding the cattle and ii) non-edible oilcakes- not suitable for feeding the cattle.

Edible oilcakes are normally fed as concentrates to the cattle, however some are also applied by the farmers to the soil. The quantity of nitrogen varies with type of oilcake. It is from 2.5 per cent in *mahua* cake to 7.9 per cent in decorticated safflower cake. Oilcakes also contain small percentage of P_2O_5 (0.8 to 2.9%) and potash (1.2 to 2.2%). Though insoluble in water, their nitrogen becomes quickly available to the plants in about 7 to 10 days after application with *mahua* cake as an exception, as its nitrogen dose become available till about two months after application. Oil cakes are suitable for application to plantation crops and can be applied few days prior to sowing or at sowing time. It should be well powdered before application, so that they can be spread evenly and are decomposed by microorganisms.

Green Manure: Green manuring is a practice of ploughing or turning the undecomposed green plant tissues into the soil for the purpose of improving physical structure and fertility of the soil. The green manure essentially composed of green matter of plants or crops at different stage of decomposition. These crops when decompose enriches the soil with their essential nutrients. These crops are described as cover crops which are meant to serve the purpose of nourishing the devitalized soil. Generally, leguminous crops are first choice which fixes nitrogen with the help of rhizobium bacteria and thus improves the soil fertility. Green manure crops also increase organic carbon in soil, water holding capacity, improve soil aeration and other soil characteristics and also prevent soil erosion. Commonly cultivated green manure crops are *dhaincha*, lucerne, barseem, sunhemp, clovers etc. Green manuring is considered as an economical way of improving the organic matter of the soils. It improves the texture of sandy and clayey soils and enhance the productivity of poor soils. Broadly two types of green manuring are followed; green manuring *in situ* and green leaf manuring. In case of *in-situ* green manuring, the manuring crop is raised and subsequently incorporated in the same land as pure crop as intercrop with the main crop. For ideal decomposition of green manure crop the green material should be succulent and sufficient moisture must be present in the soil. This practice is commonly followed in northern India. While in the green leaf manuring system the soil is incorporated with leaves, twigs, collected from trees grown on wastelands, forests, bunds etc. this practice is common in eastern and central India.

2. Inorganic Manures

The ever increasing human population and depletion of natural resources of arable land has necessitated the utilization of inorganic fertilizers to replenish the soils and to boost crop yields per unit area. These chemical fertilizers supply specific essential elements to the plants that were not met from soils in sufficient quantity to sustain high yields of the fruit plants. Unlike organic manures the synthetic fertilizers act as the immediate source of nutrients and effects are observed soon after application. Most commonly used chemical fertilizers are given as under.

Urea: Urea is widely used fertilizer to supplement nitrogen to the agricultural crops. It is an inexpensive, easily available white crystalline solid granules containing 46% nitrogen. Although urea is naturally produced in humans and animals, synthetic urea is manufactured when carbon dioxide is reacted with anhydrous ammonia. This process happens under high pressure and temperature. Enzyme urease, hydrolyze urea to ammonium and carbon dioxide as it is applied to the moist soils. This process takes 2 to 4 days and happens quicker in high pH soils. Urea is hygroscopic and very soluble, it must be kept away from moisture until its use. As a result of the chemical reaction that takes place when urea is applied to the soil, special care must be taken to ensure that the nitrogen is not lost as ammonia by volatilization. It must be well mixed in to the moist soil immediately after application to avoid this loss. Generally urea is supplied as broadcast to the fruit plants, but it can also be used as a foliar spray to promote the vegetative growth of citrus nursery plants. Urea may be mixed with other fertilizers or may be applied on its own and is one of the top fertilizers for acidifying soils.

Ammonium Sulfate: Ammonium sulfate is also an important fertilizer for soil that is high in pH. This fertilizer is a crystalline salt varying colour from white to shades of grey and contains 21-percent nitrogen. It is the most acidifying of the nitrogen fertilizers, making it a good choice for highly alkaline soils, therefore application of ammonium nitrate is recommended for soils containing high lime. After application to damp soil, ammonium particles escape, causing an increase in the acidity level of surrounding soil. The rising acidity creates a lower pH level in the soil, increasing growth of plants and crops. This salt absorbs a little moisture and can be stored well in dry conditions. Ammonium releases upon exposure to water and can make its way into the air in the form of ammonia gas.

Calcium Ammonium Nitrate: Calcium ammonium nitrate also known as CAN, is a widely used inorganic fertilizer and may be applied on any type of soil as top dressing. Calcium ammonium nitrate is a granular or prilled fertilizer, containing 27% nitrogen % (ammonia N (N-NH3) - 13.5 %, nitrate N (N-NO3) - 13.5 %), magnesium (MgO) 4.0 %, calcium 6.0 %. The combination of ammonium nitrogen and nitrate nitrogen makes CAN a universal and excellent fertilizer, which ensure an optimal nitrogen supply for all plants. It provides plants with necessary nutrients and compensates lack of light and the damage of acid soil. It is preferred for use on acid soils, as it acidifies soil less than many common nitrogen fertilizers.

Diammonium Phosphate: Diammonium phosphate (DAP) is the most widely used phosphorus fertilizer. It is manufactured in a controlled reaction of phosphoric acid with ammonia, where the hot slurry is cooled, granulated and sieved. DAP handles and stores well. The standard nutrient grade of DAP is relatively high, at 18-46-0 (18% N, 46% P_2O_5, 0% K_2O). In addition to P, diammonium phosphate fertilizer is an source of N for plants. It is soluble and thus dissolves quickly in soil to release plant-available form- phosphate and ammonium.

Superphosphate: This fertilizer is produced by the treatment of ground rock phosphate with sulphuric acid. The reaction yields a mixture of $Ca(H_2PO_4)_2$ and gypsum ($CaSO_4$). For the manufacture of triple superphosphate, phosphoric acid is used instead of sulphuric acid and the resulting product is $Ca(H_2PO_4)_2$. Several million tons are produced annually for use as fertilizers. Triple superphosphate remains popular as a phosphorous source in granular fertilizers. Triple superphosphate, or concentrated superphosphate contain 44-52 % P_2O_5. It has a high phosphorus content of dry fertilizers that do not contain nitrogen.

Muriate of Potash: The common source of fertilizer potassium is muriate of potash which chemically is potassium chloride (KCl). Potassium in this source is readily water soluble and at excessive rates, can cause salt damage to plants. MOP contains high potassium content (60 % K_2O) and is a natural salt. It is not lost as readily as N, but move into the soil to the roots more quickly than P. It also contains sodium chloride as an impurity.

3. Biofertilizers

Biofertilizers or microbial inoculants are carrier based ready to use live bacterial or fungal formulations, which on application to plants, soil or composting pits, help in mobilization of various nutrients by their biological activity. Biofertilizer are capable of fixing atmospheric N or convert insoluble phosphate in the soil into forms available to plants. Bio-fertilizers supply nutrients through the natural processes of nitrogen fixation, solubilizing phosphorus, and stimulating plant growth through the synthesis of growth-promoting substances. These fertilizers also improve the soil health and promote healthy growth of plants. They are extremely advantageous in enriching soil fertility and fulfilling plant nutrient requirements by supplying the nutrients through microorganism and their byproducts. Several biofertilizers viz; Rhizobium, Azotobacter, Azospirilium and blue green algae (BGA) have been used in agriculture for long time. Phosphate-solubilizing bacteria such as *Pseudomonas putida* strain P13 are able to solubilize the insoluble phosphate from organic and inorganic phosphate sources. A biofertilizer provides benefits of fixing the nutrient availability in the soil, symbiotic association with plant roots could readily and safely convert complex organic material into simple compounds that can be easily taken up by the plants. Biofertilizers are cost-effective relative to chemical fertilizers. Some important N bio-fertilizers are *Rhizobium, Azolla, Azospirillium, Blue green algae*. The phosphorus bio-fertilizers are of two types; a) phosphorus solublizing bacteria (*Bacillus megaterium* var. phosphaticum, *B. subtilis, B. circulans*) or fungi (Penicillium sp.) and b) Phosphorus mobilizers (AM fungi, Ectomycorrhizal fungi)

Criteria of essentiality of nutrients

This concept was given by Arnon and Stout (1939) and they considered 16 elements essential for plant nutrition. For an element to be regarded as an essential nutrient, it must satisfy the following criteria;

1. A deficiency of an essential nutrient element makes it impossible for the plant to complete the vegetative or reproductive stage of its life cycle.
2. The deficiency of an element is very specific to the element in question and deficiency can be corrected /prevented only by supplying that particular element.
3. The element must directly be involved in the nutrition and metabolism of the plant and have a direct influence on plant apart from its possible effects in correcting some micro-biological or chemical conditions of the soil or other culture medium.

Based on the first criteria, an element is considered essential if, in its absence, a plant is unable to complete its life cycle which includes the production of viable seed. The presence of an element must ensure the formation of a seed that possesses the natural capability to germinate and develop into a mature plant under favourable conditions for growth. The second criterion specifies the irreplaceability of any essential element. Based on the third criterion, an element is essential because it is indispensable to plant nutrition and does not merely correct some unfavourable conditions of the soil or culture medium. For example, carbon, hydrogen, oxygen, nitrogen and magnesium are essential elements because they are part of the chlorophyll molecule and the presence of chlorophyll is essential in photosynthesis.

The identification of any element as essential is preconditioned upon the availability of precise analytical methods of measurement. It is possible that certain chemical elements perform essential functions in plants but in extremely small concentrations which at the moment are undetectable. These chemical elements are detectable in the dry matter content of the plant body, that part that is left after water or moisture content is completely removed.

Why to Apply Manures

The crop removes large quantities of nutrients from the soil, particularly the removal of NPK nutrients at the present level of crop production has been estimated at 125 kg/ha/annum. The excessive dependence on chemical fertilizers and the reluctance towards the conservation and use of organic sources of nutrients have not only caused the over exploitation and exhaustion of the soil of its nutrient reserves but also resulted in several soil health problems. Excessive use of chemical fertilizers which are the important inputs in modern farming creates depletion in soil fertility and pollution in surface water bodies. Moreover, Indian soils are poor in organic matter and in major plant nutrients. In the absence of organic matter, the soil is a mixture of sand, silt and clay. Organic matter induces life into this inert mixture and promotes biological activities.

Advantages of organic manures

1. Organic manure provides all the nutrients that are required by plants to complete its life cycle but in limited quantities.
2. It helps in maintaining C:N ratio in the soil and increases the fertility and productivity of the soil.
3. It improves the physical, chemical and biological properties of the soil.
4. It improves both the structure and texture of the soils.
5. It increases the water holding capacity of the soil.
6. Due to increase in the biological activity, the nutrients that are in the lower depths are made available to the plants.

When to Apply Manures and Fertilizers

Fertilizer scheduling to plants is very important to get maximum benefit and to improve the efficiency of applied nutrients. Timing and activity zones vary for different fruits and method of application should be such that applied nutrients easily reach near active root zone with minimum losses of nutrients. It is desirable to apply N at the time when the plant can absorb its best, otherwise it may leach down. But fertilizer containing N in organic form must be applied sufficiently early to allow decomposition and make the nitrogen available to the plants at the time of need. For the quick response in growth and improved fruit set, nitrogenous fertilizers are most effective when applied to orchard trees starting growth in spring. If it rains heavily, some additional nitrogenous fertilizers should be given to both orchard trees and vegetable crops to replenish the leached N. Pre-harvest application of nitrogen fertilizers should be avoided as it impairs the storage quality of fruits. For elements like P, K and other which move quite slowly in the soil and remain with rhizosphere for a long time, the time of application is not so crucial. In many fruits plants potassium nitrate is foliar spray is very beneficial during the fruit development period to improve fruit size and quality. Similarly foliar application of zinc and iron are made during active growth period to meet the plant requirement.

How to Apply Manures and Fertilizers

The method of application plays an important role in the optimum utilization of nutrients. Fertilizers should be applied near the rhizoshere of the plant to improve the contact of the fertilizer with the plant roots so that plant can utilize the nutrient with maximum efficiency. Methods of application are selected keeping in view the type of crop, the season, the nature of the nutrients and plant requirement. Fertilizers are incorporated either in to the soil or through leaves as foliar spray. There are several methods of applying fertilizer to the fruit plants.

a. Broadcasting

b. Banding/strip placement

c. Foliar application

d. Fertigation (along with irrigation).

Broadcasting: In this method, fertilizer is uniformly spread over soil surface underneath the tree canopy and later mixed in to the soil. This method raises the general fertility level of soil and is suitable for perennial plants like fruit trees with well established root systems. This method is generally practiced by fruit growers. In fruit crops broadcasting is done manually and fertilizers should be applied away from the tree trunk up to the drip line of canopy. Applied fertilizer is worked in to the soil with hand hoe. Sufficient moisture must be present in soil to minimize the loss and uniform mixing. Weeds are removed before application of fertilizers.

Banding: In this method fertilizer is applied in bands near the root zone so that new developing roots have easy approach to the nutrients. After fertilizers are placed it is covered with soil. The nutrient use efficiency is higher by banding as compared to broadcast method. In large trees two to three bands are made one feet away from the tree trunk and one feet distance is maintained among bands. Banding of P fertilizers is useful. However, in banding method there is risk of salt burn to the plants.

Foliar Spray: Foliar uptake of nutrients is much faster than root uptake. Foliar application of nutrients is recommended when

- When deficiency symptoms are noted.
- When prompt correction of deficiencies is required.
- If root activity is disturbed during the reproductive phase.

The foliar sprays of one or more mineral nutrients on fruit plants are advocated to supplement the conventional soil application of fertilizers. Foliar applied nutrient penetrate through stomata afterwards translocation of ions take place from cell to cell and through vascular channels from leave to the sites where they are consumed.

Macro elements like nitrogen, phosphorus and potassium are needed in large quantities by the plant and foliar application containing with these elements will not fulfil the plant requirement, but these can be used as a supplemental spray in addition to the soil application when needed. Whereas foliar application of trace elements is useful, especially when soil applications would result in fixation in an unavailable form. Foliar application of iron, zinc, manganese, copper, boron and molybdenum are generally done in the form of salts, more commonly on tree fruits or vines. The longer the nutrient solution remains on the leaf surface in the form of a fine film, the more efficient is nutrient uptake by the leaf tissue. The efficacy of foliar sprays depends upon mineral nutrient concentration, valence of ions, ambient temperature, thickness of the cutical, nutrient status of plants and age of the leaves. During hot and dry weather foliar spray evaporates quickly and much of the salts accumulate on the leaf surface without being absorbed. Such detrimental effects can be avoided by using low concentration solutions, applying neutral solutions, and spraying on cool cloudy days or in the afternoon. The number of sprayings required to correct deficiency depends upon the severity of problem in the plants.

Fertigation: Fertigation is latest agro-technique combining water and fertilizer application through controlled irrigation system. Water use efficiency is much higher in the pressurized irrigation system. This technique regulate irrigation water application and nutrient supply and minimize the nutrient loss over broadcast method. Fertigation reduces leaching loss of nitrate form of nitrogen and thus minimize ground water pollution. In fruit crops the fertilizer is applied in small doses over the growing season to match crop nutrient requirements. The principal advantage of fertigation is that nutrients are applied in soluble form and are immidiately available to the plants. The quantity of fertilizer applied depends upon the frequency of fertigation, nutrient requirement of plant, soil type and nutrient availability in soil. Fertigation can provide nutrients and water to those area of soil where the roots are maximum concentrated. Common used fertilizer in horticulture crops are given below.

- Poly feed (19-19-19)
- MAP (12-61-0)
- Multi –K (13-0-46)
- MKP (0-52-34) SOP (0-0-50)
- Urea (46-0-0)
- Potash (0-0-60)

How Much to Apply Manures

The amount of manures to be applied depends upon the nutrient content of manure, fruit plant type, cultivar, plant age, soil inherent health and soil moisture availability. Well rotten farm yard manure contain 0.5 – 1.5% nitrogen, 1.2 -1.8%

Fruit	FYM (Kg)	Urea (g)	SSP (g)	MOP (g)	Time of application
Kinnow	100	1920	2750	-	FYM in December, Urea: first dose February-March and second dose in April-May SPP in February-March
Pear	50	1000	2000	1500	FYM, SSP and MOP in December, Urea: half dose in February and second half in April.
Mango	100	500	1000	1000	FYM and SSP in December, Urea and MOP in FebruaryNote: apply additional dose of 500g urea during 'On' year in June.
Guava	50	1000	2500	1500	FYM in May Half dose of inorganic manures in May-June and second half in September-October
Ber	100	1000	-	-	FYM in May-June, Urea: half dose in July-August and half after fruit set
Peach	26	1000	760	830	FYM and SSP, MOP in December, Urea: half dose in January and half dose in March
Plum	36	360	570	360	FYM and SSP, MOP in December, Urea: half dose before flowering and half after fruit set
Grapes	80	1000	4500	800	FYM, SSP and half dose of urea & MOP after pruning. Remaining dose of urea and potash should applied after fruit set.
Litchi	60	1000	2250	600	FYM and SSP, MOP in December, Urea: first half in February and second in April

phosphorus and 1.2-2.0 % potassium. For citrus orchards, nitrogen is required in fairly large amounts (1-1.5 kg per tree). Of the various types of manures, FYM is considered best for orchards. If FYM is available, chemical fertilizers should be used as supplement; it is recommended that half the nutrient requirements be supplied by FYM and half by chemical fertilizers. Organic manures like farm yard manure although contains relatively low levels of nutrients but its addition to orchard soil improves the organic matter level, texture and facilitates the growth of micro-organisms. Inorganic manures on the other hand are the instant source of nutrient elements and should be applied when plants need them in maximum quantity Well rotted FYM should be applied during November and December so that it decomposes by the onset of active growth and bloom in the spring. The dose and time of application of various farm yard manure and inorganic fertilizers for full grown fruit plants in north western India are as on page 127.

Role of Nutrients in Fruit Crops

Mineral nutrition of fruit trees is vital orchard management practice. Nutritional programme to the fruit plants must be managed properly for optimum growth and production of plants. When any essential element is deficient in plant it negatively affects various metabolic plant process and thereby affects the optimum uptake, utilization and distribution of other nutrients. Similarly, if element is supplied in excess quantity, it adversely affects availability of other nutrients in soil. It is not necessary that all the essential nutrients should be included in regular fertilizer programme because (a) many nutrients are present in abundant quantities in soils (b) fruit plants recycle a considerable part of of some nutrients back to the tree before senescence. Fruit trees require nutrition according to their ability to produce the quantity and quality of fruit; bearing behaviour (regular or alternate); bears once in a year or more number of time in a year; type of bearing whether bears on spurs or on new growth and nature i.e., evergreen or deciduous etc. Fruit plants needs both macro and micronutrients for the production of good quality fruits for longer period. Application of N, P and K fertilizers in split doses gives better results than one time application.

Effect of Nutrients on Vegetative Growth, Yield and Fruit Quality

Fruit quality: External appearance, fruit size and maturity date are most important for fresh fruit consumption, but high soluble solids and juice content are desired for processing purposes. The effects of each of these elements on fruit quality are described:

Nitrogen: Increases juice volume, total soluble solids (TSS), acid content and juice color, decreases fruit size and weight. Pre-harvest application of nitrogen reduce the storage of fruits.

Phosphorus: High P application rates reduce grapefruit juice acid levels, increases number of green fruit.

Potassium: Decreases juice content, TSS and juice colour and increases acid content, fruit size and weight.

Calcium: A low level of calcium in fruit flesh has been linked with several physiological disorders in temperate fruits. The calcium deficiency results in many disorders of the fruits viz bitter pit, cork spot, internal breakdown, scald and water core in apples, cork spots of pears, end spot of avocados, cracking of cherries and prunes, soft nose of mango.

Vegetative growth: The effect of certain elements on vegetative growth of plant is described: The optimum supply of nitrogen is essential for proper tree growth. Reduction in level of N caused stunting of growth and reduction in leaf size. Tree growth as indicated by percentage increase in trunk cross-sectional area was increased slightly by heavy N fertilization. Response of P fertilization with respect to both vegetative growth and yield in fruit crops depends mainly on the native P content of the soil. With few exceptions, P application to orchards has not produced corresponding increase in plant growth or yield. K when present above critical level does not affect tree growth. The low Ca trees had a more rounded tops due to dwarfing effect of the deficiency. S deficiency reduces tree growth and cropping in fruit plants. Deficiency of zinc result in intervienal chlorosis of leaves with rosette appearance.

Fruit yield: Different elements effects fruit yield of the plant. Adequate amount of N is needed to obtain good yields of fruit crops. Increase in quantity of N applied results in increase in fruit production but the increase is not linear thus, increase in yield become proportionately smaller with addition of more N. Application of phosphorus does not have a conspicuous effect on fruit yield. The use of both leaf and soil analysis should be made to ascertain the need of P of orchards in a particular locality. The yield of fruits increases with K application by increasing fruit size. Fruit yield also decreases in trees supplied with low levels of Ca.

Nutrient Deficiency Symptoms and Their Correction

Major Elements

Nitrogen: Nitrogen is one of the most widely distributed elements in nature; even then it is the element most deficient for crop production. Nitrogen is not available to plants in the forms of nitrate and ammonium. It is an integral constituent of proteins, nucleic acids, chlorophyll, co-enzymes, phytohormones and secondary metabolites. Normal metabolic processes can continue only in presence of an optimum level of N. Nitrogen deficiency leads to low fruit production and small yellow leaves. Chlorosis caused by N deficiency typically begins in the older leaves as N is remobilized to younger leaves.Citrus trees with a low N level shed their leaves excessively during winter causing a heavy setback to fruiting during the following spring. Under optimum soil conditions weak and old trees can be reinvigorated and their production increase by supplying sufficient amount of N. The deficiency of nitrogen is prevalent in soils low in organic carbon. The deficiency of nitrogen can be corrected by soil application of nitrogen fertilizer like urea, ammonium sulphate and calcium ammonium nitrate.

Phosphorus: Phosphorus is a constituent of all plant tissues and is found especially concentrated in younger parts and in the flowers and seed. It is an essential element required in the formation of ATP and is extremely important in plant metabolism. In P deficient plants, reduction in leaf expansion and number of leaves are most obvious effects. The average length of cell division zone is decreased in P deficient plants. Leaf expansion is strongly related to the expansion of epidermal cells and this process may be impaired in P deficient plants. The shoot becomes thin and spindly. Flowering is reduced greatly which result in poor fruit yield. In phosphorus deficient plants fruit maturity is delayed. Animal and municipal wastes are excellent sources of plant available P other sources are green manures, leaf moulds, fish meal, bones of the animals etc. The P deficiency can be corrected by applying superphosphate, di-ammonium phosphate or any other readily available source of P to the soil.

Potassium: It is the third important major element and it may exists in large quantities in soils but more than 90 per cent of potassium is fixed by clay particles and other minerals thus is unavailable to the plants. K is considered important in photosynthesis, sugar translocation, nitrogen metabolism and protein synthesis, enzyme activation. In deficient plants leaf margin become chlorotic and then necrotic, leading to a marginal scorch that is particularly evident. The fruit size decreased markedly in absence of adequate supply of potassium. Correction of potassium can be done by by soil application of muriate or sulphate of potash. For improving fruit size, foliar application of potassium nitrate @ 1-1.5 per cent is used during active initial fruit developmental stage.

Calcium: Ca concentration of the soil solution is about ten times greater than K, while the uptake rate of Ca is usually lower than that of K. Ca can be absorbed only by the young root tips that are not suberized. Ca is required for cell division and elongation. It precipitates as calcium pectate in cell-wall structures and in middle lamella and makes a cementing layer between the cells. The movement of calcium in plants takes place through xylem. Since the nutrient to fruits are supplied through phloem the fruit calcium concentration is low. Ca-deficient plants show stunted growth, necrosis at the tips and margins of young leaves. The cell membranes become leaky leading to the tissue breakdown. Several calcium related disorders have been reported in fruits. The deficiency of calcium is more prevalent in acidic and hill soils due to leaching of salt. When it is deficient, cells become leaky and lose control over the import and export of nutrient elements, leading to tissue breakdown. Due to its immobility, deficiency symptoms first appear in young tissues. This sort of tissue breakdown also occurs commonly in fruit tissues, since calcium levels are naturally low. Under such a condition, addition of lime or single superphosphate will correct the deficiency. Some of nitrogenous fertilizers such as calcium ammonium nitrate and calcium nitrate would also supply enough Ca to plants to correct the deficiency of this element.

Magnesium: The main function of Mg in plants is in the formation of chlorophyll. Mg deficiency appear on the older leaves. Being one to the basic nutrients, Mg affects the uptake of K and Ca. Soil application of heavy doses of K may cause Mg deficiency. The characteristic pattern on the leaf is a marginal chlorosis, leaving an inverted green "V ' around the midrib. These symptoms appear first on the most basal leaves and develop sequentially up the shoot. As the season progresses, the chlorotic area increases and develops into necrosis. Eventually, the whole leaf is affected and then abscises from the shoot. Thus the lower part of the shoot becomes devoid of leaves.

Mg deficiency induces chlorosis of on leaves. Mg deficiency in fruit plants can be corrected by both soil application of some soluble Mg salt or by foliar spray. Foliar application of Mg would be effective in correcting the deficiency within a few days but this recovery is comparatively of short duration and several applications are needed for proper control.

Sulfur: S is required for synthesis of amino acids cystine, cysteine and methionine, which are component of protein that comprise about 90% of the S in plants. Plants absorb sulphur from soil as sulphates which are mobile in plants. This element is absorbed by leaves in gaseous form as SO_2 particularly in heavily industrialized localities. Sulfur deficiency symptoms starts on new growth the plants turn light-green or yellow, but even under severe deficiency the leaves abscise pre-maturely. The veins of the leaves become greener than tissues between the veins. Sulfur deficiency in fruit plants can be corrected by soil application of any S-containing fertilizer such as superphosphate, ammonium sulphate or elemental sulfur.

Minor Elements

Iron: It is Involved in the activity of many enzymes such as catalase, cyrochrome, ferrodoxin, hematin and cytochrome oxidase. Most of the iron is present in chloroplasts, where it has some role in synthesizing chlorophyll. Deficiency symptoms initially appear on young leaves, particularly in the summer season. In beginning, the leaf colour turns pale but veins remain green. The next stage is yellowing between veins of leaves, leaving the veins green. Leaves remain smaller than normal, growing points may also die and growth is stunted. Iron deficiency symptoms is often observed under conditions of poor soil aeration. Fe deficiency is corrected by the use of iron chelate or foliar application of $FeSO_4$.

Zinc: Zinc is involved in synthesis of tryptophan, a precursor of IAA. Fruit plants grown on sandy soils soils are more susceptible to Zn deficiency than those grown on clayey soils. Deficient leaves of fruits tree are easily identified by their characteristic interveinal chlorosis, small leaves, the rosette formation which results from insufficient synthesis of IAA and shortening of internodes. Soil application $ZnSO_4$ did not give proper control of Zn deficiency in fruit plants in high pH soils. In citrus zinc sprays are given @ 0.3 % to 0.45 % in April, June and August-September.

Manganese: Manganese assists in formation of chlorophyll. It also functions in enzymes in oxidation-reduction reactions. Manganese in the plant participates in several important processes including photosynthesis, nitrogen and carbohydrate metabolism. Symptoms first appear on the youngest leaves. Symptoms include yellowing between the veins of leaves while the veins remain green. In severe deficiency, complete yellowing and necrosis or dead areas appear on the leaves. It can be corrected by foliar spray of soluble manganese salts like manganese sulphate. Under neutral or alkaline soil conditions, Mn deficiency can be corrected by annual spray application of manganese sulphate when a major flush of leaves has expanded to at least two-thirds of normal size.

Copper: Only a small fraction of copper is required by the plants. More than half of the copper in plant is present in the chloroplasts and involves in photosynthetic reactions. Only a small amount of copper is required by the plants. Copper deficiency symptoms appear on fruit and younger leaves first, leaves drop off and twigs die back. Gumming of bark is seen in affected plants. Copper deficiency is seldom observed in fruit plants due to regular use of fungicides like Bordeaux mixture.

Boron: Boron plays important role in flowering and fruiting processes, N metabolism, hormone movement and movement of sugars. Like calcium, boron also moves through xylem and hence several boron related disorders in fruits were reported. Internally the fruit may develop corky areas in the cortex and browning in the core region. If boron is deficient, young growing tissues are discoloured, disrupted, and die; and the plants remain stunted. Young leaves become brittle, thick and discoloured. Flowering is suppressed. Boron deficiency can be corrected by application of Borax.

Molybdenum: It is essential constituent of enzymes involved in nitrogen metabolism. Mo-deficient plants show nitrate accumulation in the leaves. Molybdenum deficiency unlike other trace elements occurs under acidic conditions. The symptoms appear on older leaves first and move from older to younger leaves. In citrus Mo deficiency appears first as watersoaked areas develop in to large interveinal yellow spots with gum on the lower leaf surfaces. Mo deficiency can be managed by foliar application of sodium molybdate or ammonium molybdate.

Chlorine: Chlorine is abundant in soils and taken up easily by plants but its requirement is very small. It is essential for growth and development of tomato and sugar beets. Exact function of chlorine is still unknown. Plant deficient in chlorine show wilting, chlorosis and death of tissues with leaf bronzing.

LEAF SAMPLING TECHNIQUES

Plant life cannot be sustained without proper nutrition. The deficiency or an excess of an essential element may cause disturbance in plant metabolism and its vital functions may fail leading to a subnormal performance. Appropriate steps, therefore, need to be taken to correct this situation by diagnosing the problem.

Diagnostic Techniques

To identify the essentiality of the elements and to assess their relative requirements in growing fruit plants in an intensive orchard cultural practice, a number of methods of investigations have been devised which may be successfully utilized for fruit production. Determining the nutritional requirements is not an easy task. However, there are several diagnostic techniques, which may be used to ease the task:

1. Visual deficiency symptoms.
2. Soil analysis.
3. Tissue/Leaf analysis.

Leaf Analysis: Leaf analysis, in all stages from sampling to interpretation of results and developing the nutritional program is a relatively complicated and costly procedure. Sampling guidelines should be followed precisely to ensure that analytical results are meaningful. Leaf analysis is considered the most reliable method for determining the nutrient status of the plant. In contrast to soil analysis, which indicates total amounts of certain elements in soil, leaf analysis indicate the quantity of different elements that have been taken up by the tree and translocated to the leaves. Leaf analysis is helpful in diagnosing the hidden or incipient deficiencies. With leaf analysis, it is possible to prevent nutritional disorders before symptoms of deficiency occurs. Deficiency of some elements does not produce symptoms on leaves clearly until their content falls below threshold. However the deficiency inflicts loss to fruit growers in form of reduced fruit size and yield. So leaf analysis indicates the balance between the nutrients.

Principles of leaf analysis

1. It is the principle site for the plant metabolism.
2. Changes in the nutrient supply are reflected in the composition of leaf.
3. Changes are more pronounced at certain stages of the development.
4. Leaf nutrient content at a specific growth stage are related to the performance of the crop.

Common Guidelines for Leaf Sampling

The sampling procedure and season vary with respect to fruit crop and is given in table.

Fruit	Leaf Age	Sample size	Season	Remarks
Citrus	4-8	100	July-October	Immediately behind fruit
Peach	3-5	100	Mid-May to Mid-July	Mid shoot leaves from current season's growth
Pear	4-6	50	July-September	-Do-
Plum	3-4	100	May-July	-Do-
Guava	5-7	50	August-October	Mid shoot leaves from non fruiting terminals
Mango	5-7	30	March-April	Non-flushing and non-flushing twigs
Ber	5-7	70	November-January	From middle of shoot

In order to get appropriate information from leaf analysis, some common guidelines have to be followed for getting best information about the nutritional disorders:

To monitor changes in nutrition status based on recommendations, samples are collected from sampling sites. Select the leaves which are fully exposed to sunlight. Leafs samples must be taken at the proper time and in the proper way for the analysis to be representative of the orchard. Collect the leaves from the periphery of the tree and at height of 1.0 to 2.0 meter from the ground level. Do not mix leaves from young and old plants. Sampling must be done from several trees and collect 4-8 leaves from each direction from each plant. Not more than two leaves should be taken from an individual shoot. Leaf sampling should be carried out early in the morning and should be collected prior to irrigation and fertilizer application. Leaves that have been recently sprayed with micronutrients should not be sampled for micronutrient analysis. Samples should be placed in polythene bags, kept in cold condition and send to nearest leaf analysis laboratory as soon as possible.

Uptake of Nutrients From the Soil

Nutrient uptake is the process of nutrient movement from an external environment into a plant system. Plants absorb mineral nutrients through roots in a form of small molecules without an electric charge or in a form of positively or negatively charged ions of a range of elements. During transport throughout a plant, minerals can exit xylem and enter cells that require them. The most normal entrance portal for mineral uptake is through plant roots.

Ions may be taken up by the plant cells by two methods:

i. Passive transport

ii. Active transport

Passive Absorption- it refers to movement of molecules down the gradient from higher concentration to lower in order maintain the equilibrium. It does not require direct cellular energy. During passive absorption, the mineral salts may diffuse through cell membranes directly through lipid bilayer but mainly through trans--membrane ion-selective protein channels or trans-membrane carrier proteins. Passive diffusion depends upon concentration gradient, permeability of membrane; separating the substances, temperature and pressure. Increase in transpiration pull increases the uptake of ions by the roots.

Active Absorption of Ions: The cell membrane also contains protein carrier molecules which can use energy to actively move molecules and ions through the membrane. This active transport moves particles through cell membranes from regions of lower concentration to ones of higher concentration. Several mechanisms have explained for process of active absorption. The transport of minerals across the membranes is facilitated by the presence of carrier molecules on cell membrane.

The major hypotheses that explain the mechanism of active transport of ions are:

1. Carrier Concept – transport by a carrier protein.
2. Cytochrome Pump – transport by electrochemical gradient generated by electron transport.

LITERATURE CONSULTED

Chandra,K. 2005.Organic Manures. Published by Regional Director, regional centre of organic farming. pp 1-46.

Johnson ,R.C. and Uru, K. 1989. Mineral Nutrition. In : peaches, plums & nectarines – growing & handling (Eds: Larue JH and Johnson RC) pp 68-81.

Malik,M.N. 2010.Horticulture. Biotech books.

Nijjar,G.S. 1985.Nutrition of Fruit Trees. Kalyani Publishers, New Delhi.

Prasad,S. and Kumar,U. 2005.Principles of Horticulture , Second edition. Agrobios Publications. Jodhdpur.

Tiwari,J.P.and Srivastava,R. 2006. Integrated Crop Management in Subtropical Fruits .Agrotech Publishing Academy, Udaipur.

Tucker,D.P.H.,Alva,A,K.,Jackson,L.K.andWheaton,T.A. 1995. Nutrition of Florida Citrus Trees. University of Florida, Institute of Food and Agricultural Sciences.

http://www.slideshare.net/AbeerElhakem/uptake-of-nutrients-by-plants

http://fertilizerbrokerage.com/ammonium-sulfate.html

http://www.ib-article.com/2013/03/various-kinds-of-organic-and-inorganic.html

http://agriinfo.in/?page=topic&superid=1&topicid=364

http://www.extension.umn.edu/agriculture/nutrient-management/nitrogen/fertilizer-urea/

http://www.cropsreview.com /essential-elements.html

7

Water Need of Fruit Plants

J.S. Chandel and Chaitanya R. Belsare

INTRODUCTION

Water is one of the most important inputs essential for tree growth, development and fruit production. Fruit trees need it continuously during their life in optimum quantities. It profoundly influences photosynthesis, respiration, absorption, translocation and utilization of mineral nutrients, and cell division besides some other processes. Both its shortage and excess affect the growth and development of a plant directly and, consequently, its yield and quality. In India, however, rainfall is notoriously capricious, causing floods and droughts alternately. Its frequency distribution and amount are not in accordance with the needs of the particular fruit crops. Artificial water supplies through irrigation on one occasion, and the removal of excess water through drainage on another occasion, therefore, become imperative, if the fruit crops are to be raised successfully. Water management in India, thus, comprises irrigation or drainage or both, depending considerably on the environmental conditions, soil, crops and climate. Applying the proper amount of water at the right time is important for reducing costs and maximizing profits. It is essential, therefore, that there be a scientifically sound method for determining when to irrigate and how much water to apply for optimum crop response and efficient use of resources.

The aim of irrigation should be to obtain the maximum possible yield of marketable produce from a given amount of water supplied to the crop. In order to achieve this, a thorough understanding of the soil in the orchard and the various growth stages and water requirements of the crop are required. The three main driving variables on which irrigation decisions are based are: (1) how much water the root zone of the crop can hold, (2) how much water infiltrates into the soil and (3) how much water is the crop using? The irrigation should be, therefore planned according to soil water holding capacity, plant water use, prevailing weather conditions and quantified management decisions. The level of irrigation in an orchard will depend on environmental factors which drive evaporative demand and transpiration, salinity, and electrolyte composition in the soil solution, the resistance of the soil to root penetration and moisture transport, soil aeration, tree hydraulic architecture (including the rootstock), and crop load. Losses due for long term sustainability of perennial fruit trees, it is important to safeguard against drought at all stages, but during certain stages of development each season, water may have to managed more carefully than at other stages of growth e.g. flowering or early fruit growth. Maintaining adequate soil moisture conditions during moisture-sensitive stages of growth will have a beneficial effect on plant growth. Restricted water supply during these critical periods will negatively impact upon yield as the provision of adequate water at other stages will not compensate for the harm sustained. Irrigation scheduling involves applying the correct amount of water at the correct time to maintain optimal soil moisture conditions for the production of maximum yield and fruit quality. Recent improvement of irrigation and management methods has reduced irrigation-relatedwater losses, thus decreasing the amount of applied water per unit of irrigated land. Drip irrigation method offer higher irrigation efficiency and water saving and can be used efficiently in water scarce areas for sustainable fruit production. *In-situ* moisture conservation techniques like use of mulching, crescent bunding, V-ditch basin preparation and use of drought tolerant rootstocks have pronounced effect on growth and fruit production under rain-fed conditions.

Source of Irrigation

The source of irrigation water is primarily classified into 2 groups – natural and artificial. The natural source provides moisture to fruit plants mainly in the form of rainfall. However, atmospheric humidity only modifies the absorption pattern and seldom utilized by the fruit plants. So, also the precipitation of snow on the artic regions and mountain tops becomes available to the fruit plants when liquidified as water. The artificial supply of water is accomplished through irrigation. The artificial source of irrigation may be a reservoir, underground aquifer and a natural drainage channel or a river. Each of these has varied influence on growth and production of fruit plants with respect to quality of water. Stability of the water is also important from the practical point of view. Orchard plants require regular and steady supply of irrigation water. While, exploring the sources or constructing permanent channels in the orchard as steady and dependable source should be identified for long term operation.

Water Use Efficiency

Efficient use of irrigation water has always been an important factor in irrigation system design and operation. As demands and costs, both for water itself and for the energy to transport it increase, improved efficiency in irrigation water use will become more critical.

The term 'water use efficiency' as used by different authors may include such factors as irrigation system leakage losses, evaporative losses, and deep-percolation losses as well as plant performance. Where a cover crop is present, some water use will be accounted for by that crop and should be included in overall efficiency calculations. Horticulturally, water use efficiency is most commonly defined in terms of total or useable crop yield per unit of irrigation water applied. Alternative methods that have been used include estimates of crop water use relative to total pan evaporation plus percolation, yield per unit of evapo-transpired water and in humid regions, yield or yield increases over un-irrigated controls per unit of irrigation supplied water.

Water Relation in Fruit Crops

Water is the most abundant and, at the same time, the most limiting resource, that plants need to grow and function for the best efficiency or productivity in terms of biomass. Water is important to the physiology of plants due to its crucial role in all physiological processes and due to the large quantities that are required. Water typically comprises 80-95% of the biomass of non-woody tissues. Whereas plants incorporate more than 90% of absorbed nutrients into new tissues, less than 1% of the water absorbed is retained in the biomass. The remainder is lost by transpiration, which is the evaporation of water from plants. When plants are unable to absorb enough water to replace that lost by transpiration, water stress develops in the plant system. The results may be wilting, cessation of growth, or even death of the plant or plant parts. Therefore, if we want to maintain or increase the productivity of crops, it is crucial that we understand plant water relations and the consequences for plant growth of an inadequate water supply.

Plant water relation can be detected by measuring soil moisture status, leaf or stem water potential (øw), leaf relative water content (RWC), sap flow or trunk and fruit diameter. Monitoring soil moisture status under field conditions may be easy and quick. However, soil moisture measurements cannot be used alone to estimate plant water relation because plant water status depends more on evaporative demand of the atmosphere. Measurement of stem øw is a more integrative and stable than single leaf øw since it is influenced by all the leaves and organs of the branch (leaves, fruits, shoot tips and other organ that do not transpire rapidly). Leaf relative water content (RWC), the mass of water held in relation to the mass that can be held at full turgor, has been used to estimate plant water status in terms of cellular hydration under the possible effect of leaf øw and osmotic adjustments. However, leaf RWC measurement may be time consuming for day-to-day monitoring of plant water status under orchard condition. Sap flow through woody tissues is used to estimate total tree or stand water use. The advantage of sap flow is that it can be measured automatically throughout the day, night or season. Trunk and fruit diameter monitoring may also provide information on plant water status.

Water loss of crops is driven by solar radiation but modified by temperature, vapor pressure gradient, boundary layer of still air, leaf area and crop structure that influence light interception and stomatal opening. Leaves intercept radiation that drives transpiration. They provide the most active transpiring surfaces. As the leaves intercept radiation, the energy warms the leaves and provides the energy for the evaporation of water within the stomatal cavities of the leaves. Broad canopies and trees or orchards spaced more closely will intercept more light and use more water. Wider or larger tree forms use more water than thinner or more vertical forms.Fruits have a very high content of solutes and show increasingly negative osmotic potential as the season progresses. Fruit øw is much higher than leaf øw due to which fruits shrink (loss water) during the day as leaf øw becomes more negative and then expands again as leaf øw begins to increase. Fruit water loss could be the result of direct evaporation or water moving out of the fruit back through the pedicel in equilibration with the water tensions in the conducting system. Water movement into the fruit is largely through the phloem as fruit pedicels have low hydraulic conductance and xylem becomes non-functional in some cultivars. Fruit surface conductance varies between cultivars, being high in russeted (dark) cultivars than those with waxy skins.

Two processes in the soil-plant-atmosphere continuum, *viz.* water up-take by roots and water loss via stomata greatly influence tree water relation and hence orchard water use. Therefore, the influences of root system characteristics and stomatal conductance on apple water relations are discussed briefly.

Root System Characteristics

Two key characteristics of tree root system of relevance to water relation are the extremely low root length density (RLD) in soil and a very non-uniform root distribution. The implications of these low RLD and erratic rooting are: (1) reduced effective soil volume explored for water and nutrients (specially that of nonmobile nutrients), (2) less competition ,(3) root distribution may concentrate in the wetted zone, especially if nutrients are supplied by fertigation, (4) localized drying in the rhizosphere (interface between roots and soil) during times of high transpiration at midday, (5) greater dependence on mycorrhyzae to absorb immobile nutrients such as phosphorus. Where rooting is shallow, such as the root systems of growth-controlling rootstocks of high-density plantings, the need for irrigation will be higher than for standard, low-density plantings.

Stomatal Conductance

Stomata are pores on the leaf epidermis that play pivotal role in controlling the balance between water loss and carbon gain, i.e. biomass production. Maximum photosynthetic rates and, therefore, maximum carbohydrate production occur when the stomata are fully open. When stomata are less active or closed due to drought, both photosynthesis and transpiration may reduced as much as 40% before the leaves show any wilting and over 90% at wilting. When soil dryness is severe, stomata may not open at all to prevent water loss from the leaves. Hydraulic (leaf Ψw, cell turgor) and chemical (abscisic acid) signals cause stomatal closure. To maintain turgor and carry on photosynthesis longer into a drought, mature leaves can adjust osmotically by as much as 2 MPa or more over time as stress develops. The osmotic adjustment is due to the accumulation of monosaccharides, especially sorbitol. Water stress could, therefore, impose stomatal and/or non-stomatal limitations of photosynthesis. Stomatal control is often thought to be the first line of defense against water stress. Non-stomatal limitations of photosynthesis prompted by water stress include reductions in carboxylating enzymes, electron transport and chlorophyll content.

Factors Affecting Irrigation

Climatic factors, rainfall, temperature and relative humidity determine the pattern of water loss through evaporation. For determining frequency of irrigation, evaporation from a free water surface (e.g., USDA class A pan) is considered the best index. Soil moisture particularly in the root zone, provides fair indication of its irrigation needs. The moisture content varies depending on soil texture, rooting depth and extraction capability of the fruit species. Plant parameters, leaf water potential and relative water content of leaves indicate water status of plant, which are gauged easily and gives a fair indication for irrigation. Irrigation water requirement also depends on the type of rootstocks used and their root distribution and varies according to fruit species.

Frequency of Irrigation Water

The irrigation to be supplied in terms of dose and frequency depends on a number of factors like type of fruit crops, nature of the soil, slope of the land and the prevailing weather condition. Nevertheless, the cost benefit ratio for the operation is also serious considered by the growers. In general, irrigation is provided to the plants after adjusting the natural precipitation, but it is over simplification to advice withdrawal of irrigation during rainy season. Fruit trees should be provided with deep irrigation at an appreciable interval to favor deeper and wider root growth, and to minimize the cost of operation which otherwise may develop goose foot type of roots having poor anchorage and may cost more for the operation. But quick, growing fruits like pineapple, papaya strawberry etc. having shallow root system favour frequent light irrigation. Banana, on the other hand, needs at least 10 cm irrigation every fortnight during drier months. In all the fruit plant, the active period of growth demands steady and adequate supply of soil moisture. In fruit, crops like nuts and mango, the extent of soil moisture during flowering determines the size of the crop. In most of the tropical fruits like mango, cashew etc; adequate supply of moisture throughout the year improves vegetative growth and emergence of new shoots well in time for production in guava may be regulated by withholding irrigation during appropriate time in expectation of quality fruit production. Fruit drop or June drop may also be minimized by providing adequate water to the fruit plants during drier period. Most of the fruit plants have their own peculiarity to endure quantum and interval of irrigation under a set of agro-climatic conditions. In other words, the plants have the capability to withstand wet or dry conditions. Most of the fruit trees cannot stand submergence for a prolonged period where orange trees are most susceptible. In case of temperate fruits like apple and pear it is important to understand the growth stages of the crop, so that irrigation can be tailored and optimized throughout the season these critical stages are bud burst and flowering, beginning of rapid shoot growth and beginning of fruit fill. The type of rooting system will influence the volume and frequency of irrigation required. Generally dwarf rootstocks for apple have shallow root systems (down to 60 cm) that can be highly fibrous and dense. The shallow rooted and dense nature of dwarf species means irrigation is required more frequently, to compensate for the increased evaporation at shallow soil depths and higher uptake of water.

Water Requirements of Fruit Crops

Fruit crops have variable degree of soil moisture stress tolerance limit hence water requirement of different fruit crops will vary under similar soil and climate conditions. The rainfed fruit crops have different degree of drought tolerance or drought resistance.

Table 1: Water requirement of fruit crops

Fruit crop	Water requirement (cm)
Aonla	60
Apple and pear	114
Banana (wetlands)	222
Ber	26
Citrus	66-104
Custard apple	12
Date plam	132
Grape	48
Litchi	60
Mango	36-54
Papaya	87
Pineapple	36
Pomegranate	54-78

Direct Method: Water requirements of crops (WR) are decided either by direct or indirect methods. The direct method is classified as lysimetric technique or soil moisture or soil water balance method. The lysimetric technique is precise for shallow root system, close spaced, short duration vegetable crops.

WR = P + IR + ÄSW - (r + PW)

P = Precipitation, cm

IR = NIR, cm

ΔSW = Soil water contribution, mm

r = Surface runoff, mm

PW = deep percolation, mm

The soil moisture monitoring to a desired depth either by nucleic technique or neutron probe are necessary in case of widely spaced, long duration horticultural crops. Water requirement of mostly rainfed fruit crops such as mango, sapota, pomegranate, citrus are difficult to predict due to their deep root system and ground water contribution.

Water requirement is also determined by the field experimentation method.

CU = ER + IR + ASW

CU = Seasonal consumptive use, cm

ER = effective rainfall, cm

IR = net irrigation water, cm

ASW = soil water contribution, cm

Estimation of effective rainfall under this method is very crucial. There are other methods of estimating water requirement like soil water depletion method and inflow - out flow method but precision in soil moisture monitoring is very important in estimating the requirement.

Indirect Method: Evapotranspiration rates of various crops are estimated by pan evaporation multiplied by a pan factor (Kp) and crop coefficient factor (Kc). The crop coefficient values increase with age of the crop approaching grand growth period nearly stabilizes at the grand growth reaching its maximum at flowering and then declines with senescence.

Net irrigation requirement of horticultural crops by using daily pan evaporation data is estimated and used for micro-irrigation purpose. The equations for estimations of ETc and ETca are as follows:

Eo = Ep x Kp

Where

Eo = Potential evapotranspiration, mm/day

Ep = USWB class A pan evaporation mm/day

Kp = Pan factor

Evapotranspiration (ETc) is again estimated using the crop coefficient constants (Kc)

ETc = Eo x Kc

Water Budgeting for Fruit Crops

Irrigation can be scheduled according to a water budget approach and has the following advantages: 1) no equipment requirements, (2) accuracy, (3) simplicity of use, and (4) flexibility allowing easy adaptation to use in other crops. The water budget approach is slightly different for high volume sprinkler-irrigated orchards than for those with low-volume drip or micro-sprinklers. For sprinkler irrigation the grower must first estimate the amount of available water in the root zone. This is achieved by multiplying the appropriate available water value by the average rooting depth.

Table 2: Ranges in available water capacity and intake rate for various soil textures

Soil texture	Available water capacity (mm of water cm^{-1} soil)		Intake rate (mm h^{-1})	
	Range	Average	Range	Average
Sand	0.5-0.8	0.65	12-20	16.0
Loamy sand	0.7-1.0	0.85	7-12	9.5
Sandy loam	0.9-1.2	1.05	7-12	9.5
Loam	1.3-1.7	1.50	7-12	9.5
Slit loam	1.4-1.7	1.55	4-7	5.5
Silty clay loam	1.5-2.0	1.75	4-7	5.5
Clay loam	1.5-1.8	1.62	4-7	5.5
Clay	1.5-1.7	1.60	2-5	3.5

Secondly, the allowable soil water depletion in the root zone (about 50%) should be estimated. This is the amount of water in the soil profile that can be extracted without causing adverse effects on tree growth, yield and quality and is calculated by multiplying available water by 50%. Thirdly, the water use rate of the various fruit crops needs to be calculated. This will vary according to site and time of year. Through the use of crop factors and the average weekly or average daily maximum evapotranspiration, average daily water use rates can be calculated. Fourthly, decisions must be made as to when to irrigate. This should be calculated after a thorough wetting of the soil and the amount of available water in the crop root zone must be determined by direct measurement e.g. tensiometer. Deciding when to irrigate is then determined by subtracting daily water use of crops from total available water in the root zone until the soil water has been reduced to the allowable depletion level. The amount of irrigation to apply is calculated by dividing the allowable soil moisture depletion by the irrigation efficiency. Irrigation efficiency depends on the size and uniformity of fields, the climatic conditions and the manner in which the irrigation is delivered as water may be lost through deep percolation, runoff and evaporation. The duration of the water application depends on the amount of water to be applied and the water intake rate of the soil, as you do not want runoff to occur.

Table 3: Irrigation depths for various fruit crops

Fruit crops	Depth of irrigation (cm)
Apple	90
Cherries	60
Grapes	90
Pears	60
Strawberries	30
Peaches	60

For low-volume drip or micro-sprinkler systems the daily water use rate by trees needs to be calculated as L tree-1 day-1. The frequency of irrigation is related to system design, tree daily water requirements, number of emitters per tree and the emitter flow rate. Low volume drip systems operate daily and maintain optimum moisture conditions in the root zone thereby preventing moisture stress. Micro-sprinklers on the other hand wet larger areas of the soil surface and meet the soils water requirements every 2-3 days, allowing greater flexibility in irrigation. Irrigation frequency is equivalent to total usable water in the soil profile divided by the daily water use of the tree. Total usable water is a function of total usable water per meter of soil depth, total soil root zone reservoir and area shaded per hectare. Soil type will determine the number, spacing and flow rate of emitters. In sandy soils emitters will have to be placed closer to the trees while in clay soils will require a low emitter flow rate to allow for slower infiltration. The actual irrigation amount will be calculated by dividing the water usage of each fruit tree (L tree-1 day-1) with the application efficiency and multiplying this by the irrigation frequency. The duration of the water application will depend on the amount of water to be applied and the irrigation application rate, which is calculated based on discharge per emitter and number of emitters per tree.

Critical Periods for Drought Stress in Fruit Crops

Irrigation is critical during plant establishment, which is usually from the first to the third year after planting, depending on the fruit crop. During this period plants have not established an extensive root system and are prone to drought stress. Early in the life of a fruit crop growers work at trying to develop an extensive plant root system and an above ground plant structure to produce fruit on. Lack of water can drastically reduce the time to bearing and increase production costs substantially. Weed control is particularly important in a dry season, because weeds compete with fruit crops for available water. Weeds should be eliminated beneath the fruit crop canopy and the row middles should be kept closely mowed to reduce water loss.

Systems of Irrigation

Different systems of irrigation are adopted to irrigate the fruit orchards throughout India. The systems adopted are determined by a number of factors like

- Preference by the fruit crops
- Economization of usage
- Distance and stability of source
- Prevailing agro-climatic conditions of the region.

The amount of water to be applied to crops also depends on all the above mentioned factors.

Irrigation Methods

There are 4 irrigation methods, which are widely used all over the world. These are flood, furrow, sprinkler and drip irrigation.

Flood Irrigation: It can be used only if sufficient water is available and orchard floor is flat. It is confined to valley areas of temperate fruit growing regions of India. Small bunds are made on all sides of the field. The water is flooded into the field and allowed to seep in the soil. This system is economical and feasible in areas with abundance of water. However, it causes problems like root rot and collar rot. It is also not suitable for rat infested orchards, where soil erosion can take place.

Furrow Irrigation: Furrow irrigation is suitable for many crops, especially row crops. Crops that would be damaged if water covered their stem or crown should be irrigated by furrows Furrow irrigation is also suited to the growing of tree crops. In the early stages of tree planting, one furrow alongside the tree row may be sufficient but as the trees develop then two or more furrows can be constructed to provide sufficient water. Sometimes a special zig-zag system is used to improve the spread of water. Furrow length must be on consonance with the slope, the soil type, the stream size, the irrigation depth, the cultivation practice and the field length.

Basin Irrigation: This system is a modified system of flooding and practiced in orchards where water is available in plenty. For this the land must be levelled. Ridges are raised along contour lines and water is applied to individual basins which are levelled. Since water stagnates in the basins, this system should be used with caution where root diseases are a problem.

Modern Irrigation Systems

The conventional systems of irrigation, *viz* flood, furrow or basin irrigation systems are not suitable for irrigating temperate fruit orchards in India. These require level fields and plenty of water. Most temperate orchards are planted on sloping and undulating land where water is scarce. The model irrigation techniques like sprinkler and drip are perfected in some of the advanced countries. These systems have many advantages over the conventional methods and are well adapted to tree crops in hilly terrains.

Sprinkler Irrigation: Sprinkler systems in fruit tree groves are designed to use overlapping patterns to provide uniform coverage over an irrigated area. Sprinklers are normally spaced 50% to 60% of their diameter of coverage to provide uniform application in low wind conditions. Application efficiencies of sprinkler systems are relatively low at less than 80%. Because networks of pressurized pipelines are used to distribute water in these systems, the uniformity of water application and the irrigation efficiency are more strongly dependent on the hydraulic properties of the pipe network. Thus, application efficiencies of well-designed and well-managed pressurized sprinkler systems are much less variable than application efficiencies of gravity flow irrigation systems, which depend heavily on soil hydraulic characteristics. Therefore, during water applications, sprinkler irrigation systems lose water as a result of evaporation and wind drift. More water is lost during windy conditions than calm conditions. More is also lost during high evaporative demand periods (hot, dry days) than during low-demand periods (cool, cloudy, humid days). Thus, sprinkler irrigation systems usually apply water more efficiently at night (and early mornings and late evenings) than during the day.

Application efficiencies of micro-irrigation systems are typically high because these systems distribute water near or directly into the crop root zone; therefore, water losses resulting from wind drift and evaporation are typically small resulting in a highly efficient (90% to 95%) water delivery system. The advantages of micro-irrigation over sprinkler include: reduced water use, ability to apply fertilizer with the irrigation, precise water distribution, reduced foliar diseases, and the ability to electronically schedule irrigation on large areas with relatively smaller pumps.

Water source: Lift irrigation is too costly for small orchards. To ensure a continuous supply during the period when sprinklers are functioning it is advisable to maintain a storage tank at a higher point to maintain an even flow of water. A tank of 1.5x1.5x1.5m, having capacity of 4000 litres, will be adequate emergency storage for a 2 hectare plot.

Most of the sprinkler systems are usually run on electric or diesel pumps. As a rule of thumb, 0.5 horsepower is sufficient to create a pressure of 2.8 kg per cm^2, on which most sprinklers usually function and can irrigate about 0.4 hectare. The head created by gravity on slops is very useful power source for sprinkler irrigation. To create pressure of 2.8 kg per cm^2, a head of 28 m will be required.

Rate of water application: The rate of water application should not be higher than basic infiltration rate of soil. The water application rate will vary with sprinkler design, nozzle size, operating pressure and spacing of sprinklers.

Fertilizer application through sprinklers: It is easy to apply chemical fertilizers through sprinklers. Fertilizers are also supplied through sprinkler equipment. Urea can be applied easily. The use of pesticides and herbicides through sprinkler is not recommended.

Drip Irrigation: Nearly all species of fruit/nut bearing trees, at any stage of growth, will benefit from irrigation applied through a drip or low-volume irrigation system. Considering that fruit trees are the longest living and largest growing planted by orchardist. Drip irrigation is the most efficient method of irrigating. While sprinkler systems are around 75-85% efficient, drip systems typically are 90% or higher. What that means is much less wasted water. For this reason drip is the preferred method of irrigation in the water scarcity regions. But drip irrigation has other benefits which make it useful almost anywhere. It is easy to install, easy to design, can be very inexpensive, and can reduce disease problems associated with high levels of moisture on some plants.Drip irrigation (sometimes called trickle irrigation) works by applying water slowly, directly to the soil. The high efficiency of drip irrigation results from two primary factors. The first is that the water soaks into the soil before it can evaporate or run off. The second is that the water is only applied where it is needed, (at the plant's roots) rather than sprayed everywhere. While drip systems are simple and pretty forgiving of errors in design and installation, there are some guidelines that if followed, will make for a much better drip system. Drip irrigation has the greatest potential where, water is expensive or scarce and where the soils are sandy, rocky or difficult to level and high value crops are produced. To encourage maximum root density, drip irrigation should be put into operation towards the end of a stress season rather than at the beginning. It is widely recommended that minimum size of the wetting pattern should be such that 25% of the root system is supplied with water.

Advantages of drip irrigation

- Economy of water use through efficient application and distribution or high water use efficiency
- Keeps down the weed growth
- No seepage or evaporation losses
- Fertilizers, insecticides and fungicides can be applied with irrigation water
- Land leveling is not essential
- Frost protection through hot water application
- Water can be precisely applied in the root zone.

Rate of water application: The size of the wetted zone in the soil depends upon the total volume of the water applied, water application ratio, infiltration capacity, hydraulic conductivity and rate of evaporation and transpiration characteristics of the soil.

An experimental result shows that if depth of wetting is more, then application rate is low and wetting duration is longer as compared to higher application rate for a shorter duration. Horizontal wetting is practically same for both.

Quantity of water to be applied: The water requirement will vary with age and size of the fruit trees. For every m of water loss, 0.375 mm water will have to be applied. The calculations for one hectare mm daily evaporation loss per emitter with 270 trees in a hectare can be worked out as follow:

1Hectare mm water = 10,000 liters

0.375 hectare water = 3750 liters

If 270 emitters are used per hectare, water to be applied per emitter is 3750/ 270 = 13.81 or say 14 liters.

Thus, only 14 liters of water per tree will be required to compensate for 1mm evaporation loss. It is divided by the number of hours that the system is to be operated per day. If weekly evaporation loss of 38 mm or a daily loss of 5.43 mm for June is to be made up, the water required to be applied per day per emitter is 14x 5.43= 76 liters. This can be provided at the rate of 6 liters for 13 hours, 9.3 liters for 8 hours or 13 liters for 6 hours.

Water pressure: Drip irrigation system can be run either on force of gravity or by pumping water. The system works on a low pressure, usually between 0.5 and 3 kg per cm^2. For breaking pressure force, a small tank of 50 cm x 50 cm can be installed, and water is allowed to flow into it. This tank can be installed a few meters above tree to be irrigated to create required pressure.

Layout of a drip irrigation system: Since drip-irrigation system on the concept of wetting one fourth of the root system of the trees, the emitters or outlets should be located in root zone of the trees. For large trees, 2 to 3 outlets may be necessary. In a smaller area, one main line may adequately supply water to laterals (Fig. 1).

Various types of drippers and micro-sprinklers are available in the market. A 100 m pipe will normally cater to 15 or 17 trees at 6 to 7 m distance. A storage tank should be used to obtain a continuous flow of water. It is necessary to fix a filter at the initial entry point of the water to main line or storage tank so as to prevent slit or other particles from choking emitters. The emitters are usually fitted with self-cleaning filters. Where micro-tubes are used instead of emitters, fitted with filter, it will be necessary to install additional filters on every lateral.

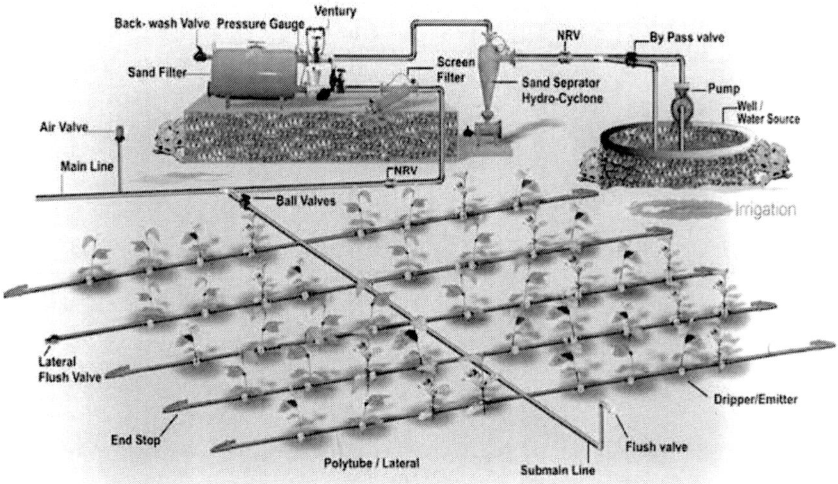

Fig. 1: Layout of a drip irrigation system. *Source*: Horticulture Water Initiative Drought Information, http://horticulture.com.au/drought

Water Conservation

Water is becoming limited due to limited resources of good quality water and increasing demand from various sectors. It is therefore, necessary that what so ever water comes in the form of precipitation is properly conserved to avoid both floods as well as the effects of drought. Conservation techniques need to be practices intelligently. These could be:

- In-situ water harvesting in macro as well as micro-catchment
- Mulching
- Close mowing of sod
- Minimum tillage
- Farm ponds and recycling
- Earthen dams across the gullies
- Use of antitranspirants on plant.

LITERATURE CONSULTED

Allen, R., Pereira, L.A., Raes, and Smith, M. 1998. Crop Evapotranspiration. FAO Irrigation and Drainage Paper 56, Rome, pp. 293.

Behboudian, M.H. and Mills, T.M., 1997. Deficit Irrigation in deciduous orchard. In: *Horticultural Reviews*, Jules Janick (ed) 21:105-131.

Chadha, K.L. and Shikhamany, S.D. 1990. Integrated Nutrient Management inHorticultural crops in soil fertility and Fertilizer use Vol. IV. Nutrient Management and supply system for sustaining Agriculture in 1990s (Ed. Kumar, V. Shrotriya, G.C. and Kaore S.V.) Indian Farmers Fertilizer Cooperative Limited, New Delhi. pp. 357 + xiv.

Doorenbos, J. and W.O. Pruitt, 1975. *Crop water requirements F.A.O.,* Rome.179 pp.

Fereres, E. and Goldhamer, D.A. 1990, 'Deciduous fruit and nut trees', from *Irrigation of Agricultural Crops*, ASA-CSSA-SSSA,United States.

Girona, J., del Campo, J., Mata, M., Lopez, G. and Marsal, J. 2011. A comparative study of apple and pear tree waterconsumption measured with two weighing lysimeters. *Irrigation Science.* 29:55-63.

HAL. 2006, *Managing Drought for Horticulture: Partial Budgeting Guide*, http://www.horticulture.com.au/delivering knowhow/Environment/Drought/grower_resources.asp

Horticulture Water Initiative Drought Information, http://www.horticulture.com.au/drought

Jones, H.G. 2004. Irrigation scheduling: Advantages and pitfalls of plant-based methods.*Journal of Experimental Botany*, 55, 2427-2436. doi:10.1093/jxb/erh213

Lamm, F.R., Ayars, J.E. and Nakayama, F.S. 2007. *Microirrigation for Crop Production.Volume 13: Design, Operationand Management*. Elsevier Publications. 642p.

Orloff, A. 2006. Soil Moisture Monitoring. How and why? In a NutshellMay/June 2006, SAMAC [see http://www.samac.org.za].

Stiles, W.C., T.L. Robinson, and W.S. Reid, 1995. Fertilization of Apple Orchards Through Drip Irrigation Systems (Fertigation). NY Fruit Quarterly 3(3):7-10.

Shirgure,P.S., Srivastava, A.K. and Singh, S. 2001. Effect of drip micro jet and basin irrigation methods on the growth, soil and leaf nutrient changes in acid lime. *Indian. J. Soil Cons.*, 29:229-234.

Sadler, E.J., Evans, R.G., Stone, K.C.and Camp, C.R. 2005.Opportunities for Conservation with Precision Irrigation. *Journal of Soil and Water Conservation*, 60:371-379.

Tan, C.S. 2007. Irrigation Scheduling For Fruit Crops, Agriculture Canada;R.E.C. Layne - Agriculture Canada,online:http://www.omafra.gov.on.ca/english/crops/facts/90-069.htm last accessed 23July 2007.

Tan, C.S. 1980. Estimating crop evapotranspiration for irrigation scheduling. Agriculture Canada vol. 25(4): 26-29.

Verma, A.K. 1998. Watershed management - A strategic approach for development of agriculture in NEH Region.*Indian Journal of Hill Farming*,11(1&2):1-6.

Westwood, M. N. 1993, *Temperate-Zone Pomology: Physiology and Culture*, 3rd edn, Timber Press Inc, Oregon.

Worthington, J.W., Coordinator, 1994. Micro Irrigation for Fruit Crops. Southern Cooperative Series Bulletin 378. Texas AES, Texas A&M University, College Station,Texas. 35pp.

8

Physiology of Fruit Plants

J.S. Bal and K.S. Gill

INTRODUCTION

Plant growth regulators (PGR) are bioactive molecules that modulates the physiological processes in plants and are effective at ultra-low concentrations. Minute alteration in the concentration and bioactive forms of PGRs' throughout the plant development phases regulate the flowering, fruit set and development. The use of plant growth substances to control fruit set, size, shape, and maturation has become important in agriculture today because they have the ability to increase fruit size, colour, and shape, thereby increasing the marketability. By hastening or delaying maturation the grower can utilize peak demands, avoid unfavourable environmental conditions, and extend the market period. PGRs are applied in fruit crops at all developmental stages for various purposes as discussed below:

Root induction : Auxins and cytokinins induce profuse root formation in the treated cuttings and layers of the fruit crops e.g. peach, guava, fig, pomegranate,etc. IBA is the most commonly used growth regulator followed by NAA for promoting root formation. GA induces cell elongation but it has an inhibitory action on root formation in cuttings.

Seed germination : GA enhances the germination of seeds by enhancing the activity of á-amylase that break down the complex carbohydrates into simple sugars and provide energy to the growing embryo. Dip treatment in 10-20 ppm solution of GA for 12 hours before sowing, significantly enhance the germination in papaya.

Canopy management : Application of growth retardants like cycocel, ethrel, SADH, paclobutrazol manages the excessive vegetative growth and cause floral induction in pear, peach, lemon, apple, litchi, apricot, plum and mango.

Flowering regulation : Both natural (NAA) and synthetic formulations (Planofix and Celmone) of auxins are effective in induction of flowering in pineapple. Ethrel or ethephon application is also very effective for the same as ethylene is the active principle responsible for flowering in pineapple. CCC treatments increased flowering in grapes and lemon. Paclobutrazol application enhances flowering and fruiting in mango and grape. SADH promotes flowering in apple, pear, peach, lemon, and blue berry and manage the excessive shoot growth.

Fruit set and development: Lack of caprification in figs can be compensated with 4 CPA treatment for improving fruit set. To increase fruit set in strawberry, peach, plum and cherry the application of GA, 4-CPA or B-NOA is very effective. Application of CPPU, a synthetic cytokinins at fruit set enhances the berry size in grapes and kiwifruit.

Prevention of fruit drop : Exogenous application of auxins (2, 4-D, 2, 4, 5-T and NAA) is used to manage fruit drop in citrus. Application of 50-100 ppm of NAA a week prior to harvest reduces the post-harvest berry shattering in grapes. The fruit drop in mango can be managed by application of 10-20 ppm NAA or 10 ppm 2, 4-D after fruit set.

Fruit ripening : Ethephon treatment (1000 ppm) prior to storing ensures post-harvest degreening and attractive yellow colour in citrus fruits especially lemons. Climacteric fruit such as banana, mango etc., can be ripened better and uniformly with ethylene treatment. Spray of ethephon during fruit development in apple, grapes, strawberry and cranberry advanced fruit maturity and induced uniform ripening.

The chapter on physiology of fruit plants includes information about types of fruits, flowering in fruit plants, pollination, fruit bud formation, fruit-setting and development and maturity of fruits.

A. FRUIT AND ITS TYPES

Fruit: Besides being nutritionally important for human consumption, fruits are of immense value for seed dispersal. Fruits differ in their morphology and constitution to a considerable extent. The fruit development depends upon the physiological changes occurring in the ovules following pollination. Ripened ovary alongwith whatever intimately attached to it at maturity is referred to as 'Fruit'. The fruits that have well developed carpels as the major edible portion are termed as true fruits i.e. orange, persimmon, avocado etc. However, the fruits developed from the epigynous flowers such as apple, pear etc. have the fleshy thalamus/receptacle as the edible portion such fruits are called 'pseudofruits'. The fruits are further classified into 3 categories *viz.* simple, aggregate, and multiple fruits based on morphological and botanical characteristic.

1.Simple fruits: Consists of ripened ovary either inferior or superior, dry or fleshy with one pistil. A fruit developing from inferior ovary consists of other parts of flower besides the ovule in contrast to the fruits arising out of superior ovary consisting of ovary alone. Simple fruits may be of 2 types : a) dry, b) succulent or fleshy

a. Dry fruits: Dry fruits i.e. follicle, legume, capsule, achene, caryopsis, samara and nuts are characterized by leathery, papery or woody fruit wall. These fruits may be indehiscent *viz.* achene, caryopsis or grain, samara, schizocarp and nut or dehiscent including legume, follicle, capsule, and silique.

i. **Indehiscent fruits :** Indehiscent fruits does not split open at maturity. Fruits developed from pistil of superior bi or polycarpellary ovary where the pericarp is hard and woody are known as *nuts.* These are dry, one-chambered and generally one-seeded fruit (sometimes two-seeded). e.g. cashewnut, chestnut, walnut, etc. These have structures formed from bracts arising from the base of one or more closely associated fleshy flowers.Nuts develop from epigynous flower and may be enclosed in fleshy or dry receptacle.

ii. **Dehiscent fruits :** It consists of fruits in which the pericarp bursts open to liberate the seeds at maturity.

b. Simple succulent fruits : These fruits develop with a fleshy fruit wall at maturity.

i. **Berry** : It is a fruit developed from a superior (sometimes inferior) ovary in which the entire ovary wall turns into a fleshy, juicy and edible pericarp. Berry is an indehiscent fruit with many-seeded flesh or pulp. It develops from a single carpel or from a syncarpous pistil with axial or parietal placentation e.g. date palm, guava, grapes, etc. The date palm is a single-seed berry. Berry remains succulent. The berry may or may not have seeds and may have, one or several carpels. True berries are distinguishable from false berries by the persistence of remnants of flower at the top of the fruits of the former. In cranberries, several sepals persists at the apical end. It is also evident in banana because the fruits have scars from where the floral parts have fallen off.

ii. **Modified berries**:

Balusta : It is an inferior, many-celled and many-seeded fruit with leathery pericarp; the carpels are placed in two rows one above the other with thin yellow papery partition walls; the seeds have succulent seed coat which forms the edible part e.g. pomegranate.

Amphisarca : A superior, many-celled and many-seeded fruit with woody pericarp; the placenta and the inner layer of the pericarp forms the pulpy edible part. e.g. wood apple

Hesperidium : Many-celled fleshy fruit which develops from a polycarpellary superior ovary with axial placentation, the epicarp and mesocarp are fused together forming the rind. Endocarp remains thin and papery forming the outer covering of compartments and the edible part is the inner juicy succulent hairs of endocarp e.g. citrus fruits.

iii. **Drupe**: A drupe is a fleshy fruit, with one or more-chambers and one or seeds developing from a monocarpellary or syncarpous pistil. Olive is an exception as the flowers consist of 2 carpels with 4 ovules but only a single carpel develops. The pericarp developed from a superior ovary is easily distinguishable into an exocarp, mesocarp and endocarp. The exocarp or epicarp constitute the thin skin of the fruit, the mesocarp is the fleshy pulp and endocarp is represented by the stone e.g. mango, peach, apricot, plum, coconut-palm, cherry etc. In.almond, leathery exocarp and mesocarp are not edible, only the shell containing the edible seed is marketed. Fleshy drupes with one carpel are mango, olive, almond, plum and peach, those with two carpels include coconut. The hard covering lying over the edible portion, (the endosperm) is the endocarp in coconut. The outer fibrous covering is the mesocarp covered by the outermost smooth epicarp.

iv. **Pome:** The pome is an inferior, two or more-celled fleshy, syncarpous fruit ·surrounded by the thalamus. The fleshy edible part is composed of thalamus while the actual fruit is engulfed within. Quince, loquat, apple and pear and fleshy fruits classed as pomes with the pericarp differentiated into epicarp, mesocarp and endocarp.

2. **Aggregate fruits**: An aggregate fruit is a collection of simple fruits or fruitlets' developing from apocarpous pistil of a flower.

Etaerio: An aggregate of simple fruits borne by single flower having apocarpous pistil (free carpels) e.g., Annonaceous fruits. The ovaries of respective carpel mature into fruitlets clustered together on the pedicel. The simple fruit of raspberry has small parts, the druplets which may separate out as independent units from the dry, dome-shaped receptacle. In blackberry, the fruit is a little different from that of raspberry where the elongated receptacle is also fleshy and forms a part of the fruit. Both blackberry and strawberry fruits are classified as accessory as well as aggregate. The strawberry fruit has simple pistils developed into tiny achenes. In the strawberry the edible portion is the fleshy and much elongated receptacle, a modified stem.

3. **Multiple fruits**: In some fruits the whole inflorescence contributes in the formation of fruit it leads to development of a multiple or composite fruit. In this, every flower develops into a fruit 'which remains together as a single mass at maturity.

i. **Syconus**: A fruit type which is developed from a hollow, pearshaped, fleshy receptacle enclosing a number of minute, male and female flowers and the receptacle grows, becomes fleshy and edible. e.g. fig.

ii. **Sorosis**: A multiple fruit developing from a spike or spadix where the flowers fuse together by their succulent sepals. The axis bearing the flowers grows and becomes fleshy or woody, and as a result the whole inflorescence forms a compact mass. e.g. pineapple, jackfruit, mulberry.

B. FLOWERING IN FRUIT PLANTS

The flowering in fruit plants is determined by the environmental factors like temperature and atmospheric humidity. The morphological features of flowers are very important and one should be aware about types and bearing habits of fruit trees. In different fruit plants, three sex types of flowers are common. The position of the flower on the inflorescence is a criteria for determining fruit set in many plants. The flowering period is variable in different fruit plants on the basis of the climatic conditions. The information on types of fruiting and bearing habit, inflorescence, types of flowers and flowering period in fruit plants is given below:

Types of Bearing Habit in Fruit Plants

The flower buds are found in a certain definite position. In developing bud, they are differentiated when they unfold into a leafy structure and the subsequent formation of a fruit bud at a new growing point. Fruit buds are borne terminally on long or short growth, laterally in the axils of the current or past season leaves or adventitiously from any point on the exposed-bark of limbs, trunk etc. Fruit bud weather it is terminal or lateral, when it unfolds may give rise to any of the three distinct types of flower bearing structure i) it may contain flower parts only and develop a single flower or a flower cluster without leaves ii) it may be mixed bud and develop short or long leaf shoots terminating inflorescence and iii) it may be mixed and develop a short or long leafy shoot bearing flowers or in some of its leaf axils. There are six main types of fruiting habits in fruit plants.

Bud position	Terminal fruit plants	Lateral fruit plants
Flower bud containing flower parts only	Mango, loquat, litchi	Peach, plum (*domestica*) apricot, almond, cherry, date palm, citrus, walnut (staminate catkins), pecannut (staminate catkins)
Flower bud mixed, Flowering shoot with terminal inflorescence	Apple (principally), pear (principally) walnut (pistillate flowers), pecannut (pistillate flowers)	Grape, cashewnut, raspberry, apple (occasionally), pear (occasionally)
Flower bud mixed, Flowering shoot with lateral inflorescence	Guava, pomegranate, olive (partly)	Ber, fig, mulberry, persimmon, avocado

Bearing Habits of Fruit Plants

The bearing habits are found different in various fruit plants. There are nine distinct bearing habits noticed in fruit plants. The classification is based on location of fruit buds and type of flower bearing structure. The various cultural operations like pruning, heading back and thinning out are easily performed if bearing habit of plants is known.

Bearing habit I: Fruit buds borne terminally containing flowers parts only. It gives rise to inflorescence without leaves e.g. Mango, loquat, litchi.

Bearing habit II: Fruit buds borne terminally and unfolding to produce leafy shoots which terminate to flower clusters. e.g. apple, pear, citrus. The fruit buds are mixed and invariably give rise to very short growth and a few short internodes, leaves of ordinary size and a lateral branch arising in the axil of the leaves.

Bearing habit III: Fruit buds borne terminally and unfolding to produce leafy shoots with flower or flower clusters in leaf axils e.g. guava, pomegranate. This might be called as an incomplete terminal bearing habit. The flower buds are terminal and for the following year's production or new buds may develop from lateral leaf buds.

Bearing habit IV: Fruit buds born laterally, containing flower parts only and giving rise to inflorescence without leaves or leaves present but are reduced in size e.g. peach, almond, plum, date palm, citrus.

Bearing habit V: Fruit buds borne laterally, unfolding to produce leafy shoots that terminate in flower clusters. Fruit bud mixed e.g. grapes, blackberry, raspberry, cashewnut.

Bearing habit VI: Fruit buds borne laterally, unfold to produce leafy shoots with flower clusters in leaf axil. Fruit bud mixed e.g. ber, mulberry, fig, persimmon, avocado.

Bearing habit VII: Fruit buds in the axils of leaves. These buds unfold and develop their flowers and fruits very soon after the flower parts are differentiated e.g. papaya.

Bearing habit VIII: Fruit buds borne both terminally and laterally. Inflorescence is generally terminal, e.g. walnut and pecannut. The terminal bud may give rise to a short leafy shoot ending in a female inflorescence. Male flowers are borne on leafless inflorescence arising from lateral buds not far below the terminal.

Bearing habit IX: Fruit buds adventitious and lateral. The flower buds appear first on the trunk as the trees grow and then on the branches e.g. cacao. Its bearing habit resemble with bearing habit of IV and V.

Inflorescence in Fruit Plants

The arrangement of flowers on the stem or axis is referred as inflorescence. It may or may not bear leaves or bracts or may or may not have stalk. Inflorescence is a flower bearing shoot either determinate or indeterminate in growth. It may terminate in a shoot or may be borne in the axil of leaf. The branched inflorescence usually but not always bear bracts or modified leaves. There are several types of inflorescence or arrangement of flowers in a cluster.

1. **Solitary Flower:** A flower may be borne singly in the leaf axils at the terminal part of growth or on a specialized stalk. It is found in guava, peach, apricot, almond, apple, sapota, citrus, pomegranate etc.
2. **Racemose:** The axis of flower cluster (peduncle) continues to grow and new flowers buds with their subtending bracts continue to be formed near the growing point, producing a succession of new flowers. The lowest oldest flowers are first to open and blossoming proceeds upward. The most common one is in determinate type of inflorescence. Further, it has different types of inflorescence.

   i) **Raceme:** The flowers of equal length are borne on pedicels. This type of inflorescence found in walnut, pecannut, blueberry, wild cherry, currents, raspberry and gooseberry.

   ii) **Catkin:** A spike or raceme with a slender rachis, bear many unisexual, epipetalous flowers which falls as a whole when the fruit is mature. It is found in pecannut, walnut, mulberry, chestnut etc.

iii) **Umbel:** A short rachis long pedicellate flowers of equal length, arise from a common point and spreading like umbrella.

iv) **Corymb:** The main axis is elongated. Radicles are of unequal length, lower are longest and upper or central are shortest. Thus, flowers lying in horizontal plane e.g. pear.

3. **Cymose:** A flower bud is produced at the growing point, floral axis cannot be elongated further. The other flower buds are produced below this point on the axis towards the base. The flower opening is basipetal or centrifugal (from the centre toward circumference of the cluster). The solitary flower is cymose type e.g. papaya, ber, strawberry, phalsa, sapota, pomegranate. There are two types of cymose inflorescence

   i) **Fascicle:** A cyme with closely crowded flowers on a very short peduncle with equal length of pedicels e.g. ber, cherry, plum.

   ii) **Panicles:** The main axis is indeterminate. The secondary and ultimate axis are cymose e.g. mango, litchi, loquat, grape.

4. **Anamalous:** The terminal flowers opens first and referred as cyme. The basal flowers opens next and the anthesis of individual flowers then proceeds more or less upward to the apex and referred as racemose e.g. apple.

5. **Special Inflorescence:** These are of two types

   i) **Hypanthodium:** The flowers are produced on the inner surface of the receptacle which become spherical type having cavity inside. It opens to the outside with small opening. It can develop three types of flowers i.e. male, long styled female and short styled female e.g. fig (ficus), banyan.

   ii) **Spadix:** In this type, small flowers borne on fleshy stem. A large spathe (boat type) enclosing an inflorescence. The peduncle upper side bear coloured bracts and lower side bear unisexual flowers e.g. date palm (palmaceae), banana, coconut.

Types of Flower in Fruit Plants

The flower is a reproductive part of the plant which usually contains petals, a pistil and pollen bearing stamens. There are three sex types of flowers i.e. perfect or hermaphrodite, staminate and pistillate flowers. Teratological is an another type of flowers included in this group.

1. **Perfect or hermaphrodite flowers:** It contain both male and female reproductive structure (stamens and pistils). Citrus, guava, sapota, papaya, cashewnut, mango, grapes, apple, pear, peach and plum flowers are placed in perfect flower type.
2. **Staminate flowers:**These are flowers with non-functional pistils. It is a male flower, bears only stamens. Papaya, litchi, pomegranate, date palm, mango, cashewnut, coconut, walnut and grapes flowers are placed in staminate type.
3. **Pistillate flowers:** These are female flowers bears only pistils and lack stamens. Papaya, grape, walnut, pecannut flowers are placed in pistillate type.
4. **Teratological flowers:** These are abnormally modified type flowers. All polymorphic forms are called teratological representing residual phyllogenetic transitional stages between the perfect flowers and the pistillate flowers for one and another reason. Mango, pomegranate, papaya and grapes flowers are placed in teratological type.

Flowering Period in Fruit Plants

India has a different climatic conditions in various regions which includes environmental factors like temperature, atmospheric humidity, rainfall and wind velocity. Therefore, flowering period depends upon the prevailing agro-climatic conditions of a particular area. However, flowering period in different fruit plants can be categorized into three periods.

1) **February to May:** In Northern Indian conditions apple, peach, pear, Japanese plum, almond, apricot, walnut, citrus, mango, litchi, grapes, pomegranate and date palm flower during February to May. On the other hand, in Southern Indian conditions, banana, papaya, sapota, pineapple and fig flower during this period.
2) **June to September:** In North India, guava, ber and papaya flower during this period. In South India, citrus and pomegranate flower from June to September.
3) **October-January:** In North India, ber, guava, sapota and loquat flower during this period,whereas in South India, mango, sapota, guava and ber flower during this period.

Flowering in Temperate Fruits: The flowering in apple takes place in April in Himachal Pradesh. In pear, flowering starts in last week of February and continues upto third week of March under North Indian conditions. The flowering in peach starts in the first week of February and goes upto end of February in

Punjab, Haryana and Himachal Pradesh. In Japanese plum flowering starts in the second fortnight of February and last up to first week of March in plum growing areas. The flowering in almond takes place in very early season in spring occurring between January to April in Northern India. The flowering in apricot mostly occurred in the month of March. In walnut peak flowering takes place in mid-April under Himachal Pradesh conditions. The flowering in walnut prolonged till the first week of May in late flowering cultivars. In sweet cherry flowering occurred in end February to early March in Kashmir valley.

Flowering in Tropical and Sub-tropical Fruits: In many parts of North India, mango flowers late in January or in the beginning of February or as late as in March. Guava normally produce three crops in a year. In North India, three distinct flowering season i.e. summer, rainy and autumn have been occurred. Sapota flowers throughout the year with two main flowering seasons i.e. July to November and February to March. In evergreen pomegranate, the flower buds of the spring flush are borne on mature wood of one year old shoot. The flowers which appear during July-August are borne on current years growth. The flowering in deciduous cultivars of pomegranate takes place between July and August and these flowers are borne on current years growth. In cashewnut, three growth flushes takes place. In South India, the flower bud emergence takes place in mid-September and continuous till the end of February. The main season of flowering is October-November. In custard apples flowering is very long commencing from March-April and continuous upto July-August. The peak flowering is noticed in April and May. The avocado produces flowering from November to January in three main flushes in the climate of Pune (Maharashtra). Papaya flowers throughout the year in the country.

In citrus flowering occurred mostly in spring season. In kinnow, flowering takes place during end of February and continues upto early March in North India. In lime and lemons, flowering takes place almost throughout the year. In litchi, flowering occurred in February in North India. The flowering period in ber is quite lengthy which lasts for about two and a half months from September to November under Punjab, Haryana, Rajasthan and Uttar Pradesh. In grapes, flowering takes place in spring in north India. The flowering in date palm takes place during end February to early March. The flowering in phalsa starts from February-March and continues till May. The flowering period in loquat is quite lengthy which lasts from mid-July to January or some time even upto April-May. In Jamun flowering starts in first week of March and continues upto mid-April.

C. POLLINATION IN FRUIT PLANTS

Pollination: It refers to successful transfer of pollen from the dehiscing another to the receptive stigma of the same flower or of a different flower. Pollen transference is made either by agents or the anther should be placed in the flower above the stigma. In this way pollen drop directly on the stigma by sheer gravity.

Self-pollination means transfer of pollen to stigma in the same flower or to the stigma of other flower on the same plant or any plant of the same variety. Cross pollination is a transfer of pollen from anthers of a flower of one variety to the stigma of flower of another variety. Self-fruitful means when a plant is able to set and yield commercial crop with its own pollen. Self-unfruitful is referred as when plant is able to set and yield commercial crop with its own pollen. Partially self-fruitful is defined as when a variety produce some crop but less than the full crop.

Self-compatible means capacity of a variety having functional sex organs and gametes to produce fruits following self-pollination (Self fruitful). When a variety shows inability to produce fruits following self-pollination is called self-incompatibility. When a variety is used as pollinizer for another is able to produce full commercial crop is said cross fruitful. However, when one variety is used as pollinizer for another shows inability to produce full commercial crop is said cross unfruitful.

Pollination Agencies

Pollen grains are disseminated by various pollination agencies viz. wind, insects, birds, human being and water. The following terms are applied to indicate the type of pollination using a particular type of agency.

a) **Pollination by wind (Anemophily):** Papaya, date palm, strawberry, pomegranate, cashewnut, walnut, pecannut, mulberry and coconut are pollination by wind.

b) **Pollination by insects (Entomophily):** Mango, citrus, guava, litchi, ber, annona, apple, pear, peach, plum, almond, cherry and walnut are pollinated by insects.

c) **Pollination by birds (Ornithophily):** Pineapple and banana are pollinated by birds.

d) **Pollinated by human being:** Datepalm is pollinated by human being. To increase fruit set in some fruit crop and for the development of new variety, pollination is done by the human being.

e) **Pollination by water:**The moving stream of water carries the pollen grains over the surface of other stigmatic flowers.Water lily and Hyacinth are pollinated by water.

Pre-requisites of Pollination

1. For Wind Pollination

- Pollen grains to be produced in sufficient quantity
- Pollen grains should be very small and light
- The stigma of female flowers should be large so as to give more chance of landing for the falling pollen grains.
- The stigmatic surfaces should be sticky.

2. For Insect Pollination

- The plants must have showy flowers.
- The flower should have enough nectar which should attract the insects.
- The petals and odour of the flower should be attractive.

Pollination Mode in Fruit Plants

There are two common modes of pollination in different fruits plants.

1. Self-pollination (Autogamy): When a stigma is pollinated by the pollen of the same flower is referred as self-pollination or autogamy. It is further of two types.

a) **Cleistogamy:** The flowers never open before anthesis as opposed to most flowers which show chasmogamy. In this type of flowers (clesitogamous), the pollens are shed within the closed flower. The self pollination process takes place automatically. This mode of pollination is common in grape, sapota and papaya.

b) **Homogamy:** In this type, the stamens and the carpels of the flowers mature at the same time. Thus, the chance of self pollination enhances otherwise it is not obligatory. Citrus, peach and apricot are pollinated through homogamy process.

2. Cross pollination (Allogamy): The pollen of a flower pollinate the stigma of another flower located on a different plant is said to be allogamy. It is not necessary that the another flower is of the same kind. Some of the special contrivances ensuring cross pollination are very conspicuous are as follows:

i) **Dicliny or unisexuality:** Cross pollination is the rule among in unisexual (diclinous plants) flowers. These are also called imperfect flowers.

a) **Monoecious:** It is an occurrence of male and female flowers on different location on the same plant e.g. walnut, pecannut and chestnut.

b) **Dioecious:** When male and female flowers borne on different pant of the same species e.g. date palm, papaya, fig, strawberry, pistachio nut.

ii) **Dichogamy:** In this mode, stamens and carpals of bisexual flowers (hermaphrodite) mature at different times and pollination between then becomes ineffective. Sometimes when cross pollination fails,then self-pollination may take place at a later stage. This mode of pollination may be of two types.

a) **Protandry:** In this anthers mature prior to ovule. When the anthers bursts, it pollinate stigma of other flower before its own stigma yet not ripe e.g. walnut, coconut.

b) **Protogyny:** Stigma became receptive earlier than anthers shed pollen grains e.g. annona, fig, pomegranate, plum, banana and avocado.

Therefore, protandrous cultivars should be planted with protogynous cultivar to permit overlapping of male and female flowering times.

iii) **Self-incompatibility:** It is a condition when a variety show inability to produce fruits following self pollination (Self unfruitful) e.g. almond, apple, pear, apricot, cherry, mango, pineapple etc.

iv) **Heterostyly:** It is a condition when length of the style varies relative to other part of flowers. This mode involve flower of two (dimorphic) of three (trimorphic) different forms with anthers and stigma at different levels. The flowers have short styles and long filaments in dimorphic heterostyly whereas the stamens and styles are of three different length in trimorphic heterostyly e.g. almond, pomegranate, litchi, sapota.

It may have called hypostigmatic (stigma above the style), epistigmatic (stigma below the style) and peristigmatic (stigma at level with style). Heterostyly is of two types.

a) **Pin type:** Style is longer than stamens, broad towards apex and narrow towards base. Such plant type set more seed due to exposed stigma e.g. sapota, pomegranate.

b) **Thrum type:** When flowers has short style and stamens are longer. Such flowers give poor, fruit-setting because stigma remains hidden inside the corolla tube e.g. almond.

Factors Affecting Pollination

Pollination is affected by some factors directly or indirectly which ultimately affect the overall productivity of fruit crop. The common factors associated in this regard are as follow:

i) Time of pollen shedding and stigma receptivity.

ii) Species and population of vectors responsible for pollen transfer.

iii) Proportion and compatibility of pollinizers.

iv) Weather conditions affect directly. These conditions affect pollination in two ways:

  a) Weather has cumulative affect on floral development which influences as the timing and period of flowering.

  b) It effect on pollen shedding, effective pollination period and foraging behaviour of pollinators.

Fruit Crops Showing Pollination Problems

Fruit Crop	Pollination Problem
Mango	Bears two types of flowers i.e. male and perfect. Perfect flowers percentage vary from 5 to 50
Papaya	Dioecious nature
Date palm	Dioecious nature
Sapota	Non-synchronous maturation of flower parts
Pomegranate	Bear two types of flowers- hermaphrodite and staminate.
Sweet lime	Bears two types of flowers i.e. staminate and hermaphrodite
Fig	There are two types of fig: - Bear pistillate flowers only hence need pollination e.g. Smyrna - Some develop fruit without pollination/ parthenocarpy e.g. common fig.
Avocado	Non-synchronous maturation of flowers parts
Grape	Some varieties bears pistillate flowers only e.g. BanquiAbyad and Hur.Some varieties are unfruitful e.g. Pandheri Sahebi. Stamens which bear sterile pollen that are unable to fertilize the ovules of their own or any other pistillate flowers.
Apple	Cross pollinated crop. Needs pollinizers and pollinators
Almond	Most varieties are self-unfruitful. Two or more varieties should be planted
Plum	Most Japanese plum varieties are self-unfruitful and require cross pollination.

Pollen Viability

In most of the fruit crops initial fruit set is very important to get better yield and productivity. Proper pollination and fertilization of egg cell is essential to obtain the desired results. The pollen should be viable and should have capacity to germinate on stigmatic surface for elongation of pollen tube sufficiently to fertilize egg cell. Acetocarmine test is performed to study the pollen viability. The pollen grains are said to be viable if get stained. The unstained pollen grains are considered non-viable.With this test the living protoplasm gets the stain. By counting viable pollen and non-viable pollen under microscope, the relevant percentage is calculated

Pollen Germinability

Pollen germination is studied by using different concentrations of sugar solution in cavity slides. The sugar solution of 15, 20 and 25 per cent are used for testing germinability. The pollen grains are incubated in freshly prepared sugar solution of different concentrations for varying times. Pollen grains can be germinated in distilled water but to enhance the pollen tube growth sugar is added. The prepared cavity slides are then seen under microscope repeatedly. The number of germinated pollen grains are counted and percentage is worked out. The size of pollen tube is also measured with micrometre. The media like agar and Boron can also be used to enhance pollen germination.

D. FRUIT BUD FORMATION

Bud is defined as a growing point from which a branch, leaf or flower may be grown. It is an incipient or unelongated leafy or flower shoot or both.

Classification of Buds

Buds are classified on the basis of kinds of buds, point of origin of bud and arrangement of buds.

1. Kind of Buds

i) **Leaf bud:** When apex bears unexpended leaves together with the leaf primordia.

ii) **Flower bud:** When apex bears unexpended flower is called flower bud.

iii) **Simple bud:** When growing shoot bear only flowers or leaves.They are further of two types. Lateral simple buds are found in peach, plum, almond, apricot and cherry. Terminal simple buds are found in mango, litchi and loquat.

iv) **Mixed bud:** In this type both leaves and flowers appear within a single bud. Mixed terminal buds are common in apple, pear and quince whereas mixed lateral buds are found occasionally in apple and pear.

v) **Compound bud:** When three buds are often found within one large bud is called compound bud e.g. grapes.

2. Point of Origin of Buds

i) **Terminal bud:** This type of bud borne terminally to the growing shoots e.g. Mango, loquat, litchi, apple, pear.

ii) **Lateral or axillary bud:** When buds borne in the axil of the leaf are called lateral or axillary buds e.g. citrus, guava, ber, peach, plum, apricot etc.

iii) **Adventitious bud:** The buds are borne on the intermodal spaces e.g. blackberry.

3. Arrangement of Buds

i) **Single or solitary bud:** When only one bud is borne in the axil of the leaf e.g. guava, ber, peach, plum, apricot etc.

ii) **Cluster bud:** In this type, buds are very close together to appear like a cluster e.g. some varieties of Japanese plum.

iii) **Collateral bud:** Where two or more buds occur side by side on a node e.g. peach.

iv) **Superimposed bud:** Here buds are located one above the other e.g. Yellow wood.

Flower Bud Differentiation

The modification of bud initials into vegetative or flower bud is called flower bud differentiation.

Time of flower bud differentiation depends and varies according to kind, variety, environmental conditions and internal nutritional conditions i.e. mainly carbohydrates reserves of the plant. In temperate fruit plants like apple, pear, peach, plum, apricot, almond and cherry and in grapes, flower bud differentiation (FBD) takes place in the previous season (summer) before their unfolding. In different fruit plants, FBD period and anthesis in given below: -

Fruit Tree	FBD Period	Anthesis
• Citrus	February-March	New growth
• Mango	September-October (N.India)	Next spring
October-December (Punjab)		
• Guava	September-October	New growth
• Ber	August-September	New growth
• Apple, peach, cherry,grape, fig, persimmon , strawberry	June-July	Next spring
• Pear, plum, apricot, almond	July-August	Next spring
• Strawberry	September-October	Next spring

Fruit Bud Formation and C/N Ratio

Carbon is taken up by the plants in the forms of carbon dioxide from the atmosphere and possibly in the form of bicarbonate (HCo^{-}_{3}) from the soil solution. Nitrogen is taken up by the plants in the form of nitrate (No^{1}_{3}), ammonia (NH_4+) from the soil and N_2 gas from the atmosphere. Assimilation of carbon dioxide takes place in the chloroplast in the leaves which later converted into glucose + oxygen ($C_6H_{12}O_6$+O_2) in the process of photosynthesis as follow

$$6CO_2 + 6H_{20} \xrightarrow[Energy]{Light} C_6H_{12}O_6 + 6O_2$$

Both carbon and nitrogen are the major constituents of the organic material which are involved in the fundamental biochemical processes. The carbohydrates are converted into chemical energy and then into complex compound adenosine triphosphate (ATP). This energy is utilized by the plants for the formation of various plant parts i.e. root, stem, leaves, buds, flowers, fruits and seeds, their development and maintenance of health.

Nitrogen generally favours vegetative growth which in turn increase carbohydrates synthesis and lead to flower bud formation. Lack of nitrogen often leads to unfruitfulness in the tree even when carbohydrates are present. When nitrogen is supplied, stored carbohydrates are utilized in new vegetative growth of the trees. Leaves become longer and greenish and flower buds are formed. When leaves are removed or otherwise eaten by insects or damaged by disease, synthesis of carbohydrates is interrupted. Then, the vegetative growth stop, flower buds become weak and their development is checked. Both nitrogen and carbohydrates are very essential for the production of flower bud and must be present in a relative proportion. When carbohydrates and nitrogen are in requisite proportion, both fruiting and vegetative growth takes place normal.

Carbohydrates Utilization and Accumulation

When vegetative phase (leaves, stem and root development) is dominant over reproductive (flowers, fruits and seed development) phase, carbohydrates utilization is dominant over carbohydrate accumulation. The carbohydrates are utilized in vegetative phase during cell division, cell enlargement and initial stages of cell differentiation.

When reproductive phase is dominant over vegetative phase, carbohydrates accumulation is dominant over carbohydrates utilization. The carbohydrates are stored in reproductive phase i.e. after differentiation, bud undergo rest period for some time and carbohydrates are being stored and later on the same are utilized for the formation and development of flower buds, flowers, fruits and seeds. When vegetative and reproductive phase are balanced, utilization and accumulation are also balanced i.e. equal amount of carbohydrates are utilized and stored. The accumulation of carbohydrates accompanied by good supply of nitrogen are essential for good vegetative growth and fruit production.

In relationship of nitrogen and carbohydrates in apple it is reported that if carbohydrates are restricted and nitrogen is abundant or nitrogen is restricted and carbohydrates are abundant, the vegetative growth will be poor and bearing will be small. Ample nitrogen alongwith limited carbohydrates leads to luxuriant growth and small amount of fruit on the trees. Plants flower and fruit abundantly as a result of sufficient carbohydrates and nitrogenous substance.

E. FRUIT SETTING IN FRUIT PLANTS

The rapid growth of the ovary which usually follows pollination and fertilization is defined as 'Fruit Set'. Pollination and fertilization of the ovules must normally take place to produce fruit with viable seed. However, in parthenocarpy development of fruits occur without pollination or fertilization resulting in seedlessness. Parthenocarpy is of two types, vegetative and stimulative. Vegetative parthenocarpy is characterized by the development of fruit without pollination (e.g. pineapple and banana). Whereas, stimulative parthenocarpyrequires the stimulus of pollen without subsequent fertilization in order to set fruit (e.g. grape)

Problems of Fruit Set

Fruit set refers to initial and appreciable swelling of the ovary occurring shortly after the period of petal fall. It is generally accompanied by some thickening of the pedicel or peduncle. The inherent ability of a plant to produce optimum crop is fruitfulness and the inability to achieve this is referred to as unfruitfulness. It is a serious problem in many orchards and a thorough understanding of its causes is a prerequisite for obtaining economically acceptable production level. Problems of fruit set may be due to internal or external factors.

Internal Factors

There are a number of internal factors which are associated with unfruitfulness or sterility. These factors are further classified into three major categories i.e.

1. Evolutionary tendencies.
2. Genetic influence.
3. Physiological factors.

I.Evolutionary Tendencies: Various developmental events that occurred during the evolutionary process induced some imperfections in flowers which may lead to unfruitfulness. These factors are discussed under the various subheads.

a. **Imperfect Flowers (Monoecious and Dioecious nature) :**The plants with stamens and carpels in different flowers on the same plant are termed as monoecious. e.g. coconut, arecanut, pecan nut, capri fig and hazel. The monoecious fruit plants with ensured pollinator activity in general do not exhibit the problem of pollination, fruit setting and fruitfulness. However, the plants which bear male and female flowers on different plants are known as dioecious. e.g. papaya, date palm and strawberry. Pollination and fruit set is comparatively less as spatial separation of both the male and female flowers that act as a limitation in achieving the optimum fruit set e.g. profuse flowering without fruit set in ornamental pomegranate is due to their unisexuality. In figs there are two types of flower clusters namely staminate and pistillate flowers. In Capri fig staminate flowers are borne near the eye and pistillate flowers are borne near the end. To ensure good fruit set, retention of a few staminate trees (9:1) is essential as pollinizers. Similarly in papaya there are a number of sex forms which act as a limiting factor in effective fruit set

b. **Heterostyly :** Heterostyly is a form of polymorphism consisting of two or three morphological types of flowers on different plants. However, only one morphological form exists on a single plant. Heterostylous plants having two flower types, the 'Pin type' has short stamens and long pistils and vice-versa in the 'Thrum type'. Thus, styles and stigmas at different height prevent self pollination. In case of brinjal there are 4 types of flowers according to their length of style i.e. long, medium, pseudo-short and true short. Out of these pseudo-short and true-short do not produce any fruit. In delicious group of apples extreme upright positions of the stamens and fully spread petals makes it impossible for bees to do pollination while collecting nectar. When the pistils of heterostyled plants are pollinated with pollen from the same flowers or from other flowers containing stamens of an equal height the union may be fruitful but it may exhibit varying degree of sterility. Hence, to obtain optimum fruits some arrangement for cross pollination needs to be developed.

c. **Dichogamy :** The phenomenon in which the period of stigmatic receptivity does not coincides with pollen viability in monoecious plants is known as dichogamy. Therefore, self pollination is prevented in perfect flowered plants, due to maturity of two sex elements at different times (temporal separation). If the stamens ripe before the stigmas become receptive the flowers are known as protoandrous but, if stigmas become receptive before the stamens produce viable pollens it is known as protogynous. Both protoandrous and protogynous flowers result in low fruit set. Protogyny is exhibited by many monoecious plants like walnuts, hazels, etc. whereas protandry is present in many coconut varieties.

d. **Abortive Flowers or aborted pistils or ovules :** It may occur in the developing flower's pistils and stigmas of many species and may result in low or even no fruit set. Abortion of partially developed flower buds is common. Any interference in development and functioning of the gametes may lead to sterility or unfruitfulness; such things can be observed in some grape varieties and tomato varieties. The late flowers of strawberry cluster are always abortive. This is more common in indeterminate type of plants. Degeneration of pistils takes the form of abortion and it is more common in ornamental pomegranate. Certain olive varieties have 10-60% abortive embryos. It is also common in some apple varieties. Embryo sac abortion becomes a cause of seedlessness in certain instances than fruitfulness.

e. **Non-viable pollen:** Many varieties of grapes produce non-viable pollens though they appear as perfect flowers. Sterile pollen in the grape are formed due to degeneration in the generative nucleus or arrested development prior to mitosis in the microspore nucleus. This phenomenon is also common in Washington Navel orange and '*Tahiti*' lime.

II. Genetic Influences: Optimum fruit set depends upon many genetic factors like hybridity or incompatibility

a. **Hybridity :** Unfruitfulness is a condition which is governed by the inherent genome but can also develop in favourable environment. The degree of sterility depends upon the relative closeness of the crossing species in the evolutionary process because greater degree of sterility is observed in wider crosses. The cross between peach and plum produce profuse flowering but they lack pistils and have malformed stamens and thus, are sterile and barren. A hybrid between the pear and the quince was also seedless. In citranges (Sweet Orange x *Citrus trifoliata*) the female gametes are mostly sterile. Seedlessness in most of the banana and pineapple varieties is due to hybrid nature of their ancestors. Most of the triploid apple varieties produce aborted pollen. A number of hybrids between *Vitis rotundifolia* x *Euvitis* and *Vitis vinifera* x *Vitis rotundifolia* are sterile.

b. **Incompatibility :** One of the most common causes of self unfruitfulness and self sterility is due to incompatibility between the pollen and ovules of the same plant or of the same variety. Inspite of being fertile the pollen and ovules fail to conjugate. In apple, pear, plum and aonla self incompatible varieties require another pollinizer varieties for fruit setting. Self incompatibility has been reported in some of the mango varieties like

'Langra', 'Dashehari' and 'Chausa'. Self sterility and self unfruitfulness has been reported in apple, pears, plums, almond, apricot and Clementine' mandarins due to incompatibility where normal processes of fertilization fails somewhere between production of functional gametes and the fusion of sex cells. Normal fruit set can be restored by cross pollination.

III. Physiological Influences

1. **Slow growth of pollen tube:** It may be considered as a type of incompatibility due to chemotropic or hormone influences. In this phenomenon the pollen takes longer time to grow and reach ovules to effect fertilization and by the time, the ovules degenerates and no fruit is set. The rate of pollen tube growth varies with the crossed and selfed cultivars of e.g. apple, pears, cherries and certain citrus fruits. Therefore, to obtain optimum fruit set fertilization should take place within a short time before the abscission of the style and ovary pedicel or peduncle.

2. **Premature or delayed pollination:** The male and female gametes must be fully matured at time of pollination to produce fruits. Both early and late pollination leads to unfruitfulness because a mature pollen is rendered ineffective for fertilization if the pistil is immature with underdeveloped ovules e.g. persimmon, pear, plum and peach. Similarly, if pollination is delayed for long time (>2 days), the flowers abscises and there is no fruit set in both the cases.

3. **Nutritive condition of plant:** Fruit set occurs at the cost of the nutrient reserve of the tree and if the tree is nutritionally weak especially during the period of flowering, the fruit set percentage is severely affected. Nutritive condition of the plant seems to control the fruit set by affecting the pollen viability or fertility of pistils.

   - **Effect on pollen viability:** There was significant difference in germination percentage of pollen collected from old apple trees and from strong young trees of the same variety. Pollen grains from vigorous plants have higher viability and they take less time to germinate and grow through the style than weaker plants.

   - **Effect on pistils:** Over bearing, drought or poverty of soil that deplete the plant nutrient reserves leads to production of defective pistils. Over bearing weakens the fruit tree and in coming season production is adversely affected. Close correlation was reported between defective pistils and unfruitfulness in American plums. In *Vitis vinifera* carbohydrate deficiency is the common cause of flower drop.

Interestingly the fruit set per cent in the terminally bearing plants is normally less because the flowers compete for nutrients with the other fruits as well as with the branch. This positional competition takes place between fruits, branch and adjoining fruits influences the fruitfulness. Similarly the nutritional condition of spurs has positive correlation with fruit setting in apple. Spurs on vigorous limbs with large leaves set more fruits than those borne on weak limbs.

External Factors

1. **Temperature:** Optimum temperature conditions at the time of flowering ensures higher fruit set by facilitating the physiological processes as well as the pollinator activity. If the prevailing temperature is lower than the minimum tolerance limit of the species it will retard the physiological processes leading to reduced pollen germination and also reduce the bee activity. Similarly higher temperature reduces the fruit set by desiccation of stigmas and styles prior to pollination. Tomato varieties grown at high temperature do not produce any fruit.
2. **Light:** Light intensity and duration modulates the flowering and fruit set in different ways:
    - **Photosynthetic activity**: Photosynthetic efficiency depends upon the intensity of photo radiation perceived. If optimum photosynthetic flux is available to the plant the floral initiation is adequate which later on gets translated to fruit set.
    - **Photoperiodic response:** Photoperiod regulates flowering in Short day, Long day plants. However day neutral plants are less effected by the duration of the photo effective radiations.
    - **Total radiation during the booming period:** The intensity and duration of the light availability during the flowering period affects the development of floral parts in some species of plants eg.exposing the strawberry flowers to long day light results in development of stamens and pistils.
3. **Rainfall and humidity :** The distribution of rainfall is more important for fruit set rather than the intensity. Rainfall at flowering stage retard the pollinator activity and even prevent anther dehiscence thereby preventing the pollination. Higher humidity is congenial for the development of fungal diseases e.g. Powdery mildew in mango and grapes appears and sprouts better in foggy and cloudy weather. However, low humidity result in drying of stigmatic fluid and retard the pollen germination.

4. **Wind:** It is limiting factor in optimum fruit set. Winds affect the fruit set by desiccating the stigma and reduction in the activity of pollinator bees. Better set has been observed on the shielded side than on the wind ward side of the tree and secondly bee activity is maximum when air is still.

5. **Insect-Pests and diseases:** Sap sucking pests and diseases reduces the overall vigour and fruit set considerably e.g. hoppers in mango ,Psylla and white fly in citrus, Fire blight and scab in pear and anthrocnose and powdery mildew in mango and grapes.

6. **Spraying trees when in bloom:** Minimum intervention such as spraying should be done during flowering. Spraying at flowering reduces the fruit set by direct injury/mechanical effect on stigma and interfering with activity of pollinators. e.g. some fungicides (copper) prevent germination of pollen grains on stigma.

7. **Pruning and ringing:** Pruning effectively reallocates the nutrients and photoassimilates by removing the unproductive portions and increases the fruit set e.g. In Anjou pear higher fruit set was obtained by pruning. In grapes cv. hope the pruned vines produce more hermaphrodite flowers than the unpruned ones. Ringing in apple and grapes, increases fruit set by supplying carbohydrates to the flowering shoots.

8. **Locality:** Jonathan apple may be sterile in one location but self fertile in another location.

9. **Season:** Hybrid grape variety 'Ideal' is self incompatible in early season but becomes self potent later on. Late flowers of strawberry clusters are always abortive.

10. **Nutrient supply:** Darwin stated that too much of manure renders many plants completely sterile e.g. tomato. Very high doses of fertilizers induces sterility due to excessive growth which exhaust carbohydrate supply. Nitrogen supply affect in both ways. In starved plants, it increases set while when supply is excessive, it had harmful effect. C/N ratio is very important especially in mango because high N and low C leads to rapid rate of amino acid synthesis increases the growth rate and very few carbohydrates remain in reserve for the development of sex organs to functional conditions. Jonathan apple are self-sterile in rich soil becomes self-fertile in poor soils. In olives, low fertility leads to partial/complete degeneration of pistils.

F. FRUIT RIPENING

After completion of growth the fruits undergoes a period of ripening. Ripening includes softening of the fleshy fruit with associated biochemical changes in pigmentation and flavour. Pulp softening results from the changes in the pectic substances by hydrolytic enzymes of pectinmethylesterase (PME), polygalacturonase and cellulase, binding the cell walls resulting in loose cementing of the cells. Starch is hydrolysed and broken down into simple sugars that contributes to the sweetness of the fruit. Stone fruits do not produce starch, thus much of their increase in sweetness comes from conversion of sorbitol (D-glacitol) to fructose and from the reduction in acid content, which reduces sourness present in unripe fruits. Similarly, polyphenolic compounds such as tannins and phenols content decline during ripening, reducing the astringent or bitter taste associated with them. Growth substances hasten maturity. Ripening of some detached fruits can be successfully controlled by growth regulators under suitable conditions.

Changes at fruit ripening

The physical and chemical changes occurs during ripening of fruits. There is decrease in chlorophyll, starch (pome fruits), maleic acid (pome fruits), acidity (except banana), plastids degenerate, respiration and firmness of flesh. However, sugars , pigments (carotene, anthocyanin, xanthophyll), protein, starch (banana), ethyl alcohol, soluble protein and ester content increases during the fruit ripening. Colour changes occur with the alterations in pigment including chlorophyll. Most of the fruits begin to lose their green colour which also acts as an indicator signal of fruit quality as well as its stage of maturity. The green colour of fruits changes ranging from yellow to red. Red and yellow colours are due to a group of fat soluble pigments known as carotenoids. Blue, violet and pink colours are due to a group of water soluble pigments called anthocyanins.The fruits become soft either by breakdown of insoluble protopectin into soluble pectin mediated by the cell wall enzymes like pectin methylesterase , polygalacturonase and cellulose in most fruits, or by hydrolysis of starch as in banana. Some volatile compounds are synthesized enzymatically as the fruit ripens and emit characteristic aroma and flavour. The acid content declines with progression of ripening in most of the fruits except in banana. When fruit ripens, tannins polymerises into very large molecules and loose their capacity to react with the proteins. These tannins get entrapped in the cell in such a way that they are not released when cells are crushed under teeth. During ripening, starch is broken down enzymatically and forms soluble sugars viz.sucrose, glucose & fructose which impart sweet taste to the ripe fruits.

The environmental and cultural practices e.g. temperature, light, pruning and fertilization affect red pigmentation in fruits. The factors which contribute to higher level of carbohydrates in the fruit during pre-harvest period tends to increase anthocyanin pigments. Excess nitrogen and improper pruning can reduce colour by depleting CHO's and by preventing light distribution in the tree. Trees produces more dry weight in full sun than on cloudy, rainy or foggy days.

Classes of Fruits as per Ripening Behaviour

On the basis of ripening behaviour, fruits are classified into two categories.

1. Non-climacteric fruits

This type of fruits exhibit no dramatic changes in the post-harvest composition or physiology and ripens to its optimal quality on the tree. In fruits like citrus, grape, litchi, pineapple, ber, phalsa, peach, fig, cherry and strawberry, there is no appreciable lag period between maturation and ripening.

2. Climacteric fruits

Climacteric marks the beginning of fruits ripening after the harvest and ripens to their optimal quality off the tree. Ethylene synthesis takes place at a more rapid rate after the fruits are detached from the tree. The attached fruits are under the influence of leaves which apparently produce ethylene inhibitor. In fruits like apple, pear, banana, mango, sapota, papaya, persimmon, avocado and custard apple there is a distinct lag period between maturation and ripening.

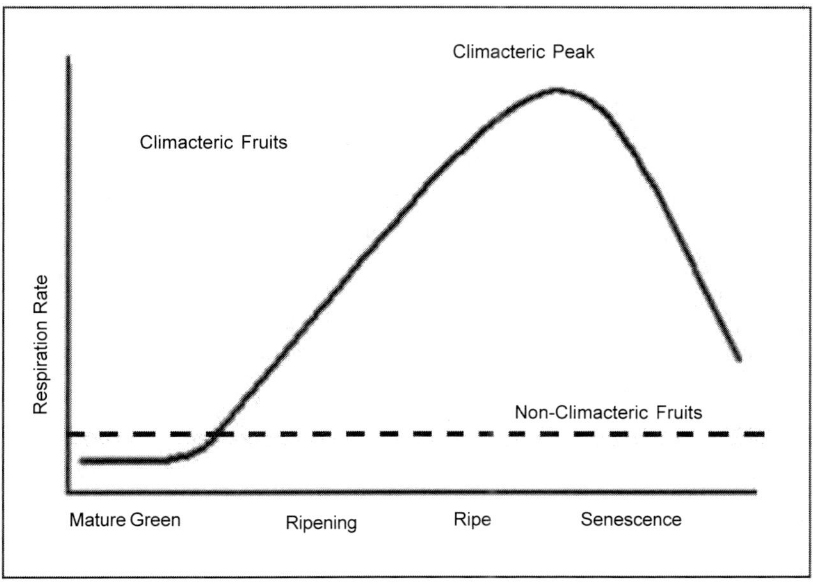

LITERATURE CONSULTED

Arteca, Richard, N.1996. Physiology of fruit set, growth, development, ripening, premature drop, and abscission. *Plant Growth Substances*. Springer US: 200-222.

Bal, J.S. 2014. Fruit Growing .3rd Edition, Kalyani Publishers, New Delhi.

Brummell, D.A. 2017. Seed development and fruit growth .http://plantsinaction.science.uq.edu.au/edition1

Chattopadhay,T.K.1994. A Text Book of Pomology,vol.1.Kalyani Publishers, New Delhi.

Imtiyaz, A. W., Bhat,M.Y., Abid, A. and Moh,. Y.2010. Unfruitfulness in fruit crops: Causes and remedies. *African Journal of Agricultural Research,* 5(25): 3581-3589,

Kumar, Rahul, Ashima Khurana, and Sharma,A.K. 2013 Molecular regulators of fruit ripening. *Stewart Postharvest Review* 9(4): 1-7.

Lescourret, F., and Genard. M. 2005. A Virtual Peach Fruit Model Simulating Changes in Fruit Quality During the Final Stage of Fruit Growth.*Tree Physiology,* 25(10): 1303–1315. Crossref. Web.

Lincoln,Tiaz and Eduardo, Zeiger, 2002. *Plant Physiology* 3rd Ed, Sinauer Associates. 690 p.

Rahul, K, and Sharma,A.K..2013. Ethylene Perception and Signalling in Ripening Fruit. Fruit ripening: physiology, signalling and genomics, 193–201.

Sandhu, A.S., Dhaliwal,G.S. and Bal ,J.S. 2004.Laboratory Manual of Fruit Growing, PAU, Ludhiana.

Serrani, J. C., Sanjuan,R., Ruiz-Rivero,O., Fos,M. and Garcia-Martinez,J.L.2007.Gibberellin Regulation of Fruit Set and Growth in Tomato. *Plant Physiology,* 145(1): 246–257.

Sharma ,G.,Sharma,O.C.and Thakur,B.S.2009.Systematics of Fruit Crops. New India Publishing Agency,New Delhi.

Singh ,J.2011. Basic Horticulture. Kalyani Publishers, New Delhi.

Singh, A.1990. Fruit Physiology and Production. Kalyani Publishers, New Delhi.

Singh, N. P.2016. Basic Concepts of Fruit Science 2nd Ed. CBS Publishers, New Delhi. 720p.

Singh,S., Krishnamurthi,S. and Katyal,S.L.1967.Fruit Culture In India. ICAR, New Delhi.

Singh,S.K., Thakur,N.and Sharma,Y .2013. Plant growth regulators in fruit and vegetable crops. *International Journal of Agricultural Sciences,* 9 (1): 433-437.

Souza, J.M.A.,Leonel,S.,Modesto,J.H.,Ferraz,R.A.and Goncalves,B.H.L.2015. Phenological Cycles, Thermal Time and Growth Curves of Mango Fruit Cultivars in Subtropical Conditions. *British Journal of Applied Science & Technology,* 9(1):100–107.

Zielinski,Q.B.1955. Modern Systematic Pomology.W.M.C.Brown Company. Dubuque, Iowa.

9

Growth and Development of Fruits

N. Sharma and Preet Pratima

INTRODUCTION

Growth and development are often used interchangeably, however, these describe separate events during the life cycle of a plant. To understand these precisely, let us take the example of multicellular organisms like annuals, monocarpic or flowering plants (angiosperms) including fruits, vegetables, flowers etc. which begins their life from a single cell fertilized egg, the zygote. The zygote develops into embryo through the cell division and differentiation. When we sow the seed in garden or pots, in a few days a young seedling emerge. With the passage of time, the seedling grow in size, the number of leaves increases, in perennial plants branching occurs for few years, and after putting forth the requisite vegetative growth plant begins flowering, produces fruits, seeds and finally the plant senesce at certain period. Generally three terms "growth", differentiation" and "development" are used to describe the different events during the life cycle of a plant.

Growth : It is defined as an irreversible increase in size, volume and mass of a cell or organ or whole organism and usually accompanied by increase in dry weight. In other words, growth is quantitative increase in plant body such as increase in the length of stem and root, the number of leaves, etc. It may be evaluated by measurements of mass, length or height, surface area or volume.

Differentiation: It is a qualitative change during life cycle of a plant and referred to the process by which new arising cells after division become specialized. During this process cell undergoes some structural changes in cell wall and protoplasm to perform different functions. The new cells differ in size and shape from the mother cell. The enlarged cell is supplied with food which may be differentiated in various ways. Inside the cambium of a woody plant, it may become "xylem" tissue or strengthening fibre or serve as food conducting tissue in the phloem; it may accumulate food reserves and become storage tissue as in the tuber of a potato or the root of carrot; it may give differentiate into floral initial and become part of the flower; it may give rise to the sex cells and be instrumental in transferring hereditary factors to the next generation. Thus each specialized cell of a plant originated from cell division, underwent a degree of enlargement and then become differentiate into its specific shape, size and function.

Development: Whole of the changes which an individual undergoes during its life cycle is referred as development. In other words, development is a sum of growth and differentiation.

The cell is the basic unit of growth. Cell divide giving rise to daughter cell, each of these enlarge and may differentiate during the process of development. Growth do not takes place indiscriminately in whole parts of plant but is restricted only to living cells at certain locations, known as 'meristem' (apical, cambial, intercalary) and at cellular level it is also accompanied by the organization of macromolecules into assemblages of membranes, plastics, mitochondria, ribosomes and other cell organelles. An important process during cell division is synthesis and replication of nuclear DNA in the chromosomes which is then passed into the daughter cells. It is accomplished by metabolic processes involving synthesis of macromolecules, such as nucleic acids, proteins, lipids and polysaccharides at the expense of metabolic energy. Thus, growth is an irreversible increase in size which is commonly accompanied by an increase in dry weight and the amount of protoplasm.

Therefore, the term growth is used to denote an increase in size by cell division and cell enlargement, together with the synthesis of new cellulose materials and the organization of cellulose organelles.

Factors Controlling Plant Growth: The two sets of factors, viz., external/ environmental (air, light, water & nutritional) and internal (genetics & hormonal), control the growth and development of plant. The raw material required for growth is supplied by nutritional factors which include the minerals, organic substances like proteins, carbohydrates, etc. Utilization of these substances for proper development of the plant is regulated by certain chemical messengers called 'plant hormones' which in minute amounts increase or modify the physiological processes in plants. The other internal factor is genetic because of this plant displays a unique trait inherited from parent.

Plant Hormone: The term plant hormone refers to compound produced within the plant which usually move within the plant from the site of production to the site of action and regulate plant physiological process in physiologically very low concentration. There are five well accepted plant hormone: auxins, gibberellins, cytokinins, abscisic acid and ethylene. However, recently several other compounds having plant growth regulating properties have been found, which include brassinosteroids (also refer as 6th group of PGR), jasmonates, salicylates, polyamines and the list is still increasing.

Plant Growth Regulators (PGRs): Refer to organic compounds other than nutrients which in small quantities promote, inhibit or otherwise modify any plant physiological process. The PGR is a broader term as it includes natural (hormone) as well as exogenous/synthetic chemicals.

Plant Bio Regulators (PBRs) encompasses plant hormones, synthetic and physiologically active growth substances, synthetic enzymes, vitamins, triacontanols, organic acids, ethylene, inhibitors, etc.

TYPE OF GROWTH

Primary Growth: The mitotic division of meristematic cells present at the root and shoot apex increases the length of the plant body and referred as primary growth.

Secondary Growth: The secondary meristem e.g. cambial increases the diameter of the plant body and it is referred as secondary growth.

Unlimited Growth- The root and the shoot system of plants grow continuously from germination stage to the death or throughout the life span of the plant. It is called 'Unlimited' or 'indeterminate' type of growth.

Limited Growth - The leaves, fruits and flowers stop growing after attaining certain size. This is called 'limited' or 'determinate' type of growth.

PHASES OF GROWTH

Whether a plant be annual, biennial or perennial, it has certain stages or period of growth. During these stages, all horticultural plants follow an established pattern of development.

Seed Germination

The process by which the dormant embryo of the seed resumes active growth and forms the seedling is referred as germination. During this process, a complex series of steps such as uptake of water, utilization of stored reserves, development and expansion of the embryonic axis takes place. Viable embryo, favourable internal and external conditions are prerequisites for obtaining a seedling. The main environmental conditions required for the seed are water, oxygen and temperature:

Water: The most important environmental requirement for germination is water availability. The water content of the seed may fall to 10 per cent during storage, but must be restored to about 70 per cent to enable full chemical activity. When water is plentiful, it is absorbed into the structure of the testa through a tiny pore called micropyle in a process called imbibation. This softens the testa and moistens the cell walls of the seed so that the next stages can proceed. The cells of the seed take up water by **osmosis**, often assuming twice the size of the dry seed. The water activates special proteins, called **enzymes**, that begin the process of seed growth. The water provides a suitable medium for the activity of enzymes in the process of respiration. As the swelling causes the seed coat to rupture, which enables the radicle to come out in the form of a primary root followed by the shoots, or growth above ground, begin to appear. The seed sends a shoot towards the surface, where it will grow leaves to harvest energy from the sun. The leaves continue to grow towards the light source in a process called **photomorphogenesis.** A continuous water supply is now required if germination is to proceed at a consistent rate, but the growing medium, whether it is outdoor soil or compost in a seed tray, must not be waterlogged, because **oxygen** essential for aerobic respiration would be withheld from the growing embryo. In the absence of oxygen, **anaerobic respiration** occurs and eventually causes death of the germinating seed, or suspended germination, i.e. **induced dormancy**.

Temperature: It is a very important germination requirement, however it is usually specific to a given species or even cultivar. Most seeds germinate at a relatively warm (21 °C) temperature. However, some seeds germinate when it is cold, such as plants in northern environments. It acts by fundamentally influencing the activity of the enzymes involved in the biochemical processes of respiration and the characteristic temperature range is between 5°C and 40°C.

Light: It may influence germination in some species, but most seeds are not affected by light or darkness. Some species found in forest settings, will not germinate until an opening in the canopy allows sufficient light for growth of the seedling. A canopy of tall deciduous plants filters out red light for photosynthesis. Seeds of species growing under this canopy receive mainly far red light, and are prevented from germinating. When the leaves fall in autumn, these seeds will germinate in response to the now available red light and to the low winter temperatures.

Typical Germination Process

Seed Dormancy

Seed is the dispersal unit of plant. One important function of most seeds is delayed germination, which allows time for dispersal and prevents germination of all the seeds at the same time. For this survival, the seed, mainly in a dry state, is well equipped to sustain extended periods of unfavourable conditions. Dormancy prevents germination during unsuitable ecological conditions, when the probability of seedling survival is low. It also prevent pre-harvest germination, therefore it is important for plant ecology and agriculture. It is a period during which very little activity occurs in the seed, other than a very slow rate of respiration. Seeds will not germinate until dormancy is broken. Seed dormancy has been defined as the incapacity of a viable seed to germinate even when the environmental conditions such as water, temperature and air.

Seed dormancy may be imposed by certain environmental factors or internal factors or genetic factors included. Broadly it is divided into three groups, exogenous, endogenous and combined.

Causes of seed dormancy: Dormancy may be caused by conditions within the **seed** that prevent under normally ideal conditions and is referred as true **dormancy** or innate **dormancy.** Often **seed dormancy** is divided into two major categories: exogenous and endogenous, which are termed as **'Quiescence' and 'Innate dormancy',** respectively. Quiescence is a condition when seed is unable to germinate because of unfavourable external condition, while in later **(innate)** the seed fail to germinate even under optimum water, temperature, oxygen and light conditions. In this seed fails to germinate is because of internal condition, it is considered as true dormancy.

Types of Seed Dormancy

Seed dormancy is either acquired during seed development/maturation (primary dormancy) or results from factors after the seed is shed (secondary dormancy).

1. Embryo Factors

i. **Immature/ Rudimentary embryo**: An undeveloped embryo in a seed is incapable of germinating until time has elapsed after the seed is removed from the parent plant e.g. tomato, palms. Some seeds such as *Acacia* are recorded to have a dormancy of more than a hundred years.

ii. **Dormant embryo/After ripening**: After their shedding, the seeds of barley, wheat, oat, etc. require an interval of ripening before they attain the power to germinate. The period of after-ripening is needed to produce the necessary growth hormones.

iii. **Inhibitory factors:** Presence of inhibitors in seeds, which do not allow the seeds to germinate. They may be present inside the fruit (e.g., Tomato juice possesses inhibitor ferulic acid), in the embryo (e.g., Xanthium), endosperm (e.g., Iris) or seed coat (e.g., Cucurbita). The chemical inhibitors are of several types like abscisic acid (ABA), phenolic acids, coumarin, short chain fatty acids, etc.

2. Seed Coat Factors

a. A thick testa prevents water e.g., Chenopodium, Trigonelta), gases (Apple, *Sinapis arvensis*) or chemicals (e.g., Xanthium) from entering the seed.

b. Mechanically resistant seed coat. The seed coats are hard and provide mechanical resistance to the growth of embryo, e.g., Lepidium, Capsella.

Methods of Breaking Seed Dormancy

1. Weakening of tough and impermeable seed coats by microbial action.
2. Physical methods by rupturing or weakening of seed coats by mechanical abrasions e.g. using sand paper or chemical treatment with sulphuric acid (collectively known as scarification) can break down the seed coat and therefore the dormancy mechanism.
3. Action of digestive enzymes present in alimentary canals of birds and other animals which happen to feed on their fruits.
4. Leaching of inhibitors present in the seed coat by soaking in water.
5. Cold temperatures cause similar breaks of dormancy in other species (stratification), the exact temperature requirement varying with the period of exposure and the plant species.
6. Hormonal treatment especially with GA is beneficial for improving germination. The use of 100 to 500 ppm GA_3 in grape, GA_3 not only substituted stratification for 40 days but also stimulated the growth of seedlings in grape. The time taken for seed germination varies from species to species. For example, mango stones require about one month to germinate whereas guava seeds require only 2-3 weeks. In jackfruit, the phenomenon of vivipary, in which seeds germinate even when they are inside the fruit/ attached to the mother plant is very common. GA are now invariably used for accelerated seed germination in citrus (trifoliate orange, sweet orange, Cleopatra mandarin and acid lime), cherry, grapes, ber, Annona, apple and peach. The treatment with GA significantly reduced/ substituted the period of seed stratification. Soaking seeds of fruits in 10-20 ppm solution of GA for 12 hours before sowing, significantly improves the yield and quality.

Physiology of Seed Germination

It is a characteristic feature of the life cycle of many organisms that at one stage a spore or seed is formed which can persist in a viable state through prolonged periods without further development. A typical seed stores carbohydrates, lipids and proteins for germination. As a result of imbibation and osmosis the embryo becomes hydrated which activates the enzymes for respiration. Digestion of food on storage centre occurs by hydrolysis and amino acids, sugars and fatty acids and glycerol are produced. Soluble products are translocated to the growing regions of the embryo. The majority of glucose is used for cellulose synthesis and other cell wall materials. Amino acids are used for enzymes and structural components. The net loss of mass of the seed occurs due to change of sugars to Co_2 and water where $C0_2$ is lost as gas than O_2 taken up by aerobic respiration. Within the embryo growth occurs by cell division and amounts of proteins, cellulose, nucleic acids steadily increase in the growing regions of the seed. The first visible sign of growth is the radicle, that emerges from the micropyle and is positively geotropic in nature. It forms the root system of the plant. The next structure that emerges is the plumule, that is positively phototropic and forms the shoot system of the plant.

PHASES OF GROWTH

A. Vegetative Phase

i. **Juvenile Phase:** It is the stage in the life of a plant after it completes germination and begins photosynthesis. During this period, a plant is characterized by rapid vegetative growth. It is the time during which a fruit tree is attaining size, framework and strength for its role of fruit production. It is the period of early growth of a tomato plant in preparation for flowering. During this period a plant will be photosynthesising as much as possible to grow as large as it can before the onset of the flowering (Generative) phase. During this phase, increase in size is due to the utilization of manufactured food for further shoot,leaf and root development.

ii. **Transition Phase**: A plant in its youth is undergoing a change from utilization of carbohydrates for growth to accumulation of carbohydrate in storage organs. The transition stage may lasts for several years in some fruits and for week or less in some annual plants. In either case, the change from juvenile to productive is not abrupt, but represent a gradual and continuous change until the flowering occurs. During the period progressive changes in morphological, anatomical and physiological characteristic takes place, which include leaf shape, thickness, and epidermal characteristics, phyllotaxis, thorniness, shoot orientation, shoot growth vigour, anthocynin pigmentation, photosynthetic production, disease and insect resistance and competence to form adventitious buds, and roots and somatic embryo. The naturally occurring plant hormones play an important role in stimulating differentiation of cells to form reproductive and storage organs. The end of transition, like the beginning is not clear cut. Some vegetative growth continues to occur even during the most productive stages of plant growth. However, a balance is reached during which carbohydrate accumulation is favoured over carbohydrate utilization and the plant is considered to be in its productive stage of growth.

B. Reproductive Phase

Depending upon the kind of horticultural plant, the productive stage results in an accumulation of food reserves in either vegetative storage organs, reproductive parts or fruits. The reproductive phase usually starts when the plant reaches maturity. Most of the apical meristems on the branches cease to initiate leaves and start producing floral parts in accordance with the traits of the species. The mechanism triggering transition from the juvenile into that of maturity phase is still not clearly understood, however, it may be due to hormonal control over the distribution of assimilates in the zone of the apical meristems.

Development of Flowers: Flower bud development involves the transformation of the vegetative apex to a reproductive structure. This reproductive development in the apical meristem is a very critical phase in the life of horticultural plants. The basic reason for flower bud formation has been viewed as something happening between the flower forming genes and the meiosis. It may be an indication of the completion of the life cycle e.g. annuals, however, in perennial plants, flowering is repeated each season. Flowers may be induced in the previously buds at the apex of the main central shoot, at the lateral branches, or on both in the form of solitary blooms or an inflorescence. The beginning of the process is often referred as induction, which is followed by differentiation of growing point and later by differentiation of flower primodia.

Initiation of Flower Buds : Floral buds are initiated after the new shoots have reached a certain diameter and a good number of their leaves have matured. In temperate fruits species such as apples, peaches, pears, cherries, or plums, flower buds are initiated between late spring and late summer during the year preceding the bloom, which followed by several months of continuing development before passing into dormancy over winter. In apple, the initiation of flower primordia occurs 3-6 weeks before bloom or after cessation of shoot growth, which coincides with embryo growth in seeds. Depending upon species, rest period begins from mid-summer to late autumn, which occurs due to accumulation of growth inhibitors like abscissic acid and the decline in the level of growth promoting substances like auxins and gibberellins because of shortening of day length. During rest period, flower as well as vegetative buds they will not open even if the plants are provided with favourable temperature, moisture, and light conditions. Flower bud development continues even after completion of rest and flowers emerge after the completion of chilling requirement during the following spring when the temperature become warm and soil moisture is at sufficient level. In tropics, citrus fruits such as oranges, lemons, and pummelos flower buds initiate and bloom throughout the year. This applies to citrus fruit. For these fruits, freezing temperature during winter is not required to break bud dormancy.

In addition to heavy crop load during the current season, canopy shade during summer will also reduce flower bud initiation and differentiation.

Flower Bud Induction: As discussed above, flowering in perennial fruit crop start after the plant has passed through the vegetative phase. The induction gives an impulse to sufficiently developed buds to transition from vegetative to generative phase. In other words, it is a qualitative change during development involving programming of strategically located meristems on plants to form flower. In fruit trees, interaction of physiological and environmental factors such as light, temperature and day length may provide proper balance of endogenous hormones to cause initiation.

Flower Bud Differentiation: After flower induction, successive steps of physiological processes occur and morphological differentiation takes place, when the apex becomes organized into a primordia of flower (or inflorescence). The first morphological change signalling transition to floral state in the meristem can be seen as broadening of the apex, which also becomes convex, followed by its lengthening into a cylindrical shape and the formation of sepal primordial in the terminal flower (king flower). After this transformation, organ primordia become differentiated laterally (Buban and Faust, 1982). The development of the floral organs occur acropetally in the order of sepals, petals, stamens and pistil. The appearance of various tissues of the flowers proceeds from the outer structures toward the inner structures.

Flower bud initiation occurs through a biochemical signal. This biochemical signal makes it possible for the tissue to change from vegetative to reproductive state in a programmed term. It occurs as a result of the balance of GA3, auxin, cytokinins, and ethylene-like hormones. Carbohydrates supply, and healthy leaves exposed to light are required. Optimum C/N ratio initiate flowering. Such practices as pruning, dwarfing, spreading of branches, nutrient control, ringing, scoring and chemical treatment may have a pronounced effect on hastening or delaying the onset of the fruiting or productive stage. Certain PGRs like Alar (SADH, now banned), chlormequat, paclobutrazol, prohexadione calcium application can induce flowering of young apple, pear and cherry trees and prevent biennial bearing in apples. These PGRs are anti-gibberellin and therefore help flowering by altering hormonal balance. Chemical blossom and fruit thinning prior to period of floral initiation lowers the GA level and increase return bloom.

Flower Bud Differentiation in Grapes

The clusters on the grapevine that make up the current season's crop actually began to form during the previous growing season in the newly formed bud on the developing shoot. The formation of clusters for next year's crop begins concurrently with the formation of leaf primordia within the compound bud. The change from vegetative primordium into reproductive primordium in the bud takes place involving in three stages, namely, anlagen formation, formation of inflorescence primordium and formation of flowers. Initially, the primary bud within the compound bud produces three or four leaf primordia and then divides into two equal or almost equal parts. The part opposite the youngest leaf primordium is the *anlage*, a undifferentiated primordia which develop into inflorescence primordia, tendril primordia or shoot primordia depending upon various factors influencing the fruit-bud differentiation. The formation of anlagen from the apex of the bud is considered to be the initiation of inflorescence axis. As the growing season progresses, successive anlagen formed toward the apex of the shoot.

The developing anlage divides into two unequal parts, the larger *inner arm* and the smaller *outer arm.* The former will give rise to the main body (*rachis*) of the cluster, while the later will give rise to either a wing or a large branch at the top of the cluster. Other branches along the inner arm will form, and these will become lateral branches along the rachis. A fully developed cluster primordium within the bud looks rather like a bunch of grapes, in which each berry like branch primordium is a mass of undifferentiated tissue. After one to three cluster primordia form, additional leaf primordia form and the bud enters dormancy.

The process of cluster primordia development is regulated by a proper balance of plant growth regulating substances. Gibberellic acid and cytokinin play roles in this differentiation. Gibberellins may promote vegetative growth as the anlagen do not proliferate resulting in the formation of either tendril primordia or shoot primordia. On the contrary, cytokinins induce the grapevine to form inflorescences from anlagen and from young tendrils. The role of cytokinin in flower formation is further strengthened by the fact that, the xylem sap (bleeding sap) of the grapevine contains high cytokinin levels during budbreak and flowering. Cluster and flower differentiation probably are regulated by the relative proportions of both of these hormones within the bud at the time of differentiation. Among the environmental factors, sunlight and temperature have the greatest potential to regulate the differentiation of anlagen into either cluster or tendril primordia. High light intensity for up to four hours per day and temperatures (20°C) favour the formation of cluster primordia. Any imbalance among the factors responsible for formation of inflorescence primordium would make the anlagen to differentiate into either tendril or shoot.

Flower Bud Differentiation in Mango

Flowering and fruiting in mango is a complex phenomenon and is influenced by many factors. In North-India, almost all the commercial varieties have tendency for alternate bearing and certain amount of shoot maturity is required to their flowering. However, regular bearing varieties grown in South India are capable to produce flower on the shoots of any size or maturity. Even in regular bearing variety, heavy cropping in one year can reduce flowering and fruiting in the following year. The time of flower bud differentiation varies with place, variety and different environmental conditions as well as endogenous conditions of the shoots. It begins within a very short period before the expansion of the terminal buds from October (Alphonso) to February.

The mango inflorescence is a much-branched terminal panicle with a few hundred to over 6000 flowers. The mango is andromonoecious, which means that each inflorescence bears both hermaphrodite and staminate flowers. Fruit yield is affected by the percentage of hermaphrodite flowers and its susceptibility to fruit drop as well as the age of the tree. Lower than 1 % of hermaphrodite flowers adversely affect cropping. The proportion of hermaphrodite flowers may even differ on the same tree from panicle to panicle depending on the location of the panicle and also period of their emergence. The percentage of hermaphrodite flowers per panicle increases with tree age. The proportion of perfect flowers can be improved by exogenous application of naphthalene acetic acid (NAA) (200 ppm) at flower bud differentiation.

Mango trees produces many growth flushes, however, early emerged shoots, which attain physiological maturity within 8 to 10 months period, accumulate sufficient reserve sugars, starch and other carbohydrates at the time of flower bud differentiation. Higher starch, total carbohydrate and C/N ratio in the shoots favour flower formation. C/N ratio can be increased in by the practice of ringing of branches in an "On year" is known to increase C/N ratio of shoots and promote flowering in "Off year". There also seems to be a possible relationship between fruit bud differentiation and amino acid, RNA, DNA, phenols, and proteins. A high level of phosphorous in leaves and flowering shoots has been reported to be favourable for flower bud differentiation.

The flowering in mango depends on environmental factors that retard growth. It begins during long dry or cool period. It is postulated that, the flowering stimulus is synthesized in the leaves is responsible for induction of flowering. However, the exact nature of 'florigen' has not been ascertain. Higher levels of auxin, zeatin riboside (cytokinin) and inhibitors (ABA like substances) appears to promote flowering, whereas, high level of gibberellins inhibits flowering in mango. Exogenous application of GA_3 just before flower bud differentiation has been shown to inhibit flower bud formation, however, response vary with cultivar and concentration of GA_3. Application of gibberellins biosynthesis inhibitor like paclobutrazol during "on year' and potassium nitrate can significantly increase flowering during the following year.

Development of Fruits: Fruits are often defined as structures derived from a mature ovary containing seeds. However, in many instances, other floral parts- the floral receptacle, calyx, even the axis of inflorescence may however, also contribute towards fruit formation. These organs may develop simultaneously with gynoecium and become an integral part of the fruit e.g. apple, pear. Such fruits are referred as false fruits. Therefore, in horticultural term, fruit is a fleshy edible part of the plant which in its development is closely associated with flower. The function of a fruit is to develop, protect, nourish, and ultimately disperse the seeds contained in it.

Initiation of Fruit Tissues/ Pre-pollination Development: In morphological terms, fruit are structures that develop from fertilized or stimulated ovules, plus associated floral parts that originate from the parent plant. As stated above, pollination, followed by pollen tube growth and fertilization, trigger fruit growth and if pollination does not occur, flowers are shed (only rare exceptions). However, fruit development often starts much before flower opening in spring. Flower primordia may have been initiated up to 6 months before a particular flower opens, and ovary development continues during flower growth with ovary tissues forming late in this process. For example in apple, flowers are initiated 4-6 weeks after bloom, starting with sepals' initiation and ending with pistil development. By autumn all parts are visible except, ovules, which some fruits get fully develop just prior to flowering bud opening in spring e.g. olive. The ovary of apple is enclosed in the receptacle and all flower parts except the petals remain attached to the fruit, which is pome. The fruit contains five carpels, each usually with two or four ovules.

Post Pollination Development: Development of fruits starts following pollination and fertilization. Growth of the ovary wall and in some cases tissues associated with receptacle is enhanced. Simultaneously shedding of floral parts takes place, however, in certain cases the calyx may persist with the fruit until its maturity, e.g. in apple, citrus, tomato, and guava. In some instances, fruit development can take place parthenocarpically i.e. without pollination and fertilization. Parthenocarpy may be vegetative or stimulative. In vegetative parthenocarpy, the fruit develops without any pollination as in tomatoes, peppers, pumpkins, and cucumbers, and the consistently seedless citrus, banana, and pineapple. In stimulative parthenocarpy, pollination occurs but fruit development is stimulated without the pollen tube ever reaching the ovule and effecting fertilization. A prominent example is seedless 'current' grape, Washington navel oranges and bananas. Fruits of triploid plants which are genetically sterile also fit into this group. In some fruits, require normal pollination and fertilization for fruit set; however subsequent abortion of the embryo cause seedlessnes, a phenomenon called Stenospermocarpy e.g. seedless grapes. Parthenocarpy may be induced in plants which are normally seeded by the application of hormones.

It is believed that gibberellins control fruit set. Pollens contains GAs, exogenous application of GA_3 can cause parthenocarpic fruit set, e.g. pear, fig. However in some horticultural plants, synthetic auxins like 4-CPA(4-chlorophenoxy acetic acid) and â-NAA can induce fruit set.

The increase in fruit volume that is associated with fruit growth is largely a result of cell division and cell enlargement. In addition, in some fruits, expansion of interstellar space may also contribute to the growth of fruit, especially during the later stages e.g. apple. The rate and duration of cell division varies among fruits and also among tissues within a fruit. Generally growth by cell division predominates in the early stages of growth, whereas growth by cell expansion predominates the later stages, but there are much varietal variations. The cell division stage involves all growing parts of the fruit and it may be short; in certain fruits completed during pre-anthesis time, however, it usually overlaps the cell enlargement stage. In tomato, cell division phase lasts approximately 7-10 days and in apple and peach, cell division ceases 3-6 weeks after bloom. In avocado, strawberry and one species of tomato (*Lycopersicum pimpinellifolium*), some cell division continue until maturity. More complicated pattern of development occurs in some fruits in which cell division ceases at different times in different parts of the fruits. The number of fertilized ovules in a fruit is correlated with both the initial cell division rate and the final size of the fruit. Also, fruits with an uneven distribution of seeds are often lopsided. Further, seeds are rich source of cytokinins and their level in developing embryos has been correlated with cell division in surrounding tissues.

There is a gradual shift from the cell division phase to cell enlargement. Cell expansion accounts for the larger increase in fruit volume, often contributing in excess of a 100 fold size increase. Gibberellins are also associated with fruit expansion and removal of the seeds from pea pods inhibited GA biosynthesis in the pericarp. Auxins from seeds may also regulate cell expansion of the pericarp, however, exogenous application of auxin does not always compensate for seed removal.

Seed Development: A seed is a reproductive unit which develops from an ovule, usually after fertilization. Both monocot and dicot seeds contain seed coat, cotyledon, endosperm and an embryo. The seed development starts with double fertilization and involves the fusion of both sperm nuclei with the egg nucleus to form zygote (embryo). The second part of this process is the fusion of the polar nuclei with a second sperm cell nucleus, thus forming a primary endosperm. During the period immediately after fertlization, the zygote is mostly inactive, but the primary endosperm divides rapidly to form the endosperm tissue. This tissue becomes the food the young plant until it has developed roots after germination. The endosperm accumulates carbohydrates, fats, proteins and growth hormones and is the nutritive tissue formed prior to embryo in angiosperm. In non-endospermic dicots, the cotyledons however, act as the storage.

The integuments of the ovule become the seedcoat of the mature seed. Seed coat may contain two distinct layers, a typically firm outer seedcoat, the testa, and a generally thin, membranous inner coat, the tegmen. The former protects the seed contents from drying out, mechanical injury, or attacks by fungi, bacteria and insects, until it is split at germination. When the seed coat forms from only one layer, it is also called the testa.

In a seed, the embryo consists of three main parts: the plumule, the radical and the hypocotyls. The development of embryo starts later and is slow. The extent of its development at the time seed maturity varies with species. In some species the embryo is still small and undeveloped when the seed is ready for dispersal, and require an additional period under suitable environmental conditions for maturation of the embryo after seed shed, before the seed can become capable of germination. If the embryo absorbs all the food reserves from the endosperm, the thick fleshy cotyledons commonly become the main organs for food storage and occupy almost the whole of the seed cavity. In fruit crop, development of more than one embryo may develop in a seed, a phenomenon called as polyembryony, e.g. many citrus fruits.

Development of seed takes place simultaneously with fruit development. The later part of fruit development is also dependent upon the plant growth substances produced in the seeds. In some species, e.g. apple, pear, the final fruit size has been well correlated with the number of seeds in a fruit. A fruit with fewer seeds is smaller, often misshapen and likely to drop pre maturely.

Fruit Maturity and Ripening

During the course of growing season fruit gradually changes in colour. Changes in fruit colour involve changes in the expression of pigment biosynthetic genes. Fruit maturation is followed by ripening, the later represents the shift from the protective function to dispersal function of the fruit. Ripening occurs synchronously with seed and embryo maturation. In dry fruits (cereals, nuts, dandelions) ripening consists of desiccation, while in fleshy fruits, ripening make the fruit appealing to animals that eat the fruit as a means for seed dispersal. Ripening involves the softening, increased juiciness and sweetness, and color changes of the fruit. Fruit softening involves a partial breakdown of cell walls. Several enzymes are known to be involved in this process. Polygalacturonase hydrolyzes bonds in pectins. The taste of the fruit changes when it ripens. At the initial stage, the fruit is a little tart or sour due to the presence of acids. When the fruit ripens, kinase enzymes turn the acidic fruit to a neutral one by converting them to neutral molecules. The fruit turns sweet when it ripens, because of the enzyme amylase that converts all the starch present in the fruit to sugars (fructose and glucose and sucrose) as it ripens. There is a general decrease in acid content as organic acids (mallic acid, citric acid, tartaric acid) are converted into sugars; protein are synthesized from amino acids, the level of later therefore decline, however, volatile organic compounds increased giving fruits their characteristic aroma. There is a steady increase in vitamin C (guava) as fruit approaches maturity.

Changes in fruit color involve changes in the expression of pigment biosynthetic genes. Pigment changes may involve loss of chlorophylls by the enzyme chlorophylases or synthesis of anthocyanin, carotenoid or lycopene. Fleshy fruits are either climacteric or non-climacteric. Climacteric fruits emit ethylene along with increased rate of respiration during the ripening process and these (apple, avocado, banana, guava, mango, papaya and pineapple) are usually harvested once they have reached maturity which then undergoes rapid ripening during transit and storage. Non- climacteric fruits do not ripen after harvest and are often harvested once they have fully ripened e.g. grapes, strawberries and oranges. Ethylene is a major regulator involved in the ripening process. This phytohormone is said to regulate the expression of several genes involved in fruit ripening so as to modulate the activity of various enzymes involved in the process of ripening. These enzymes act to soften the 'skin' of the fruit and also convert complex polysaccharides into simple sugars. Ethylene is produced and released by rapidly-growing plant tissues. It is released by the growing tips of roots, flowers, damaged tissue, and ripening fruit. Ripening process can be accelerated with exogenous application of ethylene releasing compound "Ethrel' and blocked by ethylene inhibitors such as AVG (aminoethoxy venyl glycene), AOA (amino oxyacetic acid).

Pattern of Fruit Growth

As discussed above, there is a hundred fold or more increase in fruit mass or volume from the time of fertilization to maturity. Two distinct types of growth curves are observable when the increase in such variables as the volume, fresh weight, dry weight and diameter of the fruits is plotted as a function of time after anthesis. Apple, pear, date, kiwifruit, pineapple, banana, avocado, strawberry, tomato and citrus fruits have a single sigmoid curve. The fruits such as stone fruits, fig, raspberry, blueberry, grape, and olive have a double sigmoid growth curve. In this type two periods of rapid growth are separated by an intermediate period when either less or no increase in volume occurs. There are thus three clearly defined stages of growth. In first, the ovary wall and its contents grow rapidly except, embryo and endosperm. The stage II is characterized by rapid growth of embryo and endosperm, lignifications of endocarp (e.g. pit-hardening in stone fruits) and slight growth of ovary wall. In stage III, rapid growth of the mesocarp occurs, causing the final swell of the fruit, which is followed by maturation.

Hormonal Regulation of Fruit Growth and Development

Plant bio-regulators (PBRs) are now are being increasingly used on a larger scale in propagation, development of nursery trees, control of growth and flowering of young trees, improvement of fruit set, thinning of flowers and young fruit-lets, prevention of fruit drop, regulation of fruit maturity and ripening, improvement of fruit quality and extension of post harvest life of fruits. As fruit cultivation becomes more mechanized possibilities for using inputs to enhance production increases, in which the role of PBRs becomes more vital. Infect, the use of PBR's has become an integral part of modern fruit production that involve establishing of dwarf fruit plants for early flowering and fruiting and for increased productivity.

Propagation

a). *Rooting of cuttings:* One of the oldest uses of PGRs has been to promote or accelerate the rooting of cutting. IBA is most commonly used auxin for this purpose in pome and stone fruits. Another frequently used auxin is NAA, which however, remains second choice for this purpose as it is more phototoxic if used at higher concentration. There are three different methods for applying auxins for the induction of rooting in cuttings: ***Quick dip*** -dipping the basal end of cuttings for 5-10 seconds in auxin solutions of higher concentration i.e. more than 1000 ppm; ***Prolonged soaking*** – the basal end of cuttings are soaked for up to 24 hours in auxin solution of lower concentration i.e. 10-500 ppm and ***Powder method***-in which auxin is applied in a carrier, usually talc powder to the basal end of cutting prior to placing them in rooting media. In the later method, the concentration of PGR is between 500 to 1000 ppm. Treated cutting initiate rooting earlier and more profusely, that show higher survival rate.

Clonal rootstocks for apple and pear can be rooted easily (within 2-4 weeks), when basal ends of hard-wood cuttings are dipped in IBA 1250-5000 ppm solution for 10 seconds. In stone fruits, more favourable results have been obtained by treating soft wood cutting during the active growing period than during the dormant season. Application of PGRs in plum, cherry, and apricot can accelerate rooting by two to three weeks.

Recommendation: Treating hard wood cuttings of peach in solution of IBA @ 500 ppm enhance rooting

b). *Germination:* Seed of stone fruits require certain period of stratification to break dormancy. Gibberellin treatment can replace at least a part of the low temperature. Seeds of peach and cherry require 60-100 days and 120-150 days, respectively. Seed germination can be enhanced in partially stratified seeds of peach and cherry by soaking in GA solutions at different concentration for 20-24 hours, as given below:

Crop	Stratification period	GA concentration
Peach	30-35 days	100-200 ppm
Plum (*P. domestica* L.)	35 days	20-100 ppm
Cherry	30-35 days	100 ppm

2. Tree size control and increased return bloom/yield: The use of growth retardants, paclobutrazol (Cultar) and chlormequat (CCC) is an useful aid in controlling tree vigour and to increase cropping potential in temperate fruits. One should use only recommended concentration; chemical applied at higher dose may reduce growth excessively. In apples and pears foliar spray of paclobutrazol give better response than soil application. Over dosing can prevent any terminal growth, which can have detrimental effect on production in the later years. The optimum doses are as follow:

Apple: PP_{333} @ 500 ppm as multiple (3-4) foliar spray with beginning at petal fall (PF) and subsequent spray(s) at 3 weeks intervals.

Pear: CCC @ 1000 ppm at PF and 2-3 weeks after PF

Apricot: Foliar spray @ 1000 ppm, 3 weeks after full bloom

Peaches: 2 g a.i./tree before bud-break to July Elberta peach on seedling rootstock under HDP (2500 trees/ha)

Foliar spray @ 2000 ppm at bloom

Plums: 4 litres/ ha in the end of March

Cherries: 3 litres / tree before the end of April

The use of paclobutrazol however, does encourage earlier blossoming in stone fruits, so can increase the risk of frost damage in cold springs.

Note: Soil application is made as collar line drench

Application of Ethrel, NAAand NAD to vegetative pineapple induce flowering, however, Ethrel treatment is better, in this respect. In mango, the post- harvest application of a small amount of paclobutrazol (5g a.i. $tree^{-1}$) to the soil during "on year" significantly promotes flowering and fruiting in the following year.

3. Delay bloom: Autumn application of GA_3 and Ethrel cause delaying of bloom in spring in apples and stone fruit, which can be a useful practice to increase fruit-set in frost affected areas. Another advantage of PGRs treatment is that it facilitates better cross-pollination among varieties having different (non-synchronous) bloom period. For example, blooming of Red Beaut plum can be delayed by several days with fall application of GA_3 at 200 ppm and Ethrel at 200 ppm to help in better synchronization flowering period with cultivar Santa Rosa, inter-planted for cross pollination. Concentration and time of application depend upon fruit crop and environment.

4. Fruit set: Inclement weather during flowering often limit fruit set and production in temperate fruits. Results with the use of some PGRs have been encouraging in pear. Use of GA_3 at 10-12 ppm at bloom can increase fruit set particularly in those cultivars which have tendency to bear parthenocarpic fruits e.g. Bartlett, Conference. Results are better when conditions are not ideal (low temperature) before or during bloom. In Flemish Beauty pear, combined application of GA_3 at 10 ppm + BA at 5 ppm has been beneficial, in this regard. Fruit set of apple can be improved with the bloom application GA3 at 50 ppm however, response is erratic and unpredictable. In India, a foliar application of 2, 4-D or 2, 4,5-T has been reported to be beneficial in improving fruit set and quality in mandarin. Sharma and Chopra (1978) observed increased fruit set in Pineapple and Valencia late orange following the application of 2,4,5-T at 10 to 15 ppm.

5. Thinning: Under ideal conditions, most temperate fruit tree spices often set heavier crop that cannot be adequately sized to meet market requirement for higher price. This necessitates selective removal of some of the fruits on the trees.However, manual thinning is labour intensive, time consuming and expensive. Use of chemicals including PGRs as thinning agents is gradually replacing hand and mechanical thinning in fruits. PGRs can be applied both as bloom thinner or fruit-let thinning agents.

In blossom thinning, chemicals are applied at bloom, whereas in fruit-lets thinning, chemicals are sprayed a few days after petal fall i.e. after fruit set. Blossom thinning is adopted after assessing crop potential and is always risky, while fruit-let thinning is done after observing actual crop load. However, thinning with chemicals is less satisfactory to achieve in stone in comparison to apples. Results are variable depending upon climatic conditions of the area and temperature at the time of application. A chemical may be useful for this purpose in one condition but may fail to thin satisfactorily in another area. Follow local recommendations for the type of chemical to be used, dose and time of application for better results (Table 1).

Table 1: Use of PGRs for crop thinning in stone fruits

Name	Crop/Cultivar	Concentration	Time	Remarks
Bloom Thinning				
Ethrel	Plum Cv. Santa Rosa	200 ppm	Bloom	-Increase size and return bloom
	Peach Cv. Red Haven	200 ppm	Bloom	-Increase fruit size and enhance maturity
	European Plums and Prunes	100-150 ppm	Bloom	Increase size
GA_3	Peach Cv. Red Haven	100-150 ppm	Bloom	-Increase size
Fruit Thinning				
NAA	Apples	10 ppm	7-15 Days after petal fall	-Increase size
Ethrel	Plum and Prunes	100-150 ppm	30-40 Days after Bloom	-Increase size and advance maturity
Ethrel + NAAm	Plum and Prunes	50ppm+80 ppm	30-35 Days after Bloom	-Increase size and advance maturity
Ethrel	Peaches	200-300 ppm	Pea stage (15-16 mm Fruitlet dia.)	-Increase size and advance maturity
NAA	Peaches	40 ppm	Peas stage (15-16 mm Fruitlet dia.)	-Increase size
	Apricot	25-50 ppm	Pea stage	-Increase size
	Nectarines	40 ppm	2 weeks after full bloom in snow Queen 3 weeks after full bloom in May Fire	-Increase size

Note: Doses of PGRs varies with cultivars and locations

In Himachal Pradesh, application of NAA at 350 ppm or Ethrel at 200 ppm in Kinnow mandarin during the 'On Year', 40 days after full bloom effectively control fruit drop and reduce the tendency of alternative bearing. In guava post bloom application of NAA at 80-100 ppm has been useful in reducing fruit set. This treatment can reduce more than 80 % of rainy season crop and increase flowering of the following winter crop. NAD at 50 ppm and 2,4-D at 30 ppm are also effective for de-blossoming of summer flowers. Nectarine cultivars Snow Queen and May Fire can be judiciously thinned with NAA at 40 ppm, when applied 2 weeks (Kumar *et al.,* 2017) and 3 weeks (Rimpika *et al.,* 2017), respectively after full bloom.

6. Fruit quality and post harvest life: PGRs are now increasingly used to control fruit quality (Table 2). It is a commercial practice in seedless grapes to improve berry size with the application of GA_3 at 10-20 ppm at bloom and petal fall stage. Post bloom application of forchlorfenuron (CPPU) at 5 ppm improve fruit/berry size in kiwifruit and grapes. In kiwifruit bunch dipping treatment is however, more effective than spray application. In Delicious apples, application of Promalin at 60 ppm at between full bloom to PF increase fruit size and improve fruit shape by developing oblong well lobed fruits. In marginal areas, foliar application of Ethrel (2-chloroethly phosphonic acid) at 1000 ppm (4 ml ethrel/litre of water) to mature fruits, one week before expected harvest improve surface colour. In order to control fruit drop caused by ethrel add NAA at 10 ppm (Planofix at 2.2 ml/10 litres) to the spray solution. One of the major problems with stone fruits is that they cannot endure long storage life after harvest at normal atmosphere and ambient temperature. However, Ethrel treated fruits should not be stored but only meant for fresh market. Pre harvest application of paclobutrazol and chlormequat increase fruit calcium content, firmness, and have potential to extend post harvest life. Paclobutrazol also enhances fruit colouration (Plate.1), however, it's over dose should be avoided as it decreases fruit size

Plate 1: Fruit surface colour development in July Elberta peach following the application of 250 ppm paclobutrazol, 6 weeks after full bloom (See colour version on page 421)

7. Control of pre-harvest fruit drop: Pre-harvest fruit drops coincide with lower levels of endogenous auxin and higher levels of ethylene. Under normal conditions, approximately 20% of fruit drop, if no precaution taken. In apples, foliar application of NAA at 10 ppm, 2-3 weeks before harvest or fenoprop at 40 ppm at 3-5 weeks before harvest has excellent control of fruit drop. However, fruits should not be left over the trees too long after spray application. In Chausa mangoes, NAA at 50 ppm or 2,4-D at 10 ppm increased fruit retention. Auxin, particularly 2, 4-D at varying concentrations is very effective in controlling premature/pre-harvest fruit drop in citrus. Sprays of 2, 4-D at 8 ppm at 1.2cm diameter stage in Valencia orange, and at 15 ppm in Pineapple and 2, 4, 5-T at 30 ppm in Jaffa and Mosambi are useful when applied in October. 2, 4- D have also been found to be useful in reducing fruit drop in Darjeeling mandarin. In Himachal Pradesh, application of 2,4- D at 10 in May and September controlled fruit drop in Kinnow mandarin.

Preparation of Solution

Solutions of plant growth regulators are made by dissolving their measured quantity in small amount of solvent and then diluting in water, to make an appropriate strength/ concentration. Auxins, gibberellins and cytokinins are dissolved in 95% alcohol, 90% acetone and diluted hydrochloric acid (0.1N HCl), respectively.

Commercial formulations of PGRs are also available in liquid form and solutions of desired strength of these are prepared by diluting their required quantities in water. Before applying through spray, 0.1% of surfactant (e.g. Tween 20) is added to the solution, for quick absorption by he leaves. Spray operation should be carried out in the morning hours on a clear and calm day to run off.

For making solutions of higher concentration of auxins, as in quick dip method (rooting), the quantity of alcohol to be added to the water should be up to 50 % of the total quantity of solution.

Conversion formulae for percentage to ppm solution and vice versa:

Percentage (%) = ppm/ 10000

ppm = % x 10000

Table 2: Use of PGRs for the improvement of fruit quality

Name	Crop/cultivar	Concentration	Time	Remarks
Paclobutrazol	Peach cv. July Elberta	250 ppm	6 weeks after flowering	-Increase fruit Ca content, firmness, colouration, sugar contents and shelf life.
Chlormequat	Peach cv. July Elberta	250 ppm	6 weeks after flowering	-Increase fruit firmness
Triacontanol	Almond cv. Nonpareil	10 ppm 50ppm	Pit-hardening stage	-Improve fruit quality
2,4,5-T or 2,4,5-TP	Apricots	50ppm	At the beginning of pit-hardening stage	-Increase fruit size and enhance maturity

Senescence and Death

Senescence is a process of deterioration/aging in plants. Senescence is a normal energy dependent developmental process which is controlled by plant's own genetic programme and the death of the plant or plant part consequent to senescence is called as Programmed Cell Death (PCD). It is the terminal stage of plant development causing degeneration and death of cells, tissues, organs or the whole plant. Senescence is not confined only to whole plant. It may be limited to a particular plant organ such as leaf and flowers or cells such as phloem and xylem or cell-organelles such as chloroplasts and mitochondria etc. In general, four types of senescence patterns have been found in plants:

i. **Overall Senescence**. This type of senescence occurs in annuals where whole of the plant is affected and dies.

ii. **Top Senescence.** This is represented by perennial herbs where senescence occurs only in the above ground parts, the root system and underground system remaining viable.

iii. **Deciduous Senescence.** This type of senescence is less drastic and takes place in woody deciduous plants. Here senescence occurs in all the leaves simultaneously but the bulk of the stem and root system remains alive.

iv. **Progressive Senescence.** This is characterized by gradual progression of senescence and death of leaves from the base upwards as the plant grows.

During senescence photosynthesis, starch, chlorophyll, DNA, RNA, proteins, gibberellins and auxins decrease. Senescence can best be studied in leaves or similar other organs of plants e.g. cotyledons, sepals, petals etc. or cell organelles like isolated chloroplasts.

- Senescing cells and tissues are metabolically very active and an ordered series of cytological and biochemical events occur during senescence.
- Senescence is characterized by increased respiration, declining photosynthesis and an orderly disintegration of macromolecules.
- At the cellular level, chloroplasts are the first organelles to be disintegrated. Nuclei remain structurally and functionally intact until the last stage of senescence. Meanwhile, other cell organelles and biomembranes also gradually deteriorate.
- Brilliant Colours are developed in leaves of many plants during senescence.
- Towards the end of senescence, the cells and tissues also lose respiratory control.

Studies have shown that, the programmed senescence is influenced by plant hormones. Abscisic acid, ethylene, jasmonic acid and salicylic acid promotes sencence, whereas, cytokinin has a delaying effect on senescence. Several environmental factors especially those which suppress normal plant growth also tend to enhance the rate of senescence. These are deficiency of soil nutrients, high temperatures, water deficit, darkness *etc*.

In many plants on the other hand, pruning, removal of flowers and fruits can markedly delay the senescence of leaves. Exogenous application of cytokinin and gibberellins (citrus) also have delaying effect on senescence.

LITERATURE CONSULTED

Arteca, Richard N.1996. Plant Growth Substances: Principles and Applications, Springer US, 332 p.

Basra Amarjit S. 2004. *Plant Growth Regulators in Agriculture and Horticulture*: *Their role and Commercial Uses*. International Book Distributing Co. 264 p.

Bewleyl, J.D. 1997. Seed Germination and Dormancy. *The Plant Cell*, Vol. 9, 1055-1066.

Brian J. Atwell and Colin, G. N. 1999. Flower bud differentiation and fruit set, *Plant in Action: Adaptation in Nature and Performation in cultivation*: http://plantinaction. science.uq.edu.au/ edition1 "Development of Fruit and Fruit Types." *Boundless Biology* Boundless, Retrieved ý11 ýApr. ý2017 from https://www.boundless.com/biology/textbooks/boundless-biology-textbook/plant-reproduction-32/pollination-and-fertilization-190/development-of-fruit-and-fruit-types-726-11955/

Buban, T. and Faust, M. 1982. Flower bud initiation in apple trees: Internal control and differentiation. *Horti.Rev*., 4:174-203.

Gardner, F. P., Pearce, R. B. and Mitchell, R. L. 1985. Physiology of Crop Plants. Scientific Publishers India. 327 p.

Hopkins, W. G.1995. Introduction to Plant Physiology. John Willy & Sons, Inc, New York, 464p.

Jules Janick, Buban,T. and Fau st,M. 2011. Flower Bud Induction in Apple Trees: Internal Control and Differentiation, *Horticulture Review*, Volume 4, Published Online: 8 FEB 2011.

Jules Janick and Miller Stephen S.2011.Plant Bioregulators in Apple and Pear Culture. *Horticulture Reviews,* vol.10,0nlinelibrary,wiley.com,Published Online:8 Feb 2011.

Kumar, Arvind and Purohit, S.S. 2005. *Plant Physiology: Fundamentals and Applications*, Second edition, Agrobios, India, P 776.

Kumar, Rajiv, Sharma, N., Rimpika and Sharma, D.P. 2017. Effect of hand and chemical thinning on growth and production of nectarine cv. Snow Queen. *Indian Journal of Ecology*, 44(1): 95-98.

Leopold, A. Carl and Kriedemann Paul, E.1977. MCGRAW-HILL EDUCATION; 2Rev Ed edition (January 1, 1977), 545 p.

Mango Trees: Bud differentiation. mango-trees.blogspot.com/2009/06/bud-differentiation.html, 2009.

Monselise Shaul, P. 1986. *CRC Handbook of Fruit Set and Development*, CRC Press, Inc. Boca Raton, Florida, P 568.

Nickell, L.G.1982.*Plant Growth Regulators: Agricultural Uses*. Springer-Verlag Berlin Heidelberg,176 p.

Plant growth regulators: their use in crop production 1993. North Central Regional Publication, University of Missouri, USA.

Rimpika, Sharma, N. and Sharma, D.P. 2017. Effect of chemical thinning, gibberellic acid and pruning on growth and production of nectarine cv. May Fire. *Journal of Applied and Natural Science,* 9 (1): 332-337

Robert, J. Weaver.!972. *Plant Growth Substances in Agriculture.* San Francisco : Freeman, 594 p.

Salimia, R. B. 2007. Juvenility, Maturity and Rejuvenation in Woody Plants. *Hebron University Research Journal*. 3(1):17-43.

Seed dormancy - https://en.wikipedia.org/wiki/Seed_dormancy

Sharma, S.D. and Chopra, S.K. 1978. Effect of spring and summer spays of 2,4,5-T and GA on fruit set and physical characters of some sweet orange cultivars. *Indian J. Hort.*, 35(3): 195-201.

Westwood, M. N. 2009. *Temperate-Zone Pomology: Physiology and Culture*, Timber Press; 3rd edition, 536 p.

SECTION - 3

FRUIT PLANT PROPAGATION TECHNOLOGY

10

Propagation Methods

V.K. Wali, Amit Jasrotia and Parshant Bakshi

INTRODUCTION

The propagation of plants has been one of the fundamental occupations of mankind since beginning of civilization. Plant propagation involves the application of specific biological principles and concepts in the multiplication of plants for useful purposes. The present day plant propagation industry is large and complex and involves not only the group that multiplies the plants for sale and distribution, but also a large group of industries that provide services, sell the product, is involved in regulation, provides consultation carries on research or is involved in teaching. The key person with in this complex, however, is the plant propagator who possesses the knowledge and skill either to perform or to supervise the essential propagation tasks for specific plants.

The nursery plants are said to be the foundations on which orchards are built since the ultimate successes in fruit growing depends upon the quality of the plant material used for the establishing the orchard. The use of quality nursery plants is therefore, the basic pre-requisite for the success of an orchard enterprise. In order to improve the productivity and quality of fruits, it is of paramount importance to use only genuine, true-to-type, healthy and disease free nursery plants for establishing new orchards.

PROPAGATION METHODS

Propagation of plants is defined as a controlled multiplication of plants by man to perpetuate selected individuals/or a group of plants which have specific value to him. In other words, plant propagation deals with the multiplication and propagation of plants using propagules representing a specific group of genotypes. A propagule is any plant part used to produce a new plant or a population of plants. Specific propagules include seeds, cuttings, layers, buds, scions, explants and various kind of specialized structures such as bulbs, corms and tubers.

1. SEXUAL PROPAGATION

It refers to multiplication of plants by seed. In sexual process male and female gametes are fused to produce seed. Meiosis division takes place in course of fusion and the chromosome numbers, as in parents is reduced to half, which after fertilization becomes normal. In sexual propagation during meiosis segregation, re-assortment or rearrangement of characters takes place. So, the plants thus produced may or may not be similar to their parents and the propagated plants may also be different from each other. It is called as seed propagation, since the propagation is through seed and also sexual propagation because sexes are involved. Seed is the result of fusion of male and female gametes. Seeds are fertilized ovules, containing embryos resulting from the union of a male and a female gamete during fertilization. The embryo in the seed gives rise to a new plant on germination. Plants that are produced from seeds are called seedlings.

Advantages of sexual propagation

Propagation of plants through seeds has certain advantages

1. It is the easiest and least expensive method of plant propagation.
2. Seedling trees are harder and have longer life span.
3. Plants which are difficult to propagate by vegetative method e.g. papaya, phalsa, coconut, etc. can only be propagated by seed.
4. The rootstocks for budding and grafting are obtained by means of sexual propagation.
5. Sexually propagated plants are more resistant to pests and diseases.
6. Poly-embryonic varieties (give rise to more than one seedling from one seed) can be propagated by seed.

Disadvantages of sexual propagation

Some of the limitations of sexual method of propagation are

1. Seedlings take more time to bear fruits (late bearing).
2. Quality of existing plants cannot be improved by sexual propagation.
3. Plants propagated sexually are large in size, thus the cost of manuring, pruning and spraying increases.
4. Due to cross pollination and segregation, there is no assurance about genetic purity of plant.
5. Identification of sex in seedling is not possible.

Propagation by Seed

Selection of Fruits and Extraction of Seeds

- Mature fruits should be processed for seed extraction. Fruits that are undeveloped or infected with insects and disease should be discarded. Proper extraction methods should be conducted carefully to avoid damaging seed. Normally, 2-3 days of drying is enough. Rubbing and crushing the fruits will expedite seed extraction.
- If seed extraction and processing cannot be conducted immediately, place sacks or containers of seed in a dry and cool room with good air circulation. Place the sacks or containers on a shelf or rack to facilitate air movement. Sacks may also be hung from the ceiling.
- If seed is to be sown immediately after extraction and cleaning, no further processing is required. However, if it will be stored, even for a short period, the seed of most species require drying in order to maintain viability during storage. Whether seed requires drying or not depends on its classification as recalcitrant or orthodox.

Mango: Seeds extracted from soft-pulp fruits give higher germination (76 percent) than those from firm fruits (54 percent), and seeds from firm fruits remain dormant for some time after planting.

Citrus: Take mature fruits. Carefully remove the seeds from the citrus fruit, taking care not to damage the seeds and squeezing them out gently. Rinse the seeds in water to separate them from the pulp and remove the sugar that clings to them; sugar encourages fungal growth and will jeopardize potential seedlings. Place them on a paper towel. Sort out the largest seeds; those which are more white than tan with a shrivelled outer skin are the most viable. Fruit/pod/seed collected directly from trees is of high quality because they are healthy and not been exposed to soil moisture or soil microorganisms. Selecting seed from many parts of the crown, assures broad genetic base of the seed collected.

Guava: The seeds are removed from the ripened fruits by mashing it. They should be washed well to be free of pulp, mixed with ash and stored in a cool place. It is better to sow the seeds immediately after extraction. However, guava seeds retain their vitality till one year after extraction from the fruit.

Storage of Seed: The main purpose of traditional seed storage is to secure the supply of good quality seed for a planting programme, whenever needed. If sowing time follows immediately after seed collection and processing, seeds can go directly from the processing unit to the nursery and storage is not needed. This is however rarely the case. In seasonal climates with a relatively short planting season, sowing time is normally determined by the wish to have plantable size seedling at the beginning of the planting season. Hence, seeds must often be stored during the period from harvest to sowing. That is short term storage of less than a year. Many species produce seed at long intervals, ranging from a few years to many years. To assure seed supply during the period between two good seed crops, a seed stock should be established. Even where fruiting is regular and abundant every year it may be more cost efficient to collect surplus seed to cover several years supply rather than to undertake collection every year. Hence, a seed store serves as a buffer between demand and production and has a regular turn over, seeds are stored during periods of seed availability and shipped to nurseries or other recipients when required to raise plants. A new type of seed store has arisen during the last few decades, viz. stores for conservation of genetic resources. In these so called gene banks, seed (and sometimes other propagation material) are stored for long periods at very low moisture content and temperature (cryopreservation). The techniques applied for storage at ultra low temperatures are quite different from conventional seed storage.

Eco-physiological Role of Storage: Normally seeds go through a physiologically inactive (quiescent) period from their dispersal until they can germinate. An exception is viviparous seed, e.g. seeds of several mangrove species, which often germinate before they are released from the mother tree. For other seeds, the interlude between dispersal and germination depends on the deposit site (whether favourable to germination or not) and on possible dormancy. If the deposit site is unfavourable to germination, seeds may either die or stay dormant for a period waiting for conditions to improve. The length of time seeds can stay viable in the natural environment depends on the seeds themselves and the conditions around them. Some seed types do not have the ability to stay alive for a long time. These so-called recalcitrant seeds have short physiological storability, which can only be slightly extended by sorting them under controlled conditions build up large soil seed banks, which may stay viable for decades. Soil dwelling predators, micro-organisms and a relatively high soil temperature and humidity around the seeds are the principal factors contributing to seed deterioration in the field. Only seed with special protection, e.g. hard coated legumes can survive long periods of storage under field conditions and build up large soil seed banks.

Storage Conditions: Storage conditions should be designed to prolong the viability of seeds by reducing or limiting any factor that impairs viability. The general storage conditions should therefore aim at:

1. Reducing the metabolism of seeds.
2. Keeping insects, fungi and other pathogens away.
3. Reduced general seed ageing.

The general prescriptions for seed storage are summarized below:

- Store seeds at lowest possible temperature that will not damage the seeds.
- Store seeds with lowest possible moisture content that will not damage the seeds.
- Eliminate as many pathogens as possible before storage.
- Protect seeds from pathogens during storage.
- Store in the dark.

Store orthodox and intermediate seeds with low moisture content in airtight containers. The recalcitrant seeds are stored in material which is permeable to gases but able to retain moisture.

Mango: Mango seeds remain viable for short periods and exhibit recalcitrant storage behavior. Seeds are susceptible to desiccation and are killed when moisture goes below a critical level. Further, they do not withstand chilling temperature during storage. Seeds stored under moist conditions remain viable only for a short period, from a few weeks to a few months. Reduction of seed moisture below 25 per cent in 'Alphonso' and 32 per cent in 'Totapuri' lowered the percentage of germination and seedling vigour. Fresh seeds remain viable for four to five weeks under ambient conditions, while under dry storage mango seeds remain viable for ten days. The best method is to store whole fruit at room temperature (22 to 25°C), allowing seeds to remain viable for 70 days. Storage at 5°C harms the fruit. Large-fruit seeds stored better than small-fruit seeds. Charcoal powder is a better storage medium for mango seeds than polyethylene bags, desiccators, refrigeration, or open jars. It maintains suitable moisture and protects the seeds from pathogens.

Citrus: To store the citrus seeds, place them on a moist paper towel. Keep about three times the amount of seeds that you want to plant in case some of them are not viable. Wrap the seeds in the damp towel and place them inside a plastic ziploc type bag. Place the bag in the refrigerator. Citrus seed storage in the fridge will last for several days to several months. Unlike other seeds, citrus seeds need to stay moist. If they dry out, it is very likely they will not germinate.

Testing Viability of Seeds

Viability of seeds can be determined by several tests- the standard germination, excised embryo and tetrazolium tests being the most important.

Standard Germination Tests

In the standard germination test, germination percentage is determined by the per cent of normal seedling produced by pure seeds (the kind under consideration). To produce a good test, it is desirable to use at least 400 seeds picked at random and divided into lots of 100 each. If any two of these lots differ by more than10 per cent, a re-test should be carried out. Otherwise, the average of the four tests becomes the official germination percentage. Seeds are placed under optimum environmental conditions of light and temperature to induce germination. The conditions required to meet legal standards are specified in the rules for seed, which may include type of test, environmental conditions and length of test.

Various techniques are used for germinating seeds in seed testing laboratories. Small seeds are placed on germination trays (not galvanized steel, which contains toxic salts). Plastic boxes, paraffined cardboard boxes, or covered glass petri dishes are also useful containers. The most common media used by commercial seed technology labs for germination tests are blue blotter or washed paper towels. These products ensure uniformity and reproducibility in their tests. Other media are absorbent cotton, filter paper and for large seeds, a sand, vermiculite perlite or soil (16mm; 5/8 inch) layer. Containers are paced in germinators in which temperature, moisture and light are controlled. To discourage the growth of micro-organisms, all materials and equipment should be kept scrupulously clean, sterilized when possible and the water amount carefully regulated. No water film should form around the seeds; neither should the germination medium be so wet that a film of water appears when that medium is pressed with a finger. Relative humidity in the germinator should be 90 per cent or more to prevent drying. Containers with sand should be kept tightly closed. Water should not be added during the test.

Thrilled towel test is commonly used for testing large seed like cereal grains. Several layer of moist paper toweling, about 2.8 by 3.6 cm (11 by 14 in.) in size, are folded over the seeds, then rolled into cylinders and placed vertically in a germinator.

A germination test usually run from one to four weeks but could continue for three months for some slow-germinating tree seeds with dormancy. A first count may be taken at one take and germinated seeds discarded with a final count taken later. At the end of the test, seeds are divided into (a) normal seedlings, (b) hard seeds, (c) dormant seeds,(d) abnormal seedlings, and (e) dead or decaying seeds. A normal seedling should have a well-developed roots shoot, although the criteria for a "normal seedling" varies with a different kinds of seeds "Abnormal seedlings" can be caused by age of seed or poor storage conditions; insect disease, or mechanical injury; overdoses of fungicides; frost damage; mineral deficiencies (manganese and boron in peas and beans); or toxic materials sometimes presents in germination trays, substrata, or tap water. Any germinated seed should be examined to determine the possible reason. "Hard seeds" have not absorbed water. Dormant seeds are those that are firm, swollen, and free from moulds, but do not germinate.

The Excised - Embryo Test

The excised-embryo test is used to test seed viability of woody shrubs and trees whose dormant embryo require long treatment periods to relieve dormancy before true germination will take place. In this test the embryo is excised from the seed and germinated alone.

Prior to embryo excision, seeds are soaked for one to four days of the following method until they are completely swollen: (a) in slowly running water, (b) in standing water below: 5° C (59° F) in standing water at about 25°C (68°F), with at least two changes of water daily. Storing seeds in moist peat for three days to two weeks at cool temperatures is also satisfactory in preparing seeds for excision.

The excision must be done carefully to avoid injury to the embryo. Any hard, stony seed coverings, such as the endocarp of stone fruit seeds, must first be removed. The moistened seed coats cut with a sharp scalpel, razor blade, or knife under clean but non-sterile conditions with sterilized instruments. The embryo is carefully removed. If a large endosperm is present, the seed coat may be slit and the seed covered with water and after about a half-hour the embryo may float out or be easily removed.

Procedure for germinating excised embryos is similar to that for germinating intact seeds. Petri dishes with a moist substratum, such as blotter or filter paper are used. The embryos are placed on the filter paper so that they do not touch. The dishes are kept in the light at a temperature of 18 to 22°C (64-74° F). At higher temperatures, moulds may develop and interfere with the test. The time required for the test varies from three days to three weeks.

Non-viable embryos become soft, turn brown and decay within two to ten days; viable embryos remain firm and show some indication of viability, depending upon the species. Types of responses that occur include spreading of the cotyledons, development of chlorophyll and growth of radicle and plumule. The rapidity and degree of development gives some indication of the vigour of the seed.

Tetrazolium Test

The tetrazolium test is a biochemical method of viability determined by the red colour appearing when seeds are soaked in a 2,3,5-triphenyl tetrazolium chloride (TTC) solution. Living tissue changes to an insoluble red compound (chemically known as formazon) in non-living tissue the TTC remains uncolored. The test is positive in the presence of dehydrogenase enzymes involved in respiration. The reaction takes place equally well in dormant and non-dormant seed. Result can be obtained within 24 hours, sometimes in 2 or 3 hours. TTC is soluble in water making a colourless solution. The solution deteriorates with exposure to light, but will remain in good condition for several months if stored in a dark bottle. The solution should be discarded if it becomes yellow. At 0.1 to 1.0 per cent concentration is commonly used. The pH should be 6 to 7. The TTC test is used primarily to obtain rapid result for both non-dormant and dormant seeds.

Sowing of Seeds

Sowing is done with the best seed available, to obtain the maximum number of healthy and sturdy seedlings for transplanting or field planting. Seeds are generally sown in the soil by maintaining a planting depth of about 2-3 times the size of the seed.

Pre-treatment of seed and soil before sowing

Before sowing, certain seeds first require a treatment prior to the sowing process. This treatment may be seed scarification, stratification, seed soaking or seed cleaning with cold (or medium hot) water. Seed soaking is generally done by placing seeds in medium hot water for at least 24 to up to 48 hours. Seed cleaning is done especially with fruit, as the flesh of the fruit around the seed can quickly become prone to attack from insects or plagues. To clean the seed, usually seed rubbings with cloth/paper is performed, sometimes assisted with a seed washing. Seed washing is generally done by submerging cleansed seeds 20 minutes in 50 degree Celsius water. This (rather hot than moderately hot) water kills any organisms that may have survived on the skin of a seed. Especially with easily infected tropical fruit such as litchi and rambutan, seed washing with high temperature water is vital.

In most of the fruit crops, there may be natural or chemical dormancy. The dormancy in seeds of ber, guava and walnut is due to the presence of hard seed coat which inhibits penetration of water and oxygen required for germination. Presence of chemical inhibitors (abscisic acid) is responsible for dormancy in seeds of most of temperate fruits (apple, pear, peach and walnut). Besides ABA, higher concentration of pectin, gum, tannin and amino acids are also responsible for dormancy in seeds of temperate fruits. The dormancy of seed due to hard seed coat can be overcome by softening it. It can easily be done either by scarification, stratification or by use of chemicals and hormones.

Scarification: It is the process of breaking, scratching, altering or softening the seed covering to make it permeable to water and gases. It can be achieved mechanically, or by hot water and acid. In mechanical scarification, cracking of seed with hammer, rubbing with sand paper or cutting with a file without injury to embryo is generally employed to break the dormancy (ber, peach, walnut). Impermeable seed coat of guava can be softened by hot water scarification. Seeds are placed in hot water at 77-100°C. They are immediately removed from hot water and allowed to soak gradually in cool water for 12-24 hours. Acid scarification consists of treating guava seeds with concentrated sulphuric acid for 3 minutes, ber for 5-6 hours and strawberry seeds with 0.25 per cent nitric acid or hydrogen peroxide results in higher germination.

Stratification: It is the method of handling of dormant seeds, in which, the imbibed seeds are subjected to a period of chilling to after ripen the embryo. This term originated as the nurserymen used to place seeds in stratified layers interspaced with a moist medium such as soil or sand out of door or in pits during winter. The term moist chilling has been used as synonym to stratification.

Stratification can be achieved by refrigeration of dormant seed. Dormant seeds of temperate fruits like apple, cherry, pear and apricot are generally placed in layers of sand in a box at a temperature of 1-5°C. Depending upon the seed type, treatment time varies from 1-5 months for breaking dormancy of seeds. The seed dormancy due to presence of growth inhibitors can be broken by placing the seeds in running water. It results in leaching of inhibitors. Freshly extracted seeds of strawberry and grapes, if placed in running water for 7-12 days results in increased germination.

Use of hormones: The seed dormancy can also be overcome by the treatment of growth regulators. Treatment of seeds of apple, cherry, peach, strawberry and hazelnut with 100-500 ppm of GA3 for 24-48 hours improves germination and better growth of seedlings. Ethrel (5,000 ppm) treated seeds of guava and strawberry also show better germination and growth of seedlings. Treatment of apple and peach seeds with 10-20 ppm solution of benzyl adenine (BA) is effective for higher seed germination.

Chemical treatments: Many freshly harvested dormant seeds usually respond to soaking in potassium nitrate solution. This technique is largely used in seed testing laboratories where seeds are placed in petri-dishes containing 0.2% solution of potassium nitrate. The seeds of peach and grape, treated with 5,000 ppm (5 g/litre) of thiourea show enhancement in their germination.

Raising Seedlings on Seed Beds

Seeds should be sown in lines, so as to facilitate the intercultural operations such as weeding, fertilization, etc. Shallow furrows of 0.5-1.0 cm depth should be made by fingers or using a stick, at a distance of 12-15 cm. Seeds should be placed in furrows at an interval of 2 cm. After placing the seeds, the furrows should be covered with a thin layer of the soil and pressed so as to embed the seeds. If seeds are very small, seeds are sprinkled on the top of the bed and the surface is compacted by pressing the top soil or mulched with straw to avoid the seeds from being flown away by water or picked by birds.

Raising Seedlings in Poly-bags/Pots

Poly-bags or pots are used for raising seedlings of number of species. They help to minimize the disturbance to the root system, avoid the problem of digging and are easy in transportation. Seeds planted in poly-bags or pots give a higher final survival. Generally poly-bags of 22.5 x 12.5 cm^2 size are used. Earthen pots generally 5 inch diameter are used for raising the rootstocks for the species which are multiplied by inarching. Poly-bag or pot filling mixture should contain thoroughly mixed well drained soil, FYM and clay or pond soil in 1: 1: 1 ratio. One to two seeds should be sown in each bag/pot at 0.5 to 1.0 cm depth. Gap filling in bags or pots should be carried out within two weeks of initial sowing.

2. ASEXUAL PROPAGATION

It is also known as vegetative propagation or clonal propagation. Asexual propagation is reproduction by means of vegetative parts of the plant such as roots, shoots, or leaves other than seed. In this propagation sexes are not involved–hence it is called asexual propagation.

It involves the use of any part of the plant, other than seed i.e. vegetative parts –hence called as vegetative propagation. The vegetative organs of many plants have the capacity ability of regeneration, to produce new individuals.

For instance:

(a) Stem pieces (cuttings) produce root system.

(b) Root pieces (root cuttings) develop root system.

(c) Leaves generate both roots and shoots.

Vegetative parts possess somatic cells. They divide (multiply) by mitosis it does not involve reduction in chromosomal number, but involves the duplication of chromosome structure thus the same genetic constitution is seen in the resultant plants with no variation. Whatever the characters present in the parent, the same are carried in the new plants i.e. duplicated without any change.

Advantages

1. As there is no change in the genetic makeup of the plant propagated by this method, the fruit plants propagated vegetatively are true to type, and, as a result, it is possible to get uniformity in growth, yield and quality of fruit, which makes harvesting and marketing easy.
2. Some fruits such as banana, pineapple and some guava varieties being seedless, the only way of further propagation is vegetative method.
3. Asexually propagated fruit trees come into bearing earlier.
4. Certain varieties of some fruit plants are susceptible to certain diseases. By budding or grafting them on a resistant root stock, these varieties can be grown without pest or disease incidence.
5. Hardiness to cold and other unfavourable conditions such as drought can be secured, e.g. orange do well on trifoliate stock in areas where frost occurrence is frequent.
6. Trees can considerably dwarfed by using proper rootstocks, e.g., the apple trees can be dwarfed by using Malling IX as the rootstock.
7. Methods like bridge grafting can be used for healing of the wounds caused by rodents.
8. By top working the inferior quality fruit trees can be converted into superior quality fruit trees.
9. As a fancy, it is possible to grow 2-3 varieties on the same plant, e.g., one can get 3-4 varieties of roses on various branches of the stock plant.

Disadvantages

1) No, new variety can be evolved by means of the vegetative method of propagation.
2) Vegetative propagation in many cases is more expensive than seed propagation.
3) Vegetatively propagated plants are comparatively short lived. Lack of tap root system in vegetatively propagated plants results in poor anchorage in the soil. Consequently, such plants are easily uprooted in storms and or other such severe conditions.
4) Vegetatively propagated plants are comparatively less hardy.
5) Transmit viral diseases from plant to plant.

Methods of vegetative propagation

a. Divisions or specialized organs

Rhizome: A rhizome is a specialized stem structure in which the main axis of the plant grows horizontally just below or on the surface of the ground. The stem appears segmented because it composed of nodes and internodes. The rhizome appears as a many branched clump made up of short individual sections. The rhizome tends to be oriented horizontally with roots arising from the lower side. In propagating plants through rhizome or cutting the rhizome into different sections be sure that each section has at least one lateral bud or eye. It is essentially a stem cutting. e.g., banana.

Offset: It is a short thickened horizontal branch growing out of the crown ending at the apex with a tuft of leaves and a cluster of leaves below. These are special type of branches or lateral shoots which are produced from the base of main stem of parent plant. The offset often breaks away from the mother plant and the daughter starts a new independent life. e.g., Date palm.

Suckers: It is a lateral branch developing from the underground parts of the stem or roots. The suckers arise from below the surface of the soil. There are two types of suckers.

a) **Shoot suckers:** These will arise from the base of the stem. The suckers may grow obliquely upwards and directly give raise a leaf shoot. Often it grows horizontally outwards only to certain extent but soon turn up. It strikes roots when it is still attached to the parent plant or when separated and planted. Propagation by shoot suckers can be done by separating the suckers and planting in the field. e.g., banana, pineapple etc.

b) **Root suckers:** The root suckers will arise from the adventitious buds on the roots. Propagation by shoot suckers can be done by separating the suckers and planting. e.g., Guava.

Runners: Runners are specialized aerial stems (stolons) arising in the leaf axils of plant having rosette crowns. New plants arise from nodes at interval along these runners. From these runners more new runners may arise thus developing natural clonal multiplication methods. The typical runner producing plant is strawberry which is photo sensitive with regard to its runner production. Long days favour runner production where as short days prevent runner formation.

b. Cuttings

Cutting is a portion of stem, root, or leaf, cut from the parent or stock plant and induced to form roots and shoots by chemical, mechanical and/or environmental manipulation.On the basis of plant part used and relative position on a plant, cuttings is classified in various groups as:

Stem cuttings: Soft wood cuttings,Semi Hardwood cuttings,Hardwood cuttings and Herbaceous cuttings.

Leaf and Leaf bud cuttings

Root cuttings

1. Stem Cuttings: A stem cuttings is any cutting taken from the main shoot of a plant or any side shoot growing form the same plant or stem. The shoots with high carbohydrate content usually root better. Broadly, there are four types of stem cuttings, namely hardwood, softwood, semi hardwood and herbaceous cuttings.

a) **Hardwood cuttings:** Cutting from mature and lignified stem of shrubs and trees are called as hardwood cuttings. Hardwood cuttings are prepared during dormant season, usually from one year old immature shoots of previous season's growth. The length of cuttings varies from 10 to 45 cm in length and 0.5 to 2.5 cm in diameter, depending upon the species. Usually, the cuttings of 25-30cm length, with pencil thickness are preferred. Each cutting should have at least two buds or more. While preparing the cutting, a straight cut is given at the base of shoot- below the node while a slanting cut, 1 to 2 cm above the bud is given at the top of cutting. However, in case of hollow pith species such as Kiwifruit top cut should also be close to bud to avoid drying up of top portion. This helps in maintaining the polarity of the shoot and if rain occurs, water does not accumulate on the tip of the cutting, which saves the cutting from fungal infection. A number of deciduous fruit plants like grape, hazelnut, chestnut, fig, quince, pomegranate, mulberry, plum, olive, gooseberry and apple etc. are commercially propagated by hard wood cuttings.

b) **Semi-hardwood cuttings**: Semi-hardwood cuttings are those made from woody, broad-leaved evergreen species and leafy summer and early fall cuttings of deciduous plants with partially matured wood. These types of cuttings are mostly used in evergreen fruit plants like mango, guava, lemon, jackfruit and some shrubs and shrubby plants, ornamental shrubs. The length of the cuttings varies from 7 to 20cm. The cuttings are prepared by trimming the cutting with a straight cut below a node and removing a few lower leaves. However, it is better to retain two to four leaves on the top of the cuttings. While planting 1/4th of their length should be inserted in the soil. The best time for taking cuttings in summer, when new shoots have emerged and their wood is partially matured. It is necessary that leafy cuttings be rooted under conditions that will keep water loss from the leaves at a minimum. Commercially they are rooted under intermittent mist, fog or under polyethylene sheets laid over the cuttings.

c) **Softwood cuttings:** Cuttings prepared from the soft-succulent and non-lignified shoots, which have not become hard or woody, are called as soft wood cutting. Such types of cuttings are very prone to desiccation. Therefore, proper arrangement for controlling humidity is required. Usually the cutting size is 5-5.7 cm but it varies from species to species. Usually, some leaves should be retained with this type of cuttings. The best time for preparing soft-wood cuttings is late summer. Softwood cuttings generally root easier and quicker than other types, but require more attention and sophisticated equipment. Temperature should be maintained during rooting at 23to 27°C at the base of cuttings.

d) **Herbaceous cuttings:** Herbaceous cuttings are made from succulent non-woody plants like geranium, chrysanthemums, coleus, carnation and many foliage crops. They are 7-15 cm long with leaves retained at the upper end. They are rooted under the same conditions as softwood cuttings, requiring high relative humidity. Bottom heat is also useful for initiation of rooting process. Herbaceous cuttings of some plants exclude a sticky sap (as in geranium, pineapple, cactus etc.) that interferes with root initiation process. In such cases basal ends of cutting should be allowed to dry for few hours before planting. Generally, fruit plants are not propagated by herbaceous cuttings.

Rooting Media for Planting Cuttings

Various materials and mixtures of materials are used for germinating seeds and rooting of cuttings. For good results the potting mixture should have following characteristics.

- The medium must be sufficiently firm and dense to hold the cuttings or seeds in place during rooting or germination. Its volume must be fairly constant when either wet or dry. Excessive shrinkage after drying is undesirable.
- It should be highly decomposed (preferably with C:N ratio of 20:1) to prevent N immobilization, and excessive shrinkage during production.
- It must be easy to wet (not too hydrophobic) and retain enough moisture to reduce frequent watering.
- It must be sufficiently porous so that excess water drains away, permitting adequate penetration of oxygen to the roots.

- It must be free from weed seeds, nematodes, and various pathogens.
- It must not have high salinity level.
- It should be capable of being steam pasteurized or chemically treated without harmful effects.
- It should have high cation exchange capacity (CEC) for retention of nutrients that may be applied pre-incorporated and/or in a supplementary soluble and/or slow-release fertilizer program.
- It should be readily available at affordable cost.

Propagation/potting media used in horticulture consists of a mixture of organic and inorganic components which have different but complementary properties.

Organic components include: Peat, softwood and hardwood barks, or sphagnum moss.

Inorganic component: A coarse mineral component is used to improve drainage and aeration by increasing the proportion of large, air-filled pores. These include sand, grit, pumice, scoria, expanded shale, perlite, vermiculite, polystyrene, clay granules, and rockwool.

Media components used in propagation systems

Soil: It is composed of materials in the solid, liquid, and gaseous states. These components must exist in proper proportion for satisfactory growth of the plant. The solid portion of the soil is comprised of both inorganic and organic components. The inorganic part consists of the residue from parent rock after decomposition resulting from the chemical and physical process of weathering. Such inorganic components vary in size from gravel down to extremely minute colloidal particles of clay, the texture of the soil being determined by the relative proportions of these particle sizes. The organic portion of soil consists of both living and dead organisms. The residue from decay of dead remains of animal and plant life is termed as humus which is largely colloidal and assists in holding water and plant nutrients. The liquid portion of the soil, the soil solution is made up of water containing dissolved salts in various quantities, as well as dissolved oxygen and carbon dioxide. The gaseous portion of the soil is good for plant growth. In poorly drained, waterlogged soils, water replaces the air, thus depriving plant roots as well as certain aerobic microorganisms of the oxygen necessary for their existence. The texture of the mineral soil depends upon the relative proportions of sand (0.05 to 2 mm particle diameter), silt (0.05 to 0.002 mm particle diameter), and clay (less than 0.002 mm particle diameter).

Sand consists of small rock particles, 0.05 to 2.0 mm in diameter, formed as the result of the weathering of various rocks, its mineral composition depending upon the type of rock. Quartz sand, consisting chiefly of silica complex, is generally used for propagation purposes. Sand should preferably be fumigated or steam pasteurized before use, as it may contain weed seeds and various harmful pathogens. Sand contains virtually no mineral nutrients and has no buffering capacity, or cation exchange capacity (CEC). It is mostly used in combination with organic materials.

Peat consists of the remains of aquatic, marsh, bog, or swamp vegetation which has been preserved under water in a partially decomposed state. There are three types of peat as classified by the US Bureau of Mines: moss peat, reed sedge and peat humus. *Moss peat* is the least decomposed of three types and is derived from sphagnum or other mosses. It varies in colour from light tan to dark brown. It has high water holding capacity (15 times its dry weight), has high acidity (pH of 3.2 to 4.5), and contains a small amount of nitrogen (about 1.0 percent) but little or no phosphorous or potassium. *Reed sedge peat* consists of the remains of grasses, reeds, sedges, and other swamp plants. It should not be used for propagation purposes. *Peat humus* is in an advanced state of decomposition that the original plant remains cannot be identified. It is dark brown to black in colour and has a low moisture holding capacity. It has 2.0 to 3.5 per cent nitrogen. When peat moss is to be used in mixes, it should be broken apart and moistened before adding to the mix. Peat is not a uniform product and can be a source of weed seed, insects, and disease inoculums. So peat should be pasteurized along with other media components.

Sphagnum Moss Peat: Commercial sphagnum moss peat or sphagnum peat is the dehydrated young residue or living portions of acid-bog plants in the genus Sphagnum, such as *S. papillosum*, *S. capillaceum*, and *S. Palustre*. It is the most desirable peat for horticultural purposes, but its high cost limits its commercial use. It is relatively pathogen free, light in weight, and has a very high water holding capacity, being able to absorb 10 to 20 times its weight of water. Sphagnum moss has a pH of about 3.5 to 4.0. It may contain specific fungi static substances, including a strain of *Streptomyces* bacteria, which can inhibit damping off of seedlings.

Vermiculite: It is a micaceous mineral that expands markedly when heated. It is very light in weight, neutral in reaction with good buffering properties, and insoluble in water. Vermiculite has a relatively high cation exchange capacity and thus can hold nutrients in reserve for later release. It contains magnesium and potassium, but supplementary amounts are needed from other fertilizer sources.

Perlite: It is a gray white, silicaceous material, of volcanic origin, mined from lava flows. Perlite holds 3 to 4 times its weight of water. It is essentially neutral with a pH of 6.0 to 8.0 but with no buffering capacity. Unlike vermiculite, it has no cation exchange capacity and contains no minerals. Perlite, in combination with peat moss, is a very popular rooting medium for cuttings.

Compost: It is a product of biological decomposition of bulk organic wastes under controlled conditions, which takes place in piles or bins. The process occurs in three steps: (a) an initial stage lasting a few days in which decomposition of easily degradable soluble materials occurs; (b) a second stage, lasting several months, during which high temperatures occur and cellulose compounds are broken down; and (c) a final stabilization stage when decomposition decreases, temperatures lower, and microorganisms re-colonize the material.

Coco-peat (coir pith, coir fibre pith or coir dust) is the binding material of fibre fraction of coconut husks. It is obtained as a bi-product during the processing of coconut husks for extracting coir fibres. Coco-peat can be seen as heaps near to the coir producing firms and it is available as free or at cheap rate. Potting mixture, the common medium for growing plants, is prepared by mixing soil, sand and farmyard manure (compost) in 1:1:1 proportion. Coco-peat may be added to an extent of 20 to 50 % by volume to the potting mixture. It can be used for both indoor and outdoor purposes and the pH ranges from 5.5 to 6.5. Dried Coco-peat is available commercially as bales, blocks, bricks etc. which can be re-hydrated into bulk form and easily incorporated into potting mixture. The whole sale suppliers may mix coco-peat in sufficient quantity before selling to nurseries. Nursery owners can also blend coco-peat with potting mixture according to need.

The composition of some of the commonly used potting mixtures is given below:

For seed sowing: Seeds contain their own nutrients to get going so germinate successfully in low nutrient material.

Mix: One part loam (sieved and pasteurised) + one part leaf-mould, composted bark, or coir + One part sharp sand or vermiculite.

For seedlings: Seedlings need potting on as they start growing. Transplants need more nutrients than seedlings to keep growing strongly, but must also grow sturdily, so don't need excessive nutrients.

Basic mix: One part loam + One part leaf-mould + One part home-made compost.

High nutrient mix: One part worm compost or sieved + well rotted manure + Three parts leaf-mould, composted bark or coir.

For cuttings: The mixture should have very good drainage so cuttings don't rot and should contain some nutrients for early growth.

Mix: One part home-made compost, leaf-mould + composted bark or coir + One part horticultural grit, sharp sand, or vermiculite.

For large containers: Plants growing for a long time in the same container, e.g., fruit trees and bushes, need a good balance of slow-release nutrients.

Good basic mix: One part loam (sieved if needed, not pasteurised) + One part home-made compost (substitute manure for compost for higher nutrients).

2. Leaf and Leaf Bud Cuttings

Propagation through leaf bud cuttings is partially useful in species where leaves develop root system but die because of non-development of shoot system. Leaf bud cuttings are particularly useful when planting material is scarce because the each node in leaf can be used as cutting. Leaf bud cutting should preferably be prepared during growing season. A leaf bud cutting consists of a leaf blade, petiole and shoot piece of stem with attached axillary bud of actively growing leaves. In leaf bud cutting, 1-15cm stem portion is used when propagating material is small. It is useful method of propagation in blackberry, raspberry, lemon, camellia etc.

c. Layering

Layering is the development of roots on a stem while it is still attached to the parent plant. The rooted stem is then detached to become a new plant growing on its own roots. Thus rooted stem is known as layer.

Classification of layering

I. **Ground layering:** 1) Tip layering 2) Simple layering 3) Trench layering 4) Compound or serpentine layering 5) Mound layering or stool layering.

II. **Air layering:** (Gootee or Marcottage).

1. Tip layering: It is generally followed in plants which have trailing type of shoots. It is quite similar to simple layering.

Procedure: Dig a hole 3 to 4 inches deep. Insert the tip of a current season's shoot and cover it with soil. The tip grows downward first, then bends sharply and grows upward. Roots form at the bend. The re-curved tip becomes a new plant. Remove the tip layer and plant it in late fall or early spring. Examples of plants propagated by tip layering include purple and black raspberries, and trailing blackberries.

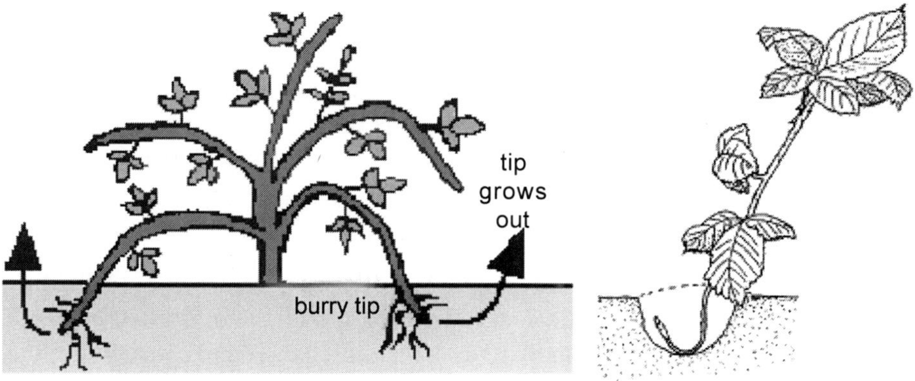

2. Simple layering: In this method, a branch is bent to the ground and some portion of it is covered by soil leaving the terminal end of the branch exposed. Root initiation takes place at the bend and buried portion. After allowing sufficient time for root formation, the rooted stem is separated from the mother plant. .

Procedure: Select a healthy, flexible and sufficiently long (50 to 60 cm) branch towards the base of the plant. The selected branch should be closer to the ground. At a distance of about 15 to 30cm back from the tip give a sharp, slanting inward and upward cut 1.5 to 2.5cm below a node and insert a small wood splinter. Bend the shoot gently to the ground so that the treated part can conveniently be inserted into the soil. Cover the treated region with soil. Peg down the shoot or keep a stone of brick on the covered soil to keep the layered shoot in place. Drive a vertical stake into the soil by the side of the layered branch and tie the terminal portion of the branch to keep it upright. Water the layered portion regularly so as to keep it moist all through till root initiation take place. After sufficient root formation separate the layer by cutting just below the rooted zone. This method of propagation is generally followed in grapes and lemon.

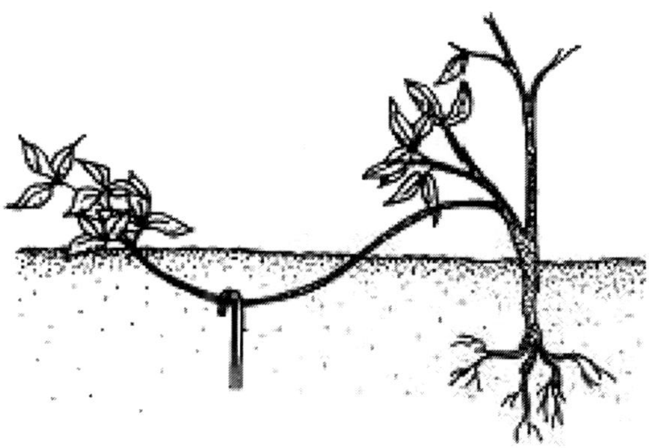

3. Trench layering: Trench layering consists of growing a plant or a branch of a plant in a horizontal position in the base of a trench and filling in soil around the new shoots as they develop, so that the shoot bases are etiolated. Etiolated roots develop from the base of these new shoots. Trench layering is used primarily for woody species difficult to propagate by mound layering. Trench layering is used for raising apple rootstocks (M 16 and M 25), cherry, quince, plum.

Procedure: Dig small trenches of about 25-30cm deep and in about 1 m wide rows. Plant rooted layers or one year old nursery – budded or grafted plants in the trenches in rows at an angle of 30 to 45°C and 50 to 10 cm apart within the row. The rows should be 1.2 to 1.5 m apart. Just before growth begins, lay the plant or a branch flat on the bottom of the trench. Plants must be kept completely flat with wooden pegs or wire fasteners. Cut back the shoots slightly and remove the weak branches. Add roots medium (sand or sawdust or peat moss) or their mixture at intervals to produce etiolating on 5 to 10 cm of the base of the developing shoots. Apply first 2.5 to 5cm layer before buds swell and repeat as shoots emerge and expand. At the end of the season, remove the medium and cut off the rooted shoots close to the parent plant. Transplant the rooted shoots in pots or suitable containers.

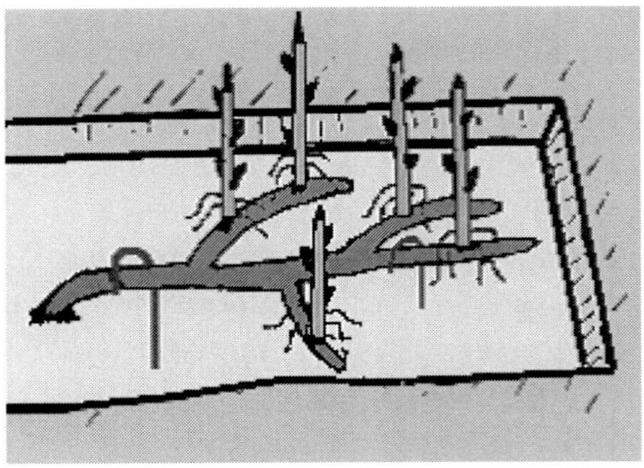

4. Compound or serpentine layering: Compound layering is essentially the same as the simple layering except that the branch is alternatively covered and exposed along its length. The branch for compound layering must be long and flexible so that it can be layered at different places along its length.

Procedure: Select a healthy, flexible and sufficiently long (100 to 250 cm) basal branch that is close to the ground. Give a sharp slanting, inward and upward cut 1.5 to 2.5 cm below a node at 30 cm interval starting from the tip leaving

3 to 3 buds in between two such cuts. Bend the shoot gently to the ground, and insert and cover the cut portions with the soil exposing the uncut portions. The remaining steps are same as in simple layering.

5. Mound (stool) layering: In this method, a plant is cut back to the ground level during the dormant season and soil is heaped around the base of the newly developing shoots. After showing sufficient time for root initiation, individual rooted layers are separated from the mother plant and panted. e.g., Apple roots stocks, Guava, Litchi, Quince,

Procedure: Select the plant to be mound layered or plant a rooted layer in a trench and allow it to grow for a year. Cut back the plant to 2.5 cm from the ground level just before growth begins. Allow the new shoots to develop. When these shoots have grown 7 to 15 cm tall, girdle them at the base and treat the girdle portion with the recommended growth regulator and draw up the loose soil round each shoot to half its height. When these shoots have are 20 to 25 cm tall add soil again to half their height. Add soil again when the shoots grow to a height of about 35 to 45 cm. Water the heaped soil regularly and allow sufficient time for the initiation of roots. A depression can be made in the centre of the heap to hold water. After sufficient root formation, remove the heaped soil and cut the rooted shoots individually to their base. Transplant the rooted shots in pots or suitable containers.

II. Air layering: In air layering roots form on an aerial shoot. The rooting medium is tied to the shoot for getting root initiation. Sphagnum moss is the best rooting medium for air layering as it holds large quantities of water till root initiation and root development. e.g., fig, guava, phalsa, pomegranate.

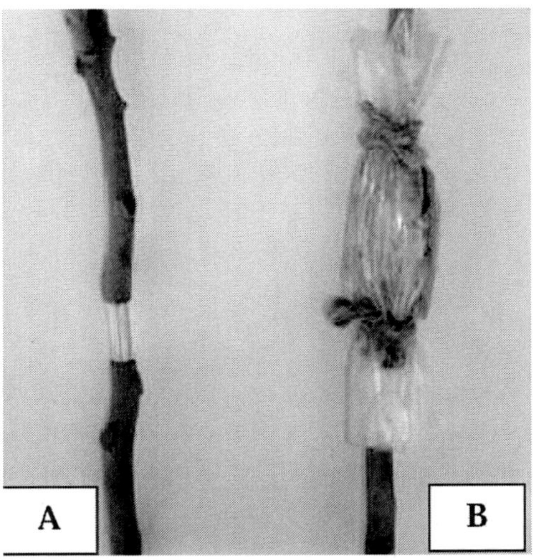

Procedure: Select a healthy branch of previous season's growth. At a point 15 to 30 cm back from the tip of the shoot make a girdle just below a node by

completely removing a strip of bark 2 to 3.5 cm wide all around the shot. Scrape the exposed surface lightly to remove traces a phloem or cambium to retard healing. In difficult-to-root species treat the girdled portion with the recommended growth regulator to induce better rooting. Cover the girdled portion with moist propagating medium *i.e.* Sphagnum moss, saw dust, vermiculite. Tie the medium around the girdled portion using a polyethylene sheet. Tying should be perfect so that no water can enter the treated part. After observing the fully developed roots through the transparent polyethylene sheet, separate the root zone and transplant the layer appropriately.

Steps showing removal (A) of outer bark of shoot, covering (B) with moss grass, wrapping with polythene & tying ends of polythene with rope.

d. Budding

Budding is a form of grafting in which one bud and a small section of bark with or without wood. In comparison to grafting, in which the scion consists of a short-detached piece of stem (short) tissue with several buds. Chip budding and T-budding are two most important types of budding for fruit crops and woody ornamentals. The type of budding methods to be adopted depend upon the bark's *slipping*, period of active growth in season and when newly formed tissues are easily torn as the bark is lifted from the wood.

Shield or T-budding: As the name indicates shield is the shape of the bud and 'T' is the shape of cut given on the rootstock. It is most common method of budding used by nurserymen worldwide. For shield budding one year old rootstock seedlings of 25-35 cm height and 2-2.5 cm thickness is selected. The bark of seedlings should slip easily. The selected bud of desired cultivar is inserted 15-20cm above the ground level and is tied with a polythene strip.

For performing budding operation, a "T" shaped cut is made on the selected portion of the stock with the help of a sharp budding knife. The incision should be given through the bark not the wood. The two flaps of bark are loosened with the help of budding knife. The healthy bud is removed from the bud wood by cutting shallowly about 5-6 mm below and 2-3cm above the bud. This shield piece containing a bud is inserted in the "T" cut made on the rootstock.

The shield should be covered by two flaps of the bark, but bud be exposed. The buds pressed firmly, fitted into the "T" cut and finally tied with polythene strip. When bud healing process is over, the bud may attain height of 15-20cm, the remaining portion of the stock is cut to about 10-15cm above the bud. Plants with thin bark, with sufficient flow of the sap like apple, pear, peach, plum and apricot, cherry, ber and citrus are propagated by this method.

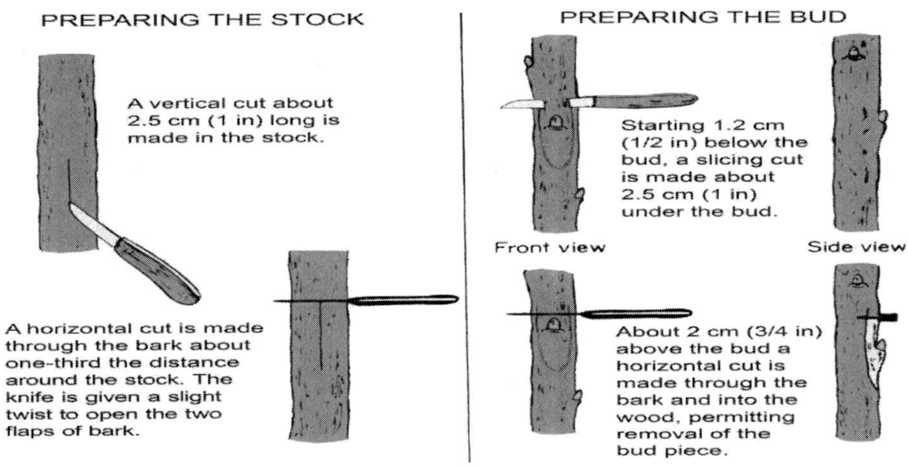

Patch budding: In case of patch budding, a rectangular patch of bark is removed completely from the rootstock and replaced with a patch of bark of the same size containing a bud of the cultivar to be propagated. It is slower and difficult to perform than T-budding. It is widely used in thick-barked species, such as walnuts and pecans and rubber tree, where T-budding gives poor results due to poor fit around the margins of the bud-particularly the top and bottom . It is usually done in late summer or early fall, but can be done in spring also.

In patch budding, the stock and scion should preferably of same thickness (20-25mm). First a rectangular piece of bark (25mm long and 10-15mm wide) is removed from the stock and a similar patch, containing a bud is removed from the scion by making two horizontal cuts above and below the bud and then two vertical cuts connecting the horizontal cut. After removing the patch, the bud should fit tightly at the top and bottom. It is then wrapped with polythene strip, keeping the bud uncovered. The wrapping material should hold the bark tightly and cover all the cut surfaces to prevent free entry of air or water. After the bud starts sprouting, the stock above the bud union may be cut off step by step. In addition to pecanut and walnut, mango, rubber plant, guava, aonla, jackfruit and Jamun are also propagated by this method.

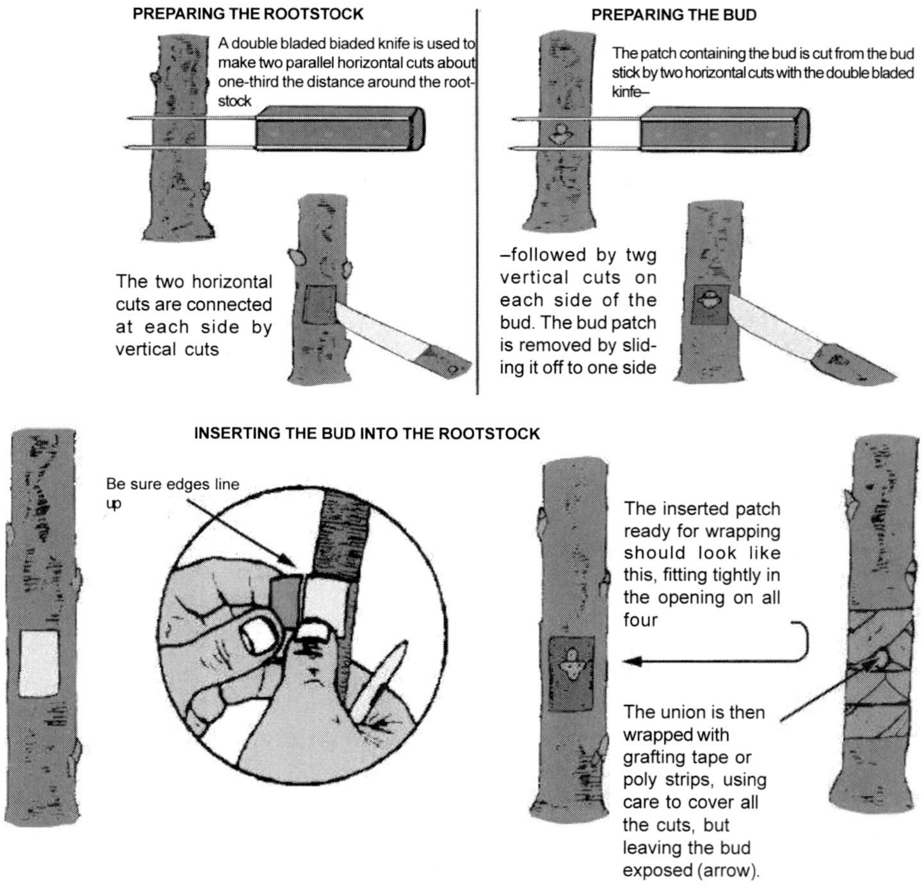

Ring or annular budding: In this type of budding, a complete ring of bark is removed from the stock and it is completely girdled. A similar ring of bark containing a bud is removed from the bud stick and is inserted on to the rootstock. The thickness of stock and scion should be same size. It has been utilized in ber, peach and mulberry because the newly emerged shoots from the heavily pruned plants are capable of giving such buds for budding, which can be easily separated. In this method since the stock is completely girdled and if the bud fails to heal in, the stock above the ring may eventually die.

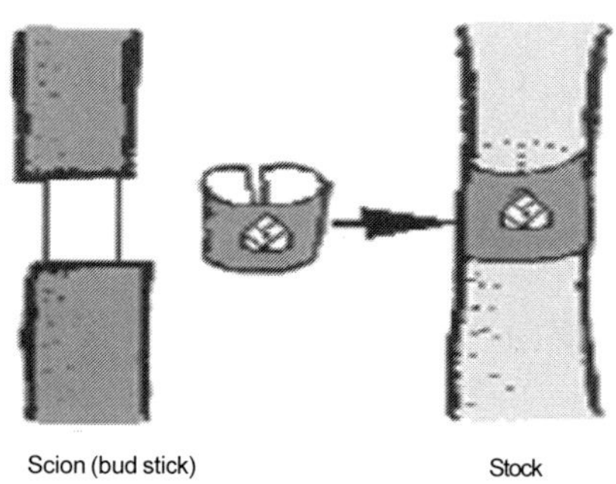

Forkert budding: In forkert budding, the stock is prepared by giving two vertical cuts and a transverse cut above the vertical cuts to join them. The bark is removed carefully along the cuts, so the flap of bark hangs down. The scion is prepared in a fashion similar to patch budding, having the size similar to cuts made on the stock. The scion is then slipped into the exposed portion of the stock and the flap is drawn over the inserted bud patch. It is then tied with a suitable wrapping material. After successful growth of bud, the portion of stock above union is removed carefully.

Flute budding: In flute budding, the bark patch from the stock is removed in such a way that it almost completely encircles the stock except with a narrow bark connection between the upper and lower cuts on the stock. A similar patch of bark is removed from bud stick containing a healthy bud. The shield containing the bud is then inserted in the vacant area of the stock; the shield should fit tightly on the stock. It is then wrapped with suitable wrapping material, leaving the bud uncovered. The other procedure is same as in patch budding. Because of presence of a narrow connecting strip of bark on the stock, it remains alive even if the bud fails to sprout.

I-budding: In I-budding the bud patch is cut in the form of a rectangle or square, just as for patch budding. With the same parallel–bladed knife, two transverse cuts are made through the bark of the rootstock. These are joined at their centre by a single vertical cut to produce the shape of letter-I. The two flaps of the bark can then be raised to insect the bud patch beneath them. A better fit may occur if the side edges of the bud patch are slanted. In tying the I-bud one should ensure, that the bud patch does not buckle outward and leave a space between the rootstock. I-budding is most appropriate when the bark of the rootstock is much thicker than that of the bud stick. In such cases, if the patch buds are used considerable paring down of the bark of the rootstock around the patch would be necessary.

Micro-budding: Micro-budding is used successfully for propagating citrus particularly in Australia. It is similar to "T" budding except that the shield (bud piece) utilized is thin and tiny like "T-budding"; the micro budding is also not done under aseptic conditions. The petiole is cut off just above the bud and then bud is removed from the bud stick by a flap cut just underneath the bud. Thus, only the buds are utilized in micro-budding. In stock an inverted "T" cut is made and the tiny shield containing the bud is inserted in it and later tied with a thin plastic tape. The tape may be removed soon (15-20 days) after the healing has taken place.

e. Grafting

Grafting is a process by which two living parts are joined together in such a manner that they would unite together and subsequently grow into a composite plant. Usually graft has two parts, the scion and rootstock.

Splice grafting: In this method, it is essential that both the stock and scion should be of equal diameter. For this , about one year old rootstock is headed back at a height of 20-25 cm from the soil and a splice(diagonal) cut 2.5 to 6 cm long) is made at the distal end of the rootstock with the help of a sharp knife. A similar slanting cut is made on the proximal end of the scion. The cut should be smooth and the cut surface of both the rootstock and scion are bound together and tied firmly with polythene strip. After the union has taken place, the rootstock above the union is loped off gradually. Sprouts arising below the grafting union must be removed at regular intervals to divert flow of metabolites for the growth of scion only. Splice grafting is done in early spring

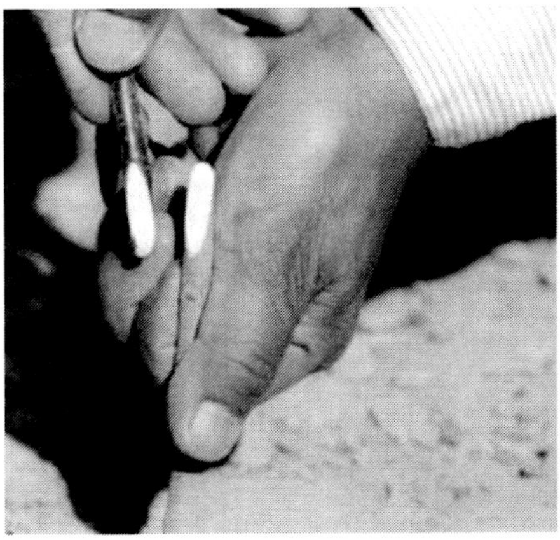

Tongue grafting: It is modified form of whip grafting. It differs from whip grafting that a reverse cut is made downward at a point about 1/3rd of the distance from the tip and should be about ½ the length of first cut. To obtain a smooth-fitting graft, the cut should not split. The rootstock and scion is then inserted into each other, with the tongues interlocking. It is important that vascular cambium layer match along at least one side, preferably alone both sides. The lower tip of scion should not overhang the stock as there is a likely hood of the formation of large callus knots. The use of scions larger than rootstock should be avoided. After the scion and rootstock are fitted together they are securely held by tying with budding/grafting tape or polythene. This method gives better success because of better cambial contact between stock and scion due to formation of tongue. Regular de-shooting of sprouts on stock is required to obtain better growth of scion. This method is performed in pear, peach plum and almond during dormant period *i.e* December-January.

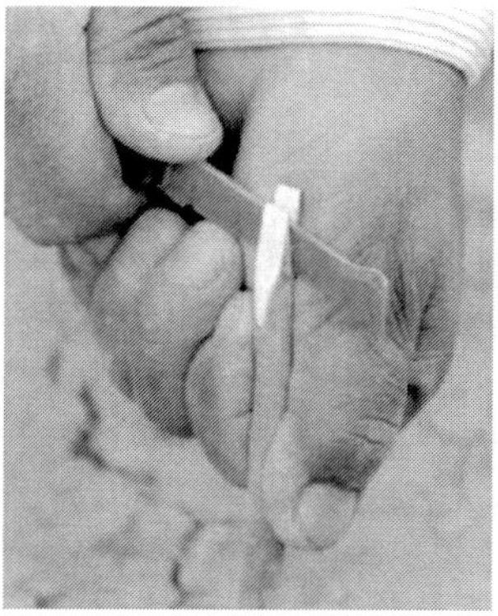
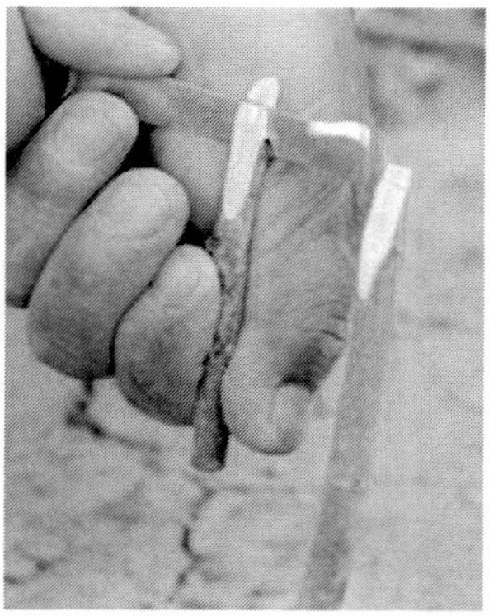
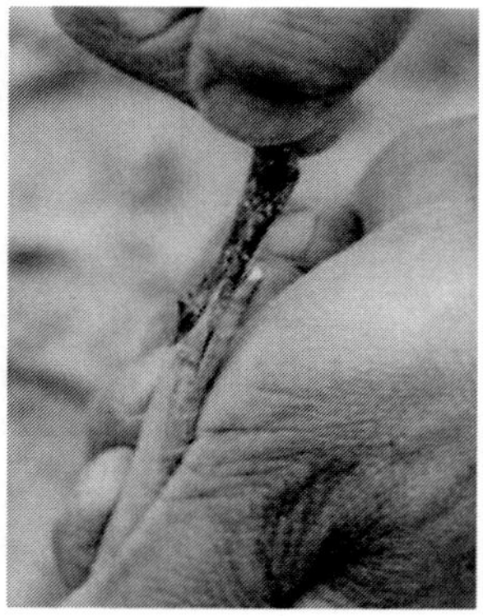

Cleft-grafting (wedge or split-grafting): It is one of the oldest methods of field grafting. It is used to top work trees, either in the trunk of a small tree or in the scaffold branches of a longer tree. In making the cleft-graft, a heavy knife is used to make a vertical split for a distance of 5 to 8 cm down the centre of the stub to be grafted. This split is made by pounding the knife in with a hammer. The branch is sawed off in such a way that the end of the stub which is left is smooth and free of knots for at least 15 cm. Two scions are inserted, one at each side of the stock where the vascular cambium layer is located. The scions should be 8 to 10 cm long and 10-15 mm in thickness and should have two or three buds. The side of the wedge which is to go to the outer side of the rootstock should be slightly wider than the inside edge. When the knife is removed, a hard wooden wedge is inserted to keep it open for the subsequent insertion of the scion. The graft should be thoroughly waxed to prevent wilting of the scion. The scion starts growing after 2 to 3 months of grafting. The right time for cleft grafting is the later part of the dormant season or just before the start of active growth. Walnut hazelnut, pecanut and grapes are propagated by this method. It is done during February- March. In pear, peach, plum and almond it is done during January-February.

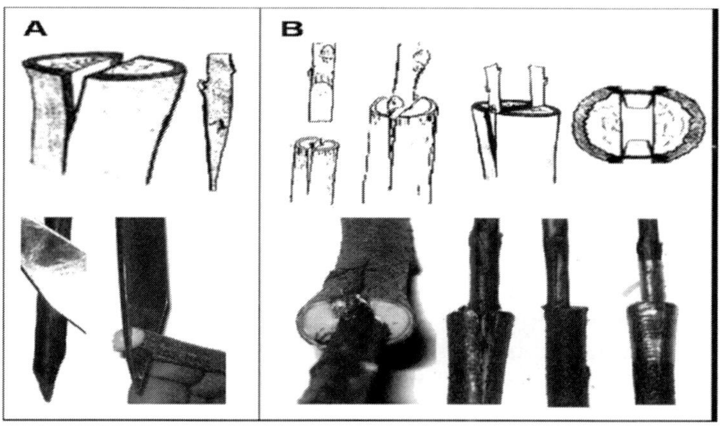

Wedge graft (saw-kerf grafting): It is performed in late winter or early spring before the bark begins to slip. A sharp, heavy, short bladed knife is used for making a V-wedge in the side of the stub or stock about 5cm long. Two cuts are made, coming together at bottom and as for apart at the top as the width of the scion. These cuts extend about 2 cm deep into the side of the stub. The base of the scion is trimmed to a wedge shape exactly the same size and shape as the opening. With the two vascular cambium layers matching the scion is tapped downward firmly into place and slanting outward slightly at the top so that the vascular cambium layers cross. After all scions are firmly tapped into place, all cut surfaces including the tips of the scion, should be waxed thoroughly. It is called raw-kerf because the cuts in the side of the rootstock can be made with a saw rather than the sharp knife.

Saddle grafting: The saddle grafting can be bench grafted by hand or machine. The rootstock and scion should be of the same size. The scion is prepared by cutting upward through the bark and into the wood opposite sides of the scion. The knife should penetrate more deeply into the wood as the cuts are lengthened. Before the scion piece, and the saddle shape is gradually formed by removing pieces of the wood. The rootstock is cut transversely and receives two upward cuts on either side. The apex of the rootstock is carved to fit the saddle so that vascular cambia of both stock and scion match. The graft is tied and cut surfaces are sealed by polythene until the graft-union is formed. It is used for bench grafting grape and rhododendron cultivars.

Side grafting: This method is used for grafting small potted liner plants such as seedling of deciduous trees, shrubs and fruit crops. A shallow downward and inward cut from 1 to $1^1/2$ inches long is made. At the base of this cut, a second short inward and downward cut is made, inserting the first cut, so as to remove the piece of wood and bark. The scion is prepared with a long cut along one side and very short one at the base of the scion on the opposite side.

These scion cuts should be of the same length and width as those made in the rootstock so that the vascular cambium layers can be matched as closely as possible. After inserting the scion, the graft is tightly wrapped with polythene sheet strips. After the union has healed, the rootstock can be cut back above the scion in gradual steps. Mango and peach and persimmon are commercially propagated by veneer grafting. July- September is ideal time of veneer grafting.

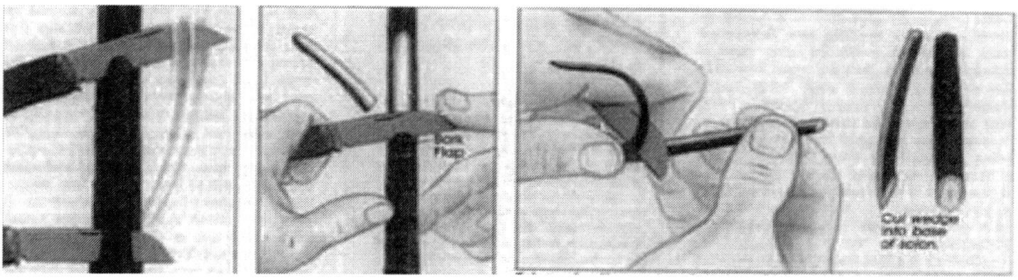

Side-Tongue grafting: In this method the cuts at the base of the scion is made as for the whip-and tongue graft. Along a smooth portion of the stem of the rootstock a thin piece of bark and wood of the same length, as the cut surface of the scion is completely removed. Then a reverse cut is made downward in the cut on the rootstock starting 1/3rd of the distance from the top of the cut. This second cut in the rootstock should be the same length as the reverse cut in the scion. The scion is then inserted into the cut in the rootstock, the two tongues interlocking, and the vascular cambium matching. The graft is wrapped tightly, to avoid desiccation. The top of the graft union is left intact for several weeks until the graft union is healed and may be cut afterwards above the scion gradually so that buds on the scion are forced into active growth.

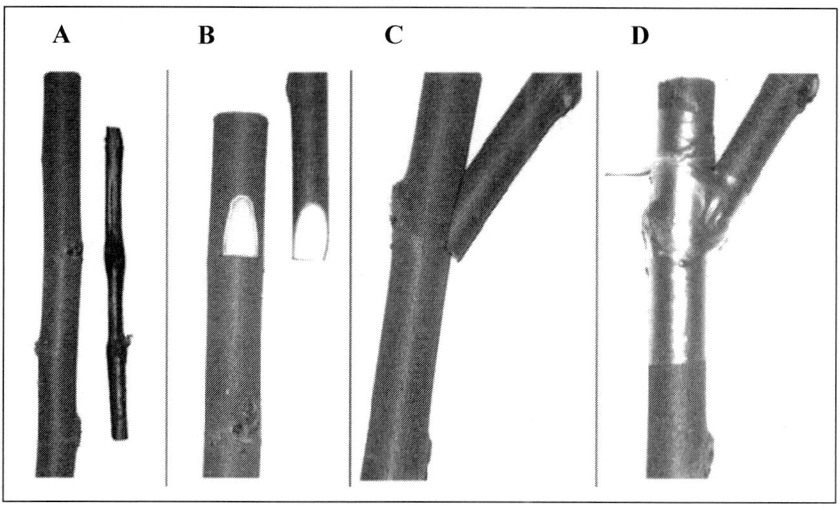

- The square cut is 2-3 cm long and 8-10 cm wide. The joining part is then tied. Nails are not used and the branch of the stock above the joining point should be cut off after the scion begins shooting. This method is most commonly used during the growing season.
- Jackfruit and mango are propagated by side grafting.

Bark grafting (Rind grafting): It is very simple and some nurserymen prefer it, because it can be done without splitting the stub, which prevents the entry of pathogens. It can be done in branches which are 25 to 30 cm long. It should be done in spring when the bark of the stock slips easily. It is important to note that the scion used in bark grafting should be dormant. The rootstock is first sawed off at a point, where bark is smooth. If the stock is thick many scions can be inserted. For each scion, bark is split downward, about 5 cm from the top of the stub. Scions of 10-12cm long containing 2-3 buds are prepared by giving a slanting cut (5cm) downward along one side of the base. On the other side a small cut is given. The scion is then inserted in the centre of split between the bark and wood of the rootstock. The longer cut of the scion is placed against the wood. The scion is hold firmly by using adhesive tape. Afterwards, all the exposed portions of the stub and scion should be covered with melted wax. The scions will start sprouting after 6-8 weeks in case of successful grafts.

Inarching: Inarching is similar to approach grafting in that both rootstock and scion plants are grafted on their roots at the time of grafting. It differs in that the top of the new rootstock plant usually does not extend above the point of the graft-union as it does in approach grafting. Inarching is used to replace roots damaged by cultivation equipments, rodents or diseases. It can be used to very good advantage in saving a valuable trees or improving its root system.

Seedlings (or rooted cuttings) planted beside the damaged tree, or suckers arising near its base are grafted into the trunk of the tree to provide a new system to supplement the damaged roots. The seedlings to be inarched into the tree should be spaced about 10-15 cm apart around the circumference of the tree if the damage is extensive. The seeding of compatible species is planted around the tree during dormant season and graft when active growth commences in early spring. Inarching will enhance growth of uninjured older trees.

A thin slice of bark (6-10 cm in length) at about 20 cm above the ground level is removed from the stock with a sharp knife. A similar cut is made in the scion. Thus, the cambium layers of both stock and scion are exposed. These cut are brought together and tied firmly with the help of polythene strip. After the successful union, stock above and scion below union are looped off gradually. In low rainfall areas, it should be done with the onset of rains, while in regions of heavy rainfall; it should be done soon after the rainy season is over, provided temperature does not fall below 15^{0}C. The method is followed in mango, guava, loquat, sapota and aonla. It is performed during end July to first week of August.

Bridge grafting: Bridge grafting is basically not a method of propagation but a form of repair grafting only in plants, which have been damaged either by frost, rodents or insects. It may, however, be kept in mind that bridge grafting is only helpful if the trunk is damaged but the root system of the plant is healthy.

In this method first, the damaged portion of the stock is cleaned and the irregular edges of the girdled area are cut evenly. The scion of desired variety is inserted in such a way that it is attached at both upper and lower ends into the living bark. Similarly, it is important that scions be inserted right side up to ensure the polarity. The exposed injured wood must also be covered otherwise; it may serve as an entry channel for decaying organisms.

Bracing: Bracing is a form of natural branch grafting that is used by fruit growers to strengthen scaffolding limbs of a tree in order to support the load of the fruit crop. In this method two strong, young lateral shoots from the limbs are to be braced. A rope or electrical cord is used to temporarily brace the larger limbs. The smaller shoots, which will naturally graft, are tied with waxed string or poly tape to keep them together.

Epicotyl grafting: Epicotyl grafting is a recently standardized method of propagation under South Indian conditions. Mango, Walnut, Jackfruit, Cashew nut are propagated by this method. It is done during August. This method is simple, economic and fast. Freshly extracted stones/seeds are sown in the nursery beds. After germination, 10-15 day old seedlings with tender stems and coppery leaves are lifted along with the stones. The roots and stones are dipped into 0.1 per cent carbendazim solution for 5 minutes after washing the soil. The seedling stems are headed back leaving 6-8 cm long stem. A 3- 4.5 cm longitudinal cut is made into the middle portion of the cut stem. A wedge shaped cut starting on both sides is made on the lower part of scion stick. The scion stick should be 4-5 months old and 10-15 cm long containing plump terminal buds. The scion stick is then inserted in the cleft of the seedlings and tied with polythene strip. The grafts are planted in polyethylene bags containing potting mixture. The bags are kept in the shade for protection from heavy rains. When the scion sprouts and the leaves become green, the grafted plants should be planted in nursery beds. July is the most suitable month for stone grafting.

Omega Bench grafting: Omega grafting was conducted using one-year old rootstocks and scions with a grafting machine. Graft unions were plunged into hot paraffin (70-80^{0}C) and then cooled in cold water. The graft unions were forced in woody boxes, filled with wet sawdust in a room under controlled conditions at 27^0 C and 80 % relative humidity.This method is also applied in walnut propagation.

Hot Callus Method: The experiment was set in an unheated green-house. Cleft grafting was carried out with one-year old rootstocks and scions. The grafted plants were tied with a plastic band and the grafting place was covered with foil. Thus prepared, the plants were placed horizontally, the place of grafting being located above an electric heating cable buried in a plastic U-shaped pipe filled with peat. After that their roots were covered with wet sand and wet foam was put over the place of grafting, pressed with weights to the plastic pipe in order to save the heat and to maintain the temperature of 27°C at the place of grafting. Periodically the roots were watered and the foam was wetted.

Green Houses for Propagation

The term Green House refers to a structure covered with a transparent material for the purpose of admitting natural light for plant growth. A greenhouse is a framed, inflated structure covered with a transparent material in which crops can be grown under at least partially controlled environment. A greenhouse is large enough to permit people to work within it and to carry out cultural operations. Various designs of greenhouse are constructed according to the need. They may be Quonset, Gutter Connected, Glass House, Plastic Film Greenhouses, Rigid Panel Greenhouses, and Greenhouse with Double-Layer Covering etc.

Types of Greenhouse

Shade Net House: A shade net nursery usually has 20 m x 10 m dimensions. It is erected using GI pipes as a support. UV stabilized HDPE shade net of 50 per cent shade intensity is used to cover the nursery area at a height of 6.5 feet. Wire grid is provided at the top of the structure as support for shade net. To prevent insect entry, 40 mesh UV stabilized nylon insect proof net is fitted on all the four sides of the nursery. Provision is also made to pull polythene sheet over the pro-trays in the event of rainfall by way of making low tunnel structure. For preparing low tunnel structure, 3/4" LDPE pipes and 400 gauge UV stabilized polyethylene sheet are used.

Plastic Film Greenhouse: These greenhouses are covered with flexible plastic films (made up of polyethylene, polyster or polyvinyl chloride). Plastic covering of greenhouse is popular because it is cheap and the cost of heating is less as compared to glass greenhouses, but such films have short life. For example, the best quality ultraviolet (UV) stabilized film can only last for four years. Quonset design as well as gutter-connected design is suitable for using this covering material.

Rigid Panel Greenhouses: Polyvinyl chloride rigid panels, fiber glass-reinforced plastic, acrylic and polycarbonate rigid panels are used as the covering material in this type of greenhouses. These panels can also be used in the Quonset type frames or Ridge and Furrow type frames. This material is more resistant to breakage. The light intensity is uniform throughout the greenhouse when compared to glass or plastic. High grade panels have long life even up to 20 years. The main disadvantage is that these panels tend to collect dust as well as to harbor algae, which results in darkening of the panels and subsequent reduction in the light transmission. There is significant danger of fire hazard.

Glass Greenhouses: The average cost per year of a Glass house is more than that of plastic film greenhouses. Several styles of glass greenhouses are designed to meet specific needs. A Glass house is preferred when the greenhouse is to be placed against the side of an existing building. It makes best use of sunlight and minimizes the requirements for roof supports. An even-span greenhouse is one in which the two roof slopes are of equal pitch and width. Glass greenhouse is seldom used today because they are not adaptable to automation. Individual greenhouses standing free of one another are well adapted to cold climates since snow easily slides from their roofs. A ridge-and-furrow design refers to two or more green houses frames connected to one another along their length. The side wall is eliminated between greenhouses, which results in a structure with a single large interior. Consolidation of interior space reduces labor, lowers the cost of automation, improves personnel management, and reduces fuel consumption because there is less exposed wall area through which heat can escape. Heating pipes are usually located beneath the gutters for this purpose.

Quonset-style Greenhouses: It is less expensive and can be purchased prefabricated or can be fabricated on the site. The trusses are constructed from water pipe that is bent to fit 180^0 and are modified for some-what more vertical sides. In a Quonset style greenhouse uses pipes which are 6 meter wide, 1.5 cm in diameter. One inch diameter pipe is used for a 30 foot wide greenhouse. Slightly larger pipe is driven into the ground into which the pipe arches are inserted for support. A 5 x 20 sq. cm wooden plant is attached to the base of the pipe arches such that it runs along the ground partially buried. This provides a basal point of attachment for the plastic film. The pipe arches or trusses are supported by pipe purling running the length of the house. Trusses are spaced 75 to 90 cm apart. The width of plastic film required to cover a Quonset greenhouse of given width can vary according to the height and shape of the trusses. A 6 meter wide greenhouse generally requires a meter wide sheet of plastic. The covering width for a 30 foot wide Quonset greenhouse varies greatly; the more common width is 12 meter. These green houses are either constructed in a freestanding style or may be arranged in an interlocking ridge-and-furrow manner. In this latter case, the trusses overlap sufficiently to place a bed of plants between the overlapping portions of adjacent houses. A single large interior thus exists for a set of houses, an arrangement that is better adapted to the movement of labour and to automation.

Gutter Connected Greenhouse: This is the second currently popular plastic film greenhouse design. The gutters can be placed at greater heights than is possible in ridge and furrow Quonset ranges. Gutters, depending upon the manufacturer can range between 3 to 10 meter intervals. Columns can be placed in greenhouses with gutters spaced 12 feet apart and the columns placed under each, every other or every third gutter. Nine meter spacing between rows of columns though expensive, greatly enhances the ease with which the shading of plants with black cloth can be accomplished. Gutter connected greenhouses greatly minimize the exposed surface area and consequently the heating cost. Only 6 meter of plastic film is required to span a 5 meter bay. Gutter-connected bays of 3, 5, 7, 8, and 9 meter widths can be covered by plastic film sheets 4, 6, 8, 10 and 12 meter wide respectively. When additions are made, the plastic film can be removed from an existing side wall and the new houses connected at that point without any resulting discontinuity. In this way, a modest initial investment, inadaptable for automation, can be developed through expansions into a structure well suited to automation. The gutter-connected green house brings us full circle to the category of permanent metal-frame greenhouses of which the glass greenhouse is a member.

LITERATURE CONSULTED

Abellanosa, A. L. and Pava, H. M. 1987. Introduction to Crop Science. CMU, Musuan, Bukidnon: Publications Office. 245 p.*Anal.*

Association of Official Seed Analysts. 1993. Rules for testing seeds. *J. Seed Tech.* 16:1–113.

Association of Official Seed Analysts.1999. Tetrazolium testing handbook, 29. *Assn. Offic. Seed* Anal.

Association of Official Seed Analysts. 2002. Seed vigor testing handbook, 32. *Assn. Offic. Seed*

Bal, J.S. 2003. Fruit Growing. Kalyani Publishers,New Delhi, ISBN 10: 9327239806, pp. 425.

Bose, T. K., Mitra, S.K., Sandhu, M. K. and Das, P. 1997. Propagation of Tropical and Subtropical Horticultural Crops. Naya Prokash, Kolkata.

Bryant, G. 1995. Propagation Handbook. Stackpole Books: Mechanicsburg, Pennsylvania.

Chandra, A.,Chandra,A. and Gupta,I.C.1994.Arid Fruit Research.Scientific Publishers,Jodhpur.

Gunjate, R.T and Tawde, A.B.1993. Propagation and rootstocks of mango. In: Advances in Horticulture, Vol. 2-Fruit crops (Eds., Chadha, K.L and Pareek, O.P). Malhotra Publishing House, New Delhi, India, pp. 515-522.

Hannah, M. 2002. Nursery stock production using the pot-in-pot technique. Department of Horticulture, Ohio State University.

Hartmann, H. T., D. E. Kester, F. T. Davies and R. L. Geneve. 1996. Plant Propagation, Principles and Practices. 6th ed. Prentice Hall: Upper Saddle River, New Jersey.

Khatiwada P. P. and Ghale, M. S. (eds.) 1998. Nepal Agricultural Research Council, Khumaltar, Second National Horticulture Research Workshop 13-15 May, 1998.

Peter, K.V., Rajan, S. and Baby L. Markose. 2006. Propagation of Horticultural Crops. (Eds.) New India Publishing Agency, New Delhi.

Rahudkar W.B., Bhujbal B.G., Sonawane, M. and Rajput, H. 2010. YCMOU, Textbook Publication No. AGR 227 Horticulture Nursery Management.

Ratha Krishnan, P., Kalia, R. K. Tewari, J.C. and Roy, M.M. 2014. Plant Nursery Management: Principles and Practices. Central Arid Zone Research Institute, Jodhpur, 40 p.

Reddy, B.M.C. and Shukla, S.K. Manual for testing of seed and planting material testing of fruit crops. ICAR-Central Institute for Subtropical Horticulture Rehmankhera, P.O. Kakori, Lucknow – 227 107

Resource Book on Horticulture Nursery Management Developed Under National Agricultural Innovation Project, Indian Council of Agricultural Research, New Delhi - 110012.

Salunkhe, D.K. and Kadam, S.S. 1995. Handbook of Fruit Science and Technology: Production, composition, Storage & Processing. CRS Press, USA, pp. 632.

Sharma, R.R. 2002. Propagation of Horticultural crops. Kalyani Publishers, New Delhi.

Sharma, R.R. and Srivastava, M. 2004. Plant Propagation and Nursery Management. International Book Distributing Co., Lucknow, India.

Westwood, M.N. 1988. Temperate Zone Pomology. Timber Press Portland, USA.

11

Propagation Techniques and Rootstocks

V.K. Wali, Amit Jasrotia and Deep Ji Bhat

INTRODUCTION

Plant propagation involves three main aspects: knowledge of mechanical manipulations and technical skills which constitutes the art of propagation; knowledge of plant growth and structures which constitutes science of propagation and knowledge of different kinds of plants and their methods of propagation. The basic objective of plant propagation is to increase the number of desirable types and to preserve the essential characteristics of the plant. Generally, sexual propagation or the seed propagation is used to raise the seedling rootstocks on which to graft or bud the desirable types. The clonal rootstocks are multiplied through vegetative means- layering, cutting etc. Various sexual and asexual methods such as use of seeds, apomicts, specialised plant parts, cuttings, layers, grafts and buds are employed to multiply plants of different fruit crops. In this chapter, the techniques used for raising rootstocks and multiplying the desirable types of different fruits have been discussed.

A. MAJOR FRUITS

MANGO

Mango is a highly heterozygous and cross pollinated crop. There are two types i.e. monoembryonic and poly-embryonic. Poly-embryonic varieties can be propagated through seed because they produce true to type seedlings where as the mono-embryonic varieties need to be propagated vegetatively. Inarching or approach grafting is the principal method of propagation followed by veneer grafting. Cuttings do not root successfully in open. There is limited success under constant growth regulators. Air layering with use of growth regulators is also reported to be of some success but did not reach commercial exploitation. Shield and patch methods of budding are practiced in Far East countries. It is not in vogue in our country. The method most commonly used in this country are inarching, side veneer grafting and epicotyl grafting.

***Raising of Rootstocks*:** Mango seedlings grown from stones of seedling trees are used as rootstocks. Stones should be collected from vigorous, disease free and high yielding trees during the harvesting season. Mango stones lose their viability very soon on desiccation. Stones are sown during in June- July. When the seedlings attain the age of 2-3 months, they should be transplanted in well prepared beds or polybags. Plant is allowed to grow as a single stem for six to eight months. Scion is ready for grafting when the plant attains pencil thickness or a little more. Now-a-days, high density planting is getting popular for which dwarf trees are demanded. To produce dwarfing effect it is necessary to select the correct type of rootstock. The rootstock such as Villaikolamban, Ambalavi are recommended for this purpose. The locally found wild mango stones may also be used as rootstock for the same purpose.

***Selection of bud wood/scion*:** The best bud wood is prepared from hardened terminal growth 6 to 10 mm in diameter. The leaves are removed, with the exception of two or three terminal ones. The buds swell in 2 to 3 weeks, and are then ready to use. If the buds are to be used on stocks older than 3 weeks, ringing the base of the shoots from which the buds are to be taken about 10 days before they are used increases their carbohydrate supply and seems to promote graft union formation. The scion should be a terminal non-flowering shoot of 3 to 4 months' maturity. The scions are defoliated 7 to 10 days before they are removed for grafting, keeping a part of the petioles attached, which helps the buds to swell. This defoliation is done from March to September in northern India.

***Methods of Propagation*:** Nurserymen in many of the mango growing areas still use inarching, traditional method of propagation. During past few decades, experimental results have shown that veneer grafting technique can be used with high success rate in North India. Stone (epicotyl) grafting is suitable for Coastal regions. Recently, softwood grafting is being used commercially for mango propagation in several parts of south India. Side grafting and soft wood grafting techniques are being used for large scale multiplication of mango all over the mango producing area.

Epicotyl grafting: Seeds are germinated in a sand bed covered with leaf mould. 8-15 day old germinated seedlings are lifted, decapitated about 5 cm above the stone, and wedge-shaped scions are inserted into a vertical split. Buds are wrapped with poly tape, and grafted plantlets are planted in poly bags or outdoors in a trench, after which they are then shifted to the field.

Softwood grafting: The technique of softwood grafting is similar to that of cleft or wedge grafting. In this case, grafting is done on 3 month to 8 month old rootstocks. In south India, the rootstocks attain graftable thickness within 3-6 months due to mild winter. In the past, this technique was used for *in situ* orchard establishment under adverse soil and climatic conditions as the grafting operation is performed using cleft/wedge method on the newly grown top portion of the plant one year after the rootstock establishment in the field. The scion shoots of the thickness equal to that of rootstocks are defoliated 7-10 days prior to grafting. The graft should be secured firmly using 1.5 cm wide, 200 gauge polyethylene strip. July and August months with high humidity and moderate temperature are the best for the success of softwood grafting.

***Rootstocks*:** A full knowledge about the effect of using of the species of *Mangifera* and other genera of the family as rootstocks is not available. Other species of *Mangifera* like *M. foetida* and *M. odorata* have been found to show promise as rootstock. Kalaypad variety is said to be dwarfing stock and a wild mango variety Pulima (Ceylon) is considered as prolific rootstock. Polyembryonic seedlings which are uniform amongst themselves are recommended as rootstocks as they reduce variability due to the rootstock. Among them Pahutan and Olour are particularly good (Table 1).

Table 1: Characteristic features of mango rootstocks

Mango rootstock	Characteristic features
Kurukkan	Salt resistant and polyembryonic.
Olour	Vigorous rootstock
Rumani,Kalepad &Vellaikolamban	Dwarfing.
Moovandan and Nekkare	Salt tolerant.
Pulima	Prolific rootstock

CITRUS

***Propagation*:** Polyembryony occurs in seeds of most citrus species used as rootstocks due to nucellar embryony. The sexual seedlings present within the embryo tend to be weak, variable, and are usually rogued out. The apomictic seedlings, which arise from the nucellus, are usually uniform and have the same genotype as the seed tree. As commercial plants, nucellar seedlings, not only have a long juvenile period but are also vigorous, thorny, upright-growing, slow to come into bearing, and are undesirable as an orchard tree. Vegetative propagation methods like budding are practiced. Shield budding or T-budding is practiced all over the world. The varieties to be propagated should be well known for their good performance and free from diseases. The rootstock should be compatible with the scions and well adapted to the soils of the regions where tress are intended to be grown.

***Raising of Rootstock Seedlings*:** The vigour and uniformity of nucellar seedlings makes them ideal as rootstocks, and most rootstock species or cultivars are nucellar. Citrus seeds generally have no dormancy but are injured if allowed to dry. Consequently, seeds should be planted immediately after being extracted from the ripe fruit. Seeds may be stored moist, in polyethylene bags at a low temperature 4°C; before storage they should be soaked for 10 minutes in water at 49°C to aid in eliminating seed-borne diseases. Treatment of the seed with a fungicide is beneficial. Freshly extracted seeds from mature fruits picked from healthy, vigorous and good yielding trees are sowed on raised beds during July-Aug. The slimy substance on the seeds is removed by mixing them with wood ash and drying them under shade. The seeds take 20-30 days to germinate and in another month, a pair of leaves emerge. After 3 or 4 months the plants reach 60cm in height. The seedlings are transplanted late the nursery beds under shade and spaced 20cm apart and the seedlings will be ready in a year.

***Rootstocks*:** The stock and scion must be compatible and should be capable of producing long lived, productive trees. The influence of the most important and widely used rootstocks and their characteristics (Table 2) should be known to the grower for choosing the right kind of stock for the locality. The most commonly used stocks for sweet orange as well as mandarins in India are Jambheri (Rough Lemon) and Karna Khatta.The selection of proper rootstock is of utmost importance.

Table 2: Characteristics of some of the important citrus rootstocks

Rootstock	Characters
Rough lemon	Suitable for light (sandy) and infertile soil, resistant to tristeza virus. But it is most susceptible to cold and induces poor quality fruit
Sour orange	Suitable for silty, clayey and poorly aerated soils , improves the fruit quality and has high sugar and acid content, improve the fruit size
Cleopatra mandarin	Suitable for heavy soils
Trifoliate orange	Cold tolerant, improves the fruit quality and has High sugar and acid content, hasten fruit maturity and the fruit size. It also induces precocity. Resistant to Citrus nematode (*Tylenchulus semipenetrans*). It is ideal for high density plantations.
Sweet lime	It gives high yield in early age followed by reduction in the yield later on. Also it is most susceptible to cold and induce poor fruit quality
Sour lime	Most susceptible to cold
Rough lemon (Jambheri)	Vigorous rootstock. Induce poor fruit quality and resistant to tristeza virus and drought and salt tolerant

Contd.

Rootstock	Characters
Sweet lime	Give high yield in early age followed by reduction in the yield
Citranges	Resistant to citrus nematode (*Tylenchulus semipenetrans*)
Rangpur lime	Tolerant to Phytophthora foot rot and resistant to Tristeza virus, Drought and salt tolerant
Sweet orange	Resistant to tristeza virus

***Selection of bud wood*:** Source selection of cultivars and rootstocks is essential in modern citrus nursery production. About 12 known viruses, a viroid causing "exocortis," and a mycoplasma-like organism causing "stubborn" disease can infect citrus. The most serious worldwide problem is *tristeza,* causing "quick decline" and/or "slow decline" in specific cultivar-rootstock combinations. Citrus sometimes produce low-producing, inferior "bud-sport limbs," so it is desirable to select bud-sticks from a single tree, avoiding any off-type "sporting" branches.

Buds should be collected from a tree or trees of the desired variety. Bud sticks are commonly selected from the next to last growth flush (the wood behind the current flush) and from the current growth flush after it has matured and hardened. Older growth flushes can be used if the bark still is green. Round twigs about the size of a pencil are preferred. The buds located in the axils of the leaves (where the leaf is attached to the wood) should be well developed, but still dormant. After the bud wood is cut from the tree, the undesirable wood and/or growth flush should be removed and the remaining bud wood should be trimmed to lengths of 20-25 cm (8-10 inches). The leaves should be cut off leaving a stub of the petiole 3-4mm (1/8 inch) long to protect the buds. Trimmed bud sticks should be labeled and used immediately or placed in plastic bags in a cool place. Include a moist paper towel to maintain turgidity and freshness. The label should include the variety, date of collection and source. Bud sticks are usually tied in bundles for ease of handling.

***Budding*:** The seedlings are ready for budding in about a year and budding along with a chip of wood gives good results. In mandarin the removal of wood seems to decrease the percentage of success. T-budding or Shield budding is easy to perform. Citrus plants are T-budded during March-April or August-September in North India

Sweet Orange: It is propagated by Shield or T- budding. The rootstocks mainly used are Jambheri and Rangpur lime. Cleopatra for Blood Red and Pectinifera for Mosambi are recommended in Punjab. In Andhra Pradesh mainly Rangapur lime is used as a rootstock.

Acid Lime: It is commonly propagated by seed all over the country because the seeds exhibit a high percentage of polyembryony. The nucellar seedlings are identical with the parent in growth and production. These seedlings are hardy compared to the budded plants. The seedlings take a little longer time than budded plants to start bearing but soon over take them in yield. If budded, Gajanimma is a prolific stock but there is scope for the spread and transmission of diseases and the budlings are short-lived.

Lemons, Grapefruit and Pummelo: Italian lemon which is seedless, it is propagated by budding on Jambheri in the Deccan region. For lemons the rootstocks found suitable in Punjab and UP are Rough lemon and Kharna Katta.Pummelo is comparatively monoembryonic and hence budded plants of pummelo are usually raised on the rootstocks of Jatti Khatti, Jambheri etc. Similar is the case of Grapefruit. Pummelo as a rootstock is found compatible with grapefruit.

Mandarins: Mandarins are largely propagated by seed all over the country except Nagpur mandarin which is the only budded variety grown on a commercial scale around Nagpur (Vidharba) in Maharashtra producing the finest mandarins of the world. Rough lemon (Jatti Khatti) is ideal rootstock for kinnow mandarin.

Micro-propagation: Citrus can be micro-propagated, but it is not a commercial method. This method is primarily used for production of virus-free clones through micro-grafting procedures.

GUAVA

Guava can be propagated both by seed as well as vegetative means. Among the asexual methods viz., cuttings, air layering, stooling, grafting and budding are important.

Raising of Rootstocks: Seeds of L-49 (Sardar)) are commonly used for raising rootstock. Portugal rootstock is most suitable for Sardar and Allahabad safeda varieties as it imparts vigour and wilt tolerance to these scion varieties. Raising rootstocks in polyethylene bags is recommended as this gives better establishment of plants in the field on account of undisturbed tap root system. Guava seeds have a hard coating over the endocarp as a result of which usually long time is required for germination. Fresh seeds should be extracted from fruits and washed thoroughly to remove the pulpy material clinging to the seeds. It should be treated with fungicide (0.03 % copper oxychloride) before sowing in the polyethylene bag to prevent damping-off of seedling. Seeds of guava are sown in polyethylene bags of 20x10 cm or 18x27 cm size during any time of the year. Polyethylene bags are filled with soil, sand and Farm Yard Manure in 3:1:1 ratio. All the polyethylene bags are covered with 100 micron (400 gauge) white polyethylene sheet soon after sowing of seed. During winter months, the polyethylene mulch conserves heat and create conducive environment (micro-climate) for rapid germination and early establishment of seedlings.

Methods of propagation in guava

Air layering: Air layering is one of the most important commercial methods in practice for guava propagation. Rainy season (preferably July-August) is the most suitable period for air layering.

Approach grafting/Inarching: Inarching or approach grafting is the oldest technique employed in many guava growing areas. This method gives very high success (85 to 95%) during rainy season. The procedure of this method is discussed in detail in Chapter 10.

Wedge grafting: The technique involves growing of seedlings in polyethylene bags, grafting, capping and hardening of grafts. Seedlings for rootstocks are allowed to grow in the nursery for approximately 6 to 8 months. When the stem diameter of seedling is of pencil thickness (0.5-1.0 cm) they are chosen for wedge grafting. In this technique, proper selection and preparation of scion sticks is very important for obtaining higher success. 15-20 cm long shoots (3-4 months old) with 3-4 healthy buds and of pencil thickness should be selected for grafting. Selected scion shoots are defoliated on the mother plant, about a week prior to detaching. At the same time, the apical growing portion of selected shoot is also beheaded. This helps in forcing the dormant buds to swell. In this way, the buds on the scion are ready to start sprouting at the time of grafting. This treatment is essential for high success of grafts. After selection of the scion, rootstock (seedling) is headed back by retaining 15-20 cm long stem above the soil level in the polyethylene bag. The beheaded rootstock is split to about 4.0 - 4.5 cm deep through the centre of the stem with grafting knife. A wedge shaped cut, slanting from both the sides (4.0 - 4.5 cm long) is made on the lower side of the scion shoot. The scion stick is then inserted into the split of the stock and pressed properly so that cambium tissues of rootstock and scion stick should come in contact with each other. Care must be exercised to match the cambium layer of the stock and scion along with full length of each component. The union is then tied with the help of 150 gauge polyethylene strip, 2 cm in width and 25 - 30 cm in length. Immediately after grafting, the graft is covered by 2.5 x 18.0 cm long white polyethylene cap which is tied with rubber band at the lower end. The scion starts sprouting after 10 to 12 days. The cap is removed after 25 days in the evening hours. The grafts are transferred to net house for hardening.

Stooling: It is the easiest and cheapest method of guava propagation which has been standardized at Indian Agricultural Research Institute (IARI), New Delhi. In this method plants are headed back to ground and shoots are allowed to develop on the beheaded stumps. When shoots attain a size of 20-25 cm, these are girdled at the base by removing a bark ring of 2- 2.5 cm. Indole Butyric acid (IBA) @ 2,500 ppm in lanolin is applied to the upper portion of the girdled shoots and shoots are left open for 2-3 cm days after IBA application.

These shoots are covered/ mounded with soil to encourage rooting in the stools. After 6-8 weeks, the roots are fully developed in the shoots and these shoots can be separated and planted in nursery. The shoots are headed back to maintain the root and shoot balance before planting in the nursery. Before separating the stools, a light irrigation should be given if soil is dry. After separating the shoots apply ample quantity of manures and fertilizers and level the field. Stooling can be done twice a year (spring and rainy season) and in this way, one can get 30-40 stools per mother plant per year. The stooling of a mother stool can be done for many years. Stooling is the easiest and cheapest method of guava propagation. The self-rooted plants (cuttings or layers) are planted 0.5m apart in the stooling bed. These are allowed to grow for about 3 years. Then these are cut down at the ground level in March. New shoots emerge on the beheaded stumps. A 30 cm wide ring of bark is removed from the base of each shoot rubbing the cambium of the exposed portion in May. All the shoots are mounded with the soil to a height of 30cm. The soil is covered with mulch to conserve the moisture. After a period of 2 months of the onset of monsoon, the shoots are detached from the mother plant at ringed portion and planted in the nursery. The shoots are headed back to maintain the root and shoot balance before planting in the nursery. By following the technique of ringing and mounding of the shoots, second time stooling is done on the same mother stools in first week of September. The rooted stool layers are detached in first week of November. Thus stooling is done twice on the same mother stools in a year. The stooling of a mother stool can be done for many years. With the advancement in its age, the number of stool layers also increases every year. The growth and development of stool layers are better than seedlings. The application of rooting hormone is not required. A semi-dwarfing rootstock for guava Aneuploid 82 has been developed

Budding: In northern India, guava is propagated by inarching, giving a very high percentage of success during rainy season. But inarching is cumbersome and, gives limited number of plants from the mother plant. Budding has been adopted only on a limited scale in some parts of the country where the atmospheric humidity is high. The main problem encountered in this method is disbudding of rootstock making it labour-intensive. Among the various methods of budding shield, patch and chip-the patch budding is ideal giving highest percentage of success. However, the best time of budding differs from locality-to-locality.

In areas of high atmospheric humidity budding could be practiced. Different budding techniques could be employed like forkert, shield and patch. Patch budding can be done during May –June with 75-80 per cent success. For patch budding seedlings of pencil thickness and about one year age with uniform active growth are selected. The trees from which buds are taken should be highly vegetative with lush succulent growth to permit easy separation of buds from the stem. It is better to take swollen and unsprouted dormant buds from leaf axils of mature twigs of the scion variety. A patch of approximately 1-1.5 cm length with a bud is removed from the scion variety. Similarly, 1-1.5 cm long patch is removed from the rootstock and bud is fitted into the remaining portion on the stock seedling. Bud should be fitted at a height of nearly 15-20 cm above the ground level. Polyethylene strip is used for keeping the buds close to the stock. When the bark adheres tightly to the wood, budding is usually successful. After about 2-3 weeks of budding the polyethylene strip can be opened to examine the success. In successful cases, about one-third shoot of the rootstock can be removed for forcing the growth of buds. The remaining two-thirds can be removed after three weeks of the first cutting, leaving about 2-3 cm above the bud.

In Odisha, air layering or goottee is the most common propagation method of guava due to better root emergence and survival rate (80-85%). Besides, it produces sufficiently large planting material which can be planted within 3-4 months. Wedge grafting is also practiced up to some extent to take the advantage of well developed root system of rootstock.

Micro-propagation: Guava can be micro-propagated with explants from mature trees.

APPLE

Apple plants are commercially raised by vegetative means, since seedling plants are not true to type. Apple is generally propagated by budding or grafting on seedlings of crab apple, Golden Delicious etc. Shield budding is done with buds of current seasons growth in June. Among grafting methods tongue grafting is the ideal method. Grafting is done in spring. Apple has been propagated mainly on seedling stocks of apple (*Malus pumila* Mill). Clonal rootstocks of M and MM series were introduced from East Malling Research Station of U.K. and M-9, M-26, M-4, M-7, MM-106 and MM-111 were identified as promising. Merton 779 was recommended as a commercial rootstock for apple for Kumaon hills of U.P. However, clonal rootstocks have not found favour among the apple growers. *In-vitro* protocols of M-9, MM-106 and MM-111 have been developed. Clonal rootstocks have been preferred only for high density orchards. Four rootstocks, namely, EMLA-111 for drought prone areas, EMLA-7 for slopy land and EMLA-106 for sloppy and less clayey soil and EMLA-9 for high density planting on irrigated deep soils have been recommended.

Propagation of Rootstocks: The apple plantations are raised on seedling rootstocks. The use of clonal rootstocks has not been commercialized despite the established superiority to raising uniform plantations, precocity and high productivity in Himachal Pradesh, Jammu and Kashmir and Uttarakhand. Non-availability of clonal material in the absence of commercial mass propagation techniques has remained the major constraint. Apples are propagated on seedlings of crab apple or self pollinizing varieties, Golden Delicious, McIntosh, Yellows Newton and Northern Spy having good seed viability, germination percentage and seedling growth. The seeds of commercial varieties from fruit juice canning units are also used by nurserymen for raising seedling stocks. However, the seedling stocks lack uniformity in tree size and productivity but show better adaptability to sloppy and shallow soils under rain-fed conditions.

***Raising of Seedling Rootstock*:** Apple seeds need stratification in moist sand at 4-7°C for 60-90 days. The water soaked seeds are placed between 2 and 3 cm thick layers of moist sand in wooden boxes or polythene bags during December. The stratification boxes of bags are placed in cool place where the required chilling temperature of less than 7°C for 1000-1500 hours is met in 60-90 days. The sand is kept moist during stratification. The stratification can be accomplished in the lower chamber of the refrigerator also. The stratification requirement is also met with in areas having very cool winters by direct sowing of seeds in nursery beds during November-December. The pre stratified seeds, as indicated by whitish tip at the micropylar end, are sown during February-March in well-developed, raised beds. Sowing is done in rows 10 cm apart with the spacing of 5cm between the seeds and 2-3cm deep. Nursery beds should be irrigated after sowing and covering with 10cm hay or pine needle mulch to protect the germinating seedlings from heavy rains and spring frost. Mulch when the seedlings are 5cm long for the straight upright growth of seedlings. One-year old seedling stocks are ready for grafting during February.

Clonal Rootstocks

Malling and Malling-Merton series: This rootstock selection program, which began in 1912 at the East Malling Research Station in England, included both a program of rootstock selection and standardization of propagation. Size of grafted cultivars range from very dwarfed to very invigorated, although size control was also a function of the scion cultivar. Fruit size was large, especially on young dwarfed trees, often larger than on standard-sized trees. The Malling rootstocks are compatible with most apple cultivars, and trees grafted to these rootstocks have been planted in varying amounts all over the world. In 1928, a second phase of this program was instigated jointly by the John Innes Horticultural Institution and the East Malling Research Station. A new series of apple rootstocks, referred to as Malling-Merton (or MM), was produced that added resistance to woolly aphids, and to provide a further range in tree vigor. Other improved cultivar characteristics associated with these rootstocks include high yield, precocity of flowering (with some rootstocks), well-anchored trees (with some rootstocks), freedom from suckering, and good propagation qualities. All of the Malling and Malling-Merton stocks are readily propagated, mostly by stool-bed layering. A number of rootstock clones, notably 'Malling 26,' 'MM 106,' and 'MM 111,' propagate well by hardwood cuttings.

In India, the size controlling Malling (M) and Malling Merton (MM) series clonal rootstocks were introduced at the Regional Horticultural Research Station, Mashobra and Temperate Fruit Research Station, Kotkhai, in Himachal Pradesh, Government Hill Fruit Research Station, Chaubattia in Uttarakhand and Fruit Research Station, Shalimar in Jammu and Kashmir during late sixties. The common method of propagation of clonal rootstocks of apple is mound layering (stooling). The rooted layers of the clone are planted in well-prepared stool beds during winter at a distance of 30cm in the row and 60cm apart. The 3-4 years old layers give rise to numerous suckers during spring. The suckers are covered with the soil before monsoon. The suckers are ringed or notched near the base during the rainy season and covered with soil to encourage rooting. The difficult-to-propagate rootstocks like M 9 need the treatment of 1000-2000 ppm IBA at the notched portion for quick root initiation. The rooted layers are separated at the onset of dormancy (December) and lined out in nursery beds for further grafting with scion varieties during February-March. The rootstock should be healthy and disease-free and should attain the thickness of 0.9-1.25cm at grafting height for proper bud-take success. The growth of the rootstock should be straight and upright with proper root development. There should be no gall, knot or injury scar at the grafting height. The important features of apple rootstocks are given below in Table 3.

Table 3: Salient features of some rootstocks of apple

Rootstock	Description	Eventual size (foot)
M 9	Very common commercial rootstock that generally should not be used by homeowners, due to the maintenance required. Trees grafted onto this rootstock require staking or support for the life of the tree. It is susceptible to fire blight	8-12
M 26	Very common commercial rootstock that, when trees are grafted onto it, requires the tree be somehow supported or staked for the life of the tree. It is generally not recommended for home use. It is susceptible to fire blight.	12-15
G 11	Produces trees 35-40% the size of a standard tree and is resistant to fire blight. Trees grafted onto this rootstock require support and availability may be limited. Of the Geneva rootstocks, this may be more adapted to home production than others	12-15
G 30	A main feature of this rootstock is resistance to fire blight. However, trees grafted onto it should receive support for the life of the tree.	18-20

Contd.

Rootstock	Description	Eventual size (foot)
M 7	Very common rootstock for home use. Trees reach 15-20 ft high and wide at maturity. Has a tendency to sucker but is resistant to collar rot. It is more tolerant of wet soil than many other rootstocks	18-20
M 106	Common rootstock at box stores and garden centers. It has a tendency to form root suckers but is moderately resistant to wooly apple aphid. Offers excellent root anchorage.	25-30
M 111	This rootstock allows trees to reach 80-90% of the size that they would on seedling rootstock. This may preclude them from use in the home orchard. It is resistant to wooly apple aphid and is reportedly more drought resistant than other rootstocks.	30-35
Seedling	This is a general term for apples grafted or budded onto rootstocks derived from apple seedlings. Trees reach 40 ft high at maturity and have delayed fruiting as compared to other rootstocks.	35-40

Promising Rootstocks

a. **Malling (M) series:** Developed at East Malling Research Station in England in 1913.

   Dwarfing : M-8, M-9, M-26, M-27.

   Semi-dwarfing : M-2, M-4, M-7.

   Semi-vigorous : M-13.

   Vigorous : M-12, M-16.

   M-7 and M-13 are tolerant to excessive soil moisture. Most suitable for use as dwarfing interstock is M-27. M-9 is sensitive to injury from very low winter temperature.

b. **Malling-Merton (MM) series:** Developed by crossing Malling rootstocks with Northern Spy to evolve wooly aphid resistant rootstocks.

   Semi-dwarf : MM-106.

   Semi-vigorous : MM-104, MM-111

   Vigorous : MM-109

   MM-111 is resistant to drought while MM-104 is tolerant to wet soils. MM-106 is the most widely used rootstock.

c. **Merton series:** Out of 4 Merton rootstocks (Merton 778, 779, 789 and 793), Merton 793 has proved a very useful rootstock. It produces trees little smaller than M-16, is resistant to wooly aphid and collar rot and is adaptable to a wide range of soil and induces early fruiting.

d. **Polish (P) series:** P-series of apple rootstocks have been developed in Poland by crossing common Antonovka with M-9. Five rootstocks (P1, P2, P16, P18 and P22) have been found promising. All these rootstocks are winter hardy and resistant to crown rot but susceptible to fire blight. P18 is semi-dwarfing while all others are dwarfing. The P2 and P22 show particular promise as dwarfing interstock.

e. **Budagovsky (Bud) series:** Bud series of rootstocks introduced by the Michurin College of Horticulture, Michurinsk, Russia. These rootstocks viz. Bud-9, Bud-490 and Bud-491 are of promise. Bud-9 and Bud-491 are dwarfing, while Bud-490 is semi-dwarfing. Bud-490 and Bud-491 are very winter hardy and Bud-9 and B-490 are resistant to crown rot.

f. **Ottawa series:** There are two Ottawa series of apple rootstock, the Ottawa Hybrid Seedlings (OH) and the Ottawa Clonal series (O).

The six OH series rootstocks (OH1 to OH6) are still under test but most of these hybrids are resistant to latent viruses commonly found in commercial apple varieties. In the Ottawa Clonal series out of the 14 rootstocks originally selected (O1 to O14) five rootstocks viz. O5, O6, O9, O10 and O11 were discarded in the course of testing because of poor performance or incompatibility. Among the others, two rootstocks namely O4 and O 8 have shown promise. These are more productive and were hardy than MM-106.

g. **Michigan Apple Clone (MAC) series**

MAC 1: Trees on this rootstock are approximately M-7 in size but do not sucker and are well anchored.

MAC 9 (Mark):Trees on this rootstock are similar in size to M-9 or slightly larger and equivalent in productivity.

MAC 24: Trees on this rootstock are semi-vigorous and in the MM-111 class. It is well anchored.

h. **Miscellaneous Rootstocks**

i. Bemali: First introduction from Balsgard, Sweden. Bemali produces trees about the size of M-26, is highly productive and resistant to wooly aphid.

ii. Jork 9 (J9): This is an open pollinated M-9 seedling released by the Jork research Station, West Germany, that has the advantages of more easy propagation with equal productivity over M-9.

iii. Alnarp 2: Introduction from Alnarp Fruit Tree Station, Sweden, is winter hardy, vigorous and induce early bearing and productive.

iv. Robusta 5: Originated in Canada, is vigorous, winter hardy, resistant to fire blight and easy to propagate.

GRAPES

Vegetative propagation in grape is practiced through cuttings, grafting, layering and budding depending on the varieties used and the growing conditions.

Hardwood Cuttings: Propagation through hardwood cuttings is the most popular method of propagation in grape. Grape cultivars and clonal rootstocks have traditionally been propagated by dormant hardwood cuttings, which root readily. Cutting material should be collected during the winter (January) from healthy, vigorous, mature vines. Well-developed current season's canes should be used; which are medium in size and have moderately short internodes. Cuttings 8 to 13 mm in diameter and 36 to 46 cm long are generally used. They are planted in well prepared flat beds, leaving two nodes above the soil surface. In North India these cuttings are planted in the nursery after allowing them to form callus by burying them in moist soil or sand for 4-5 weeks. The rooted cuttings will be ready for planting in the main field only after one year.

Leafy Cuttings: Leafy greenwood grape cuttings of *V. vinifera* cultivars root profusely under mist in about 10 days if given relatively high (27 to 30°C) bottom heat and treated with IBA. Scarce planting stock (such as virus-indexed material) can be increased very rapidly using one-node stem cuttings, then consecutively taking additional cuttings from the shoots arising from each node on the rooted cutting. Several consecutive cycles of cuttings can be taken to increase the supply of rooted cuttings to very high numbers in a short period.

Grafting: In India grape is planted on its own roots. However, use of resistant rootstocks is necessary under infestation by nematodes and other pests and diseases and also for saline soils. The scion variety can be chip budded on suitable rootstocks. Budding or grafting onto resistant rootstocks is necessary to increase vine life, plant vigor, and yield specifically where pests, such as phylloxera *(Dacylosphaera vitifoliae)* or root-knot nematodes (*Meloidogyne* spp.) are present. Bench grafting is widely used where scions are grafted either onto rooted or unrooted disbudded rootstock cuttings. Methods include the whip graft, or better, machine grafting. Grafting is done in late winter or early spring from completely dormant scion and rootstock material. The stocks are cut to 31 to 36 cm with the lower cut just below a node and the top cut 2.5 cm or more above a node. All buds are removed from the rootstock to prevent subsequent suckering. Scion-wood should have the same diameter as the stock.

***Rootstocks*:** Dog ridge is one of the commercially important rootstock and is known for its drought tolerance. It is a more vigorous variety and Thompson Seedless grafted onto this rootstock is known to produce more vegetative growth at the cost of reproductive growth, thus reducing the yield per unit area. Another rootstock that is gaining popularity in India is 110 R, which is known for its drought and salt tolerance, apart from improving bud fruitfulness due to its moderate vigour. St. George is one of the important rootstocks for phylloxera resistance. It is moderately tolerant to drought on lightly textured soil and highly tolerant of moisture stress in deep soils.

B) COMMON FRUITS

SAPOTA

Among the vegetative methods, the most important are air layering, approach grafting, inarching and soft wood grafting. However; commercially followed method is inarching and now this method is replaced by softwood grafting. For grafting sapota various rootstocks are used. Use of proper rootstock is important in grafting. Khirni and Rayan are the two important rootstocks of sapota being used for raising grafted plants of sapota in India. Since the slow growth of the seedling rootstocks become a limiting factor for its rapid multiplication.

Air layering is carried out in the month of June. In this method a 1-2 year old, healthy, vigorous, mature shoot of 45-60 cm in length and pencil thickness is selected. A circular strip of bark about 3 cm wide just below a bud is completely removed from the selected shoot. Rooting hormones like IBA & NAA 50mg each in Lanoline paste are applied over this portion. Moist sphagnum moss is packed around this portion and tied with polyethylene sheet to prevent the loss of moisture. Application of such hormones promotes early rooting. Light brown roots are visible through the polythene wrap within 4 months. The rooted shoot is slowly detached by giving 2-3 successive cuts over a period of week before finally detaching from the parent plant. These are planted in pots and kept in nursery under shade. Top of the shoot is cut back to maintain a proper ratio of leaves: roots. Sapota plants prepared from air-layering are susceptible to damage by strong winds and cannot withstand drought, as their root system does not penetrate very deep into the soil. Approach Grafting is also used for propagating sapota plants. In this method 2-3 year old 'Khirni' or 'Rayan' plants raised in polythene bags are used as rootstocks due to their strong root system. These seedlings should be at least 45-60cm in height and 1 cm in diameter at the time of grafting. Approach grafting should be preferably done in the beginning of rainy season or during February-March. However grafting during February-March is more economical as the grafts will have to be maintained in nursery for a shorter period of time. The success rate of the grafts is also high as compared to those obtained in rainy season. The mother plants selected for approach grafting should be vigorous true-to-type and between 10-15 years old. The lower most branches are usually bent and tied to a peg near the ground. 1-1½ year old branch (scion) of pencil thickness is selected for grafting. The seedling of 'Khirni' (rootstock) is brought closer to the branch near the ground. The cambium layers of one side of this plant at 10-15cm from the base of the plant is exposed by taking a 5-6cm long and 0.5cm wide cut with a sharp knife. Similar cut is taken on the selected branch of the mother plant. The exposed portion of both the plants are brought together and secured firmly with polythene strip. Care should be taken to avoid gap between the joints. The union of the scion and the stock takes place in about 2-2/months. However, the scion should not be separated for at least 3-4 months. During this period the 'Khirni' plant should be watered every day for the first 15 days and thereafter, at an interval of 3 days till the graft is separated from the mother plant. The rooted graft is slowly detached by giving 2-3 successive cuts 2cm below the union over a period of month before finally detaching from the parent plant. The grafts are maintained in a nursery under shade for few days.

Soft-Wood grafting: The technique of soft-wood grafting is similar to that of cleft or wedge grafting. The scion should be a terminal non-flowering shoot of 3-4 months maturity with the same thickness as the rootstock. Rootstock of 'Ryan' is mostly used for soft-wood grafting. The selected scion should be defoliated while on the mother plant at least 7 days before grafting, keeping a part of petiole intact on the selected shoot. Defoliation prior to grafting helps in swelling of dormant bud of the scion resulting in greater success. In this method, a wedge shaped cut is made on lower part of the scion stick. Similarly 4-6cm long cut is made in the middle portion on the rootstock. The scion stick is then inserted in the cleft of the rootstock and tied with polythene strip. Vigorous sprouting of the scion indicates the successful union of the graft joint. The grafts are further maintained in the shade for sometime before transferring them in the field.

BER

Among the different methods of budding adopted, the most common method is shield budding on a suitable rootstock. For raising a budded plantation, it is considered best to sow ber seeds in the field itself at proper distances and use the seedlings thus raised for budding in *situ*. The commercial method of propagation of ber in India is by budding.

Rootstock: Two species of Zizyphus namely *Zizyphus rotundifolia and Z.nummularia* are growing wild in India and are known for their extensive root system and hardiness. The ber plants should be budded on

Zizyphus mauritiana (Elongated Dehradun), *Zizyphus rotundifolia* for higher fruit yield. The seedlings of *Z.nummularia* are slow growing and become buddable after a long period than *Zizyphus rotundifolia. Zizyphus mauritiana* (Coimbatore) is a semi vigorous rootstock. On the rootstock *Z.nummularia*, budding forms inverted bottleneck formation.

***Raising of Rootstock*:** Rootstock seedlings are raised by sowing seed kernels extracted by breaking the stone (endocarp). These germinate in about one week. The seed stones can also be sown as such but take nearly one month to germinate. Germination of seed stones can be improved by soaking them for 48hr in water or for 6hr in concentrated sulphuric acid or in 200ppm Gibberellic acid. Seeds of any locally adapted and vigorous ber trees can be used for raising rootstocks. Seeds should be sown in a well-prepared nursery bed at 30cm × 30cm spacing and at 2cm depth during March–April. These seedlings could either be transplanted in the field during July– August for in-situ budding or can be budded in the nursery beds. The buddlings can be transplanted in bare rooted condition during January–March after treatment with 12 % Waxol or after defoliation. This is, however, possible only in irrigated areas. In rainfed areas, seeds should be sown in 300 gauge polythene tubes of 25cm length and 10cm diameter, filled with a 1:1:1 mixture of farmyard manure, sand and clay. Sowing is done during April in north India so that the seedlings become buddable during July. The buddlings become ready for transplanting 1–2 months after budding. The buddlings raised by this technique retain their deep rooting tendency and prove the most suitable under the low rainfall drylands. In drylands, ber orchard can also be raised by transplanting tube-raised ber seedlings with the onset of monsoon, leaving them to grow in the field until the forthcoming summer for budding *in situ*.

***Propagation*:** Most common method of propagation of ber is by I or T (shield) budding in the month of July. The *in-situ* raising of ber plantations in rainfed areas is advisable. Budding should ideally be carried out in the field, either with field-grown rootstock plants or transplanted from the nursery. Budding should be carried out as close to the ground as possible to stop sprouting from the rootstock portion. Lopping and topping of the rootstock from 8 to 15 days after budding in the nursery beds has been found to ensure 90 to 100% success. Just before in situ budding in the field, rootstock seedlings should be lopped (cut) to 30-45cm height. Budding can be carried out by different methods such as I or T (shield), ring, patch, chip, flute and forkert; but shield or patch budding are the most commonly used methods for ber.

Buddlings prepared in the nursery beds should be lifted with large earth balls at about 9 to 12 months after budding for transplanting in the field. A number of these plants may be lost due to root damage while lifting, packaging, transport to the field and transplanting. Buddlings prepared in polytubes become ready for transplanting at about 30 days after buddling. The polytubes with the budlings are removed from the nursery and kept in the shade for a week. These can then be easily transported or transplanted in the field with over 90% survival since there is little stress or root damage during lifting, transport and planting. The roots of the plants do not coil and therefore retain the drought hardy character and vigour almost similar to plants raised *in situ*.

AONLA

Aonla is propagated by Seeds as well as by vegetative methods. Since seed multiplication brings variability, vegetative methods should only be adopted. However for raising rootstock seedlings are required to be raised through seeds. Seeds are collected from fully matured fruits, in the month of February, and sinkers are used for sowing and floaters discarded. There is no seed dormancy, therefore, fresh seeds give almost 100% germination. Process of germination can be quickened by soaking dried seeds in 500 pm GA_3 solution for 24 hours. Seedlings can be raised in seedbeds or polythene bags. It will take them about 4 months to attain buddable size.

Both budding and inarching methods can be used for raising plants but budding is more successful. Among three methods of budding, forkert and patch budding done from June to September, gave maximum success. However, shield or T-budding is the commercial method of Aonla propagation and gave 80 percent success in rejuvenation of old orchards. One year old seedlings with a girth of about 1 cm should be shield budded in early June with healthy and plump buds from new growth.

Soft wood grafting can also be used successfully for which seedlings are raised *in-situ* and headed back severely in the month of May for forcing new vigorous side shoots. On these new shoots soft wood grafting method can be employed by way of wedge grafting. It gives 70 percent success.

Budding: Use of poly-bags of 25x15 cm size either filled with field sandy soil + pond soil + compost (1:1:1) or field sandy soil + FYM (1:2) give more than 80 per cent success after patch budding in the month of June, which induces more than 45 cm of buddling growth in the month of July. The plants are ready for transplanting in the month of August under hot arid ecosystem. The plants are also feasible for transportation due to low weight and uniform in size.

CASHEWNUT

Cashewnut is mainly propagated through seed leading high degree of phenotypic and genotypic variability. Air-layering and propagation through cuttings is also done with reasonable degree of success. Chip budding and wedge grafting have been successfully employed for propagation of cashewnut. Two grafting techniques, namely side grafting and wedge grafting are practised with success. Grafting should commence as soon as possible (seedlings of 3-4 months old) and planted out in the orchard to prevent the tap-root from bending.

***Cuttings*:** This method is cheap and easy to produce if the species has the potential to develop roots. Cuttings from young seedlings (10 to 20 cm high) and cut just above the cotyledons. The leaves are cut in half to reduce transpiration and put the stems into a mixture of 50% peat and 50% sand (without growth regulators). They are covered with a transparent plastic sheet and watered regularly. Rooting percentage of 90-95% can obtained after 8 weeks. The decapitated seedlings developed 2 new shoots that can be used again for making cuttings. Rootability after the third cycle declines to 40-60%. 78% rooting of cuttings can be achieved when cuttings are dipped in 1% IBA and grown in a vermiculite medium kept under 7 hours mist per day. Best rooting media is 1:1 mixture of vermiculite and peat.

***Grafting and budding*:** Grafting is a widespread and well known technique. When rootstocks and scions are of the same diameter, whip grafting is the easiest grafting method. It does not require any capital investment other than a sharp knife and strips to fix and protect the scion on the rootstock. August is the best time for grafting cashewnut. Cashewnut seeds are used for rootstock purpose

PEAR

Pears are commercially propagated by Tongue/Cleft grafting or Shield (T) budding. The rootstocks generally used are the wild Himalayan pears i.e. *Pyrus pashia*, *P. pyrifolia*, *P.kashiana*. To produce dwarf trees Quince-D rootstock is used. Some commercial varieties are not compatible with Quince-C rootstock. By double working using an intermediate rootstock like Old Home, the incompatibility can be overcome. Quince A is a vigorous rootstock. Tongue and cleft grafting are performed during December-January. T-budding is practised from May to July. Pear plants can also be propagated through hardwood cuttings.

Rootstocks

Seedling rootstock : Kainth (*P. Pashia*)

Dwarf: Quince-C, Oregon 211,

Semi dwarf: OH x F-230

Vigorous: Oregon 260

Rootstocks for Pear

In pears, there are not many clonal rootstocks as in apple and generally seedling rootstocks are the most widely used rootstocks. Although different *Pyrus* species such as *P. communis, P. pyrifolia, P. pashia, P. serotina (shiara),P. calleryana, P. betulifolia, P. ussuriensis, P. elaeagrifolia* are used as pear rootstocks, yet the species *P. communis, P. betulifolia* and *P. calleryana* offer the most useful traits for a general use rootstock. Although less easy to propagate clonally than Quince, they offer winter hardiness and hot climate tolerance, compatibility for all scion cultivars, tolerance to wet soils, reduced susceptibility to high pH and resistance to many diseases. Root suckers (*Pyrus pyrifolia*) are used as rootstock in Punjab and Jammu region.

Clonal rootstocks for pears have been limited primarily to selections of Quince that were developed since 1930 at East Malling. Of the five selections made from East Malling Research Station, viz. Quince A, B, C, D, and E only three i.e. Quince A, B and C are important from commercial point of view. Rootstock breeding programme has been successful in the development of OH x F series of clonal rootstocks, some of which are identified to have disease resistance with high yield potential, besides having a range of size controlling effect.

a.Quince rootstocks

Quince A: Vigorous and most important.

Quince B: Intermediate, produces smaller trees than Quince A.

Quince C: Most dwarfing, precocious and susceptible to winter injury.

C.132 selected from HRI-East Malling and Cts 212 selected from Pisa, Italy is more dwarfing than Quince C. *Pyrodwarf,* new rootstock selected from Germany is more dwarf than Quince A.

b. Pyrus communis rootstocks : Old Home (OH) x Farmingdale (F) series.

OH x F 51: Dwarf

OH x F 34, 69, 87, 230, 233: Semi-dwarf.

OH x F 217, 267, 361: Semi-vigorous.

OH x F 18, 97, 112, 198: Vigorous.

These rootstocks are highly tolerant to winter cold and bacterial canker and moderately tolerant to crown gall and collar rot but difficult to propagate except by micro-propagation.

c. *Pyrus betulifolia* rootstocks: Oregon 260, 261 and 264 : Vigorous, very tolerant to wet and dry soil and highly tolerant to bacterial canker, crown gall, collar rot and root lesion nematodes.

d. *Pyrus calleryana* rootstocks: Oregon 211 and 249: Very dwarfing, highly tolerant to wet soil and crown gall and collar rot diseases and root lesion nematodes

STONE FRUITS

Stone-fruit and nut-producing trees include sweet cherry (*P. avium*), sour cherry (*P. cerasus*), apricot (*P. armeniaca*), almond (*P. dulcis*), peach and nectarine (*P. persica*), European plum (*P. domestica*), and Japanese plum (*P. salicina*). In addition, several hundred related species and species hybrids exist worldwide.

Seed propagation is important for breeding and rootstock production. *Prunus* seeds have a hard protective endocarp (shell, pit, stone) that surrounds the seed. The embryo is non-endospermic and exhibits embryo dormancy that requires one to three months' cold stratification to germinate. Seeds of different species and sometimes individual cultivars have specific time requirements. In addition, the stone may be inhibiting either because of substances or because of hardness and/or lack of water uptake (cherry). Presoaking, stone removal, leaching in running water, or warm stratification preceding cold stratification may be helpful.

The most important method of propagation of stone fruits is to bud or graft cultivars to rootstocks in the nursery and transplant trees to the orchard. Seedling rootstocks are prepared either by stratifying the seeds during the winter and planting early in the spring, or planting in the fall and allowing natural chilling in the ground to bring about the moist chilling process. Seedling liners that have been produced during the previous year may be lined-out in the nursery row for later budding. T-budding is the usual practice and may be done in the fall, early spring or late spring, or summer depending on the length of the growing season. Chip budding may also be utilized in areas with a relatively short growing season. Trees are produced within 1, 2, or 3 years, depending upon the nursery site and conditions, the species, and the desires of the propagator.

PEACH

Peach is commercially propagated by grafting and budding. Peach seedlings are generally used as rootstocks, though plum, apricot and almond seedlings can also be used. Some times Bhemi (*P.mira*) is also used as rootstock. Since wild species produce more vigorous and hardy seedlings than the cultivated varieties, wild seedlings are preferred as rootstocks. Peaches are commercially propagated by tongue or cleft grafting and – T budding or ring budding. T-budding can be done from April to September and grafting during December-January.

Some peach cultivars can be propagated by leafy, succulent softwood cuttings taken in spring or summer, treated with auxin, and rooted in a mist-propagating bed. Also, in areas with mild winters, some peach cultivars can be started from hardwood cuttings if they are treated with a 4,000 ppm IBA quick-dip, then set out in the nursery in the fall.

***Budding*:** Peaches are propagated by T-budding or chip budding on seedling rootstocks. Seeds planted in early fall and allowed to stratify over winter germinate in early spring (March) to produce vigorous large plants by May.

Rootstocks

Seedling rootstocks: Wild peach, Sharbati and Flordaguard

Dwarf: *Prunus besseyi*

Semi dwarf: St. Julian Hybrid no. 2, Damas GF.1869, (resistant to high pH and water logging); GF 677,and GF-557, Rubira.

PLUM

Plum cultivars are propagated by T-budding or chip budding in the fall on seedling rootstocks or, with certain rootstocks, on rooted cuttings or layers. Budding can also be done in the spring. Plums tend to be easier to root than other species, being propagated by hardwood cuttings and some by leafy, softwood cuttings under mist. Bench grafting in winter and planting the grafts in the nursery in spring gives good results using the whip graft. Plum is raised on seedling rootstocks of wild apricot (Zardalu) and Myrobalon B plum clonal rootstocks. Tongue grafting, Chip budding and "T" budding are the important methods employed commercially for propagation of plum.

Rootstocks

Seedling root-stock: Wild peach and apricot, Kabul Green Gauze and Sloh (Almond X Peach)

Dwarf : Pixy

Semi dwarf: St. Julian Hybrid-2, Damas-1869, Mirabi

Vigorous : Mariana GF8/1, Myrobalan 27.

APRICOT

Apricot are propagated commercially by T-budding or chip budding on various seedling, *Prunus* rootstocks. Fall budding is the usual practice, but spring and June budding may be used. Bench grafting also has been successful.

Rootstocks

Seedling rootstock: Wild apricot

Dwarf: *Prunus besseyi* hybrid P-2038

Vigorous: Myrobalan GF-31 and Marianna GF-8/1.

C) MINOR FRUITS

JAMUN

Jamun is propagated both by seeds as well as vegetatively. Narendra Jamun 6 is usually multiplied by seeds. The seeds have no dormancy; hence fresh seeds can be sown (within 10-15 days) 4-5 cm deep at a distance of 25cm x15 cm. The seeds germinate in10-15 days after sowing. The seedlings become ready for transplanting in spring or next monsoon. Its seeds show polyembrony up to 30-40%. Therefore nucellar seedlings can be used to produce true –to-type plants. Seeds may also be sown in polythene bags as it facilitates in easy handling of rootstocks and grafted plants.

Seedling plants bear fruits of variable size and quality. Therefore, vegetative method is desirable for propagation of improved or selected types. Budding is most successful. It is done on one-year old rootstock having about 10 mm thickness. Patch and forkert methods of budding give more than 70% success if performed in March. In low rainfall area, July –August is ideal time for budding.

***Soft wood grafting*:** Soft wood grafting method proved to be better for *in-situ* grafting under rainfed conditions. Soft-wood grafting has been found successful in the month of July-August under semi-arid environment of western India.

LOQUAT

Several propagation methods both sexual as well asexual can be used to propagate loquat. Plants propagated from seed, take long time to bear fruits, may not possess characteristics of mother plant. Though used in some parts of the world for raising plants but is not recommended for raising commercial plantations.

Major vegetative methods of propagating loquat include layering, cutting, budding and inarching. Layering, especially air-layering is satisfactory as well as easier method in loquat. However, budding or inarching are preferred over other methods of propagation. These methods are easier to perform, ecomical and satisfactory.

***Rootstocks*:** Several members of loquat family (Rosaceae) such as apple, plum, pear, *Sorbus* are used as rootstocks in different countries. Quince is also used as rootstock to produce dwarf trees. In India, seedling loquats are most satisfactory rootstocks for grafting loquats.

CUSTARD APPLE

Propagation: Both sexual and asexual methods are employed for propagating the annona plants. However to maintain the genetic uniformity for obtaining higher yield of better quality fruits, the clonal propagation is a must. Different methods of vegetative propagation are-cuttings, grafting and budding.

***Cuttings*:** A very high rooting success (90%) has been recorded when custard apple cuttings were etiolated 15 days before planting treated with NAA 5000ppm and rooted under intermittent mist. Root cuttings of custard apple have given only 2-5% success.

Budding: Custard apple can be successfully propagated through budding. Among the budding methods shield, patch, modified forket and chip budding, shield budding is most successful. Sithaphal on its own rootstock and on cherimoyahas given 60-80% success. Budding is done in early spring when the sap starts flowing or in the autumn. For taking bud wood, dormant, last year‘s shoots should be preferred. The seedling rootstocks, however exhibit wide variability in growth and yield of scion.

***Grafting*:** It can be propagated successfully by grafting. Among the grafting methods like tongue, cleft, veneer, inarching, splice, inarch grafting on *A.cherimola* and *A.reticulata* is more successful. The latest technique of custard apple propagation is by veneer grafting. This method is being adopted on a commercial scale at FRS Sangareddy.

AVACADO

Avacado is propagated by seeds or vegetative methods.

Seed propagation: Avocado is commonly propagated by seed but seedling lack true to type characteristics. Seed propagation is useful for raising rootstocks. Vaibility of avocado seed is short (2-3 weeks) and can be improved by storing seeds in dry peat or sand at 5^0C. Removal of seed coat before sowing helps to speed up germination. If there is a shortage of rootstocks, seed can be split lengthwise into four or even six parts, leaving a piece of embryo on each. Seeds of varying size are of equal value for propagation. Seeds are sown in nursery beds or in bags. If in nursery bed, a spacing of 35-45 cm between seeds and 60-75 cm between rows are given. For shoot emergence, it takes 55 to 95 days which was positively correlated with residual dry matter content of cotyledons.

Seeds are immersed in hot water at 49 to 52°C (120 to 125°F) for 30 minutes before planting to eliminate infection from avocado root rot. Germination of the seeds is hastened by removing the brown seed coats and cutting a thin slice from the apical and basal end of each seed before planting. The seed coats can be removed by wetting the seeds and allowing them to dry in the sun. Seeds are then sown in nursery rows or into containers.

Vegetative propagation

***Cuttings*:** Depending upon races and cultivars, rooting capacity of avocado cutting varies greatly. Mexican race is relatively easy-to-root whereas West Indian race is quite difficult and Guatemalan race is intermediate in rooting ability. Cutting from young plants root more readily than from old trees. Leaf cutting root first and leaf survival on cutting under mist was essential for rooting to take place. Basal ends of cutting are soaked in IBA solution along with use of bottom heat (constant 26^{0}C) in cutting beds, to improve rooting of normally slow root cultivars. Also spraying of NAA and cytokinin, once a week, prevented leaf shedding and thus improved rooting of cutting. Similarly etiolatiion and use of root promoting substances like IBA usually promoted rooting on cuttings. Base etiolated cuttings rooted faster and had higher percentage. At early stages, these rooted cuttings grew slowly but eventually made satisfactory growth.

***Budding and grafting*:** Budding or grafting can be employed for propagation avocado. Shield budding, forket budding, side grafting, approach grafting or layering can be successfully followed. In West Bengal, chip budding was successful while in Tamil Nadu, layering and approach grafting gave up to 75% success. In Puerto Rico, 95% success was obtained with cleft grafting. Grafted plants become ready in six months and planted in field after another six months. Rootstock affected yield, tree efficiency (yield/unit area) and alternate bearing and choice of appropriate rootstock may overcome some of disadvantages associated with avocado cultivars. West Indian rootstocks are preferred in warmer regions or where salinity is a problem. Purple avocado serves as a good rootstock for avocado. Guatemalan rootstocks are more sensitive to cold and also prove more susceptible to high pH, chlorosis and verticillium wilt. Mexican race rootstocks are generally in Isreal and in South Africa. A number of immune or highly resistant Persea species are known, but unfortunately all have proved incompatible for hybridization and grafting with *P. americana. P. schiedeana* is a rootstock to induce vigour or for use where soil conditions are limited.

Whip grafting: The best grafting technique for small diameter (1/4 to ½ inch) rootstock is whip grafting. Whip grafting should be done in the fall or spring. Although whip grafts use more scion wood than budding does, they allow the grafted plant to develop more rapidly.

Bark grafting: The best grafting technique for large-diameter trees or branches is bark grafting. To make a bark graft, first cut off the rootstock (the trunk or branch to be grafted) just above a crotch where smaller branches sprout out. If possible, try to retain one branch of the original plant as a nurse branch. The nurse branch will provide the scion nutrition and support from wind (the nurse branch will eventually be removed).

ALMOND

Almonds are propagated through T-budding performed during spring.

Rootstocks

Seedling root-stocks: Bitter almond and wild peach Kabul green gauze and Sloh

Semi dwarf: Ishtara

Vigorous: Hansen 2168, Hybrid GF-677 (replant area) and GF. 557

WALNUT

Walnuts are propagated by cleft grafting or patch budding. Hot callusing technique has also been used with a reasonable degree of success. *Juglans regia* is propagated by patch budding, T-budding, or whip grafting on one-year-old seedling rootstocks. Patch budding in the spring or summer works well when top-working smaller sized material.

Rootstocks

Seedling rootstock: Katha;

Dwarf: *Juglans microcarpa*

Vigorous: Paradox (resistant to crown gall, very susceptible to oak root, black line, coloured kernels are produced).

CHERRY

Cherry is propagated commercially by T budding or Inverted T budding. The rootstocks mostly used are *Prunus cerasoides*, *P. mahaleb*, *P. padus* etc. Different states are using different rootstocks depending on their merits and demerits

Rootstocks

Seedling root-stock: Sweet/Sour cherry

Dwarf: Gisela 10, (tree size is 40-60% of F12/1)

Stockton Morello, Cab-6p, Cab-11E & Weihroot

Semi dwarf: Charger, Colt, M x M 14.

F12/1: This clone of *P. avium* was selected because it is vigorous and has high resistance to bacterial canker (Pseudomonas).

Colt (*Prunus avium* x *P. pseudocerasus*): A semi dwarfing clonal cherry 'Colt' rootstock was released in 1958 at the East Malling Research Station. This rootstock, used for both sweet and sour cherry cultivars, is easily propagated by hardwood cuttings as well as by mound layering. Tree size is about 60 to 70 per cent of trees on F12/1 and 80 per cent of trees on mazzard seedlings. It is resistant to bacterial canker (Pseudomonas), stem pitting virus, and replant problems. It is susceptible to crown gall (Agrobacterium).

Mazzard x Mahaleb Clones (*P. avium* x *P. mahaleb*): M x M 14: Semi-dwarfing clone, appears to be the most favored and is used in Europe.

Gisela Series: Gisella 1 (17 to 45 per cent of 'F12/1'); Gisella 5 and Gisella 10 (40 to 60 per cent of 'F12/1'). Trees on these rootstocks appear to be precocious, heavy bearing, well anchored, and hardy.

D.PROPAGATION TECHNIQUES IN OTHER FRUITS

A) DECIDUOUS FRUITS

POMEGRANATE

The common method of propagation of pomegranate is by rooting of hardwood cuttings (6-8 inch in length and 0.5 inch in diameter) from fully mature wood. It is globular grafting in local language known as "Gooti Kalam". Cutting obtained from one-year old shoots are universally used for pomegranate propagation. 9-10" long cuttings are planted in such manner that 2/3rd of their basal portion are buried in soil. Cuttings taken from main shoot or from basal suckers root easily. Best result are obtained when basal portion of cutting are treated with Seradix hormonal powder and planted so that 4-6 nod are buried in the soil. Cutting should not be obtained from those shoot which are less than six months old or more than 2 year old. Leaves are removed from cutting before planting. In most part of South India cuttings are planted during monsoon season and in Central and Western India they are planted during monsoon rains. Root initiated occurs within fortnight after planting and will be ready for potting in 9 months.

One-year-old rooted cuttings are best for planting. Pomegranate can also be propagated by ground layering, air layering or root suckers. In hard wood cutting, low rate of survival and unbalanced root : shoot ratio are the problems. The use of IAA and IBA is superior to NAA in stimulating rooting in different type of cutting. The survival rate of as high as 80-90% has been obtained under mist, especially after treatment of cuttings with IBA 3000 ppm.

Air-layering: Air layering is carried out in the month of May/June. In this method a 1-2 year old, healthy, vigorous, mature shoot of 45-60 cm in length and pencil thickness is selected. A circular strip of bark about 3 cm wide just below a bud is completely removed from the selected shoot. Rooting hormones like IBA & NAA 50mg each in Lanoline paste are applied over this portion. Moist sphagnum moss is packed around this portion and tied with polyethylene sheet to prevent the loss of moisture. Application of such hormones promotes early rooting. Light brown roots are visible through the polythene wrap in the month of July-August. The rooted shoot is slowly detached by giving 2-3 successive cuts over a period of week before finally detaching from the parent plant. The polythene sheet is removed before planting them in pots. They are planted in pots and kept in nursery under shade. Top of the shoot is cut back to maintain a proper ratio of leaves:roots. The grafts can be transplanted in the field in the month of September/October.

Micro-propagation: Several studies have been conducted on micropropagation of pomegranate trees over the past several years. Protocols have been developed for regeneration of *P. granatum* L. plantlets *in vitro* through either organogenesis from callus derived from leaf segments, cotyledons anthers or through embryogenesis from various seedling explants, petals and immature zygotic embryos.

Reliable and reproducible protocols to get healthy and well formed plants from nodal explants of the pomegranate (*Punica granatum* L) cv 'Bhagwa' has been developed. Nodal segments were cultured on two different media at full strength Murashige and Skoog (MS) and Woody Plant Medium (WPM). The media was prepared as a basal medium supplemented with 0.2 to 2 mg/L 6-benzylaminopurine (BAP), 0.1 to 1 mg/L 1-naphthalacetic acid (NAA), 0.5 to 2.5 mg/L silver nitrate ($AgNO_3$) and 10 to 50 mg/L adenine sulphate for establishment stage. For proliferation stage, 0.1 to 0.5 mg/L BAP and NAA was tested. For rooting stage 0.0, 0.25 and 0.5 mg/L 3-Indolebutyric acid (IBA) and NAA on MS and WPM medium were tested. The nodal explants grown on MS medium containing 1.8 mg/L BAP, 0.9 mg/L NAA, 1 mg/L silver nitrate and 30 mg/L adenine sulphate had the highest proliferation rate (10 to 15 shoots/ explants) in establishment stage. The same trend was found concerning the maximum leaves numbers (15 to 20 leaves/explants) on proliferation medium containing 0.4 mg/L BAP and 0.3 mg/L NAA.

The plantlets grown on MS medium were found to have better survival compared to WPM medium. 0.5 mg/L NAA and 0.5 mg/L IBA showed equal rooting response in both the medium, whereas thick root formation was observed in the medium containing IBA.

STRAWBERRY

Strawberry can be propagated through sexual (seed) and asexual (vegetative) means. Propagation by seed is not considered as a viable method of propagation for cultivated strawberry as the plant raised through seeds do not come true-to-type. However, strawberry is usually propagated through runners. Nowadays, its large-scale propagation by tissue culture is being used widely.

Vegetative propagation: The arial stems (stolon), is produced in the leaf axil and grows out from the parent plant during summer. At the second node, a runner plant is formed and a new stolon arises on the runner plant to continue the runner train, which often branches. Initially, the runner plant produces a few roots and thereafter makes excessive fibrous roots thus, the strawberry naturally propagates itself by vegetative method of runner production after the runner plant has acquired sufficient growth and roots it may be separated from the mother planted elsewhere. Although, propagation by runners perpetuates all the character of the mother plant, viral diseases are quite often transmitted through runners. Thus, for runner production, a separate bed should be used. The site and soil where strawberry had not been grown for the last 3-4 year should be selected for the runner production. Soil should be tested for viruses and other soil-borne pathogens. Virus-free or tested clonal material of the required variety should be used. The planting for runner production should be done at 1.2mx1.2m or at 1.8m x1.8m distance. All other cultural practices like irrigation, nutrition, weeding and mulching should be strictly followed till the runners are lifted from the bed. Usually, 30-40 runners from a mother plant is considered satisfactory rate of production. Runner's production is not a problem in June-bearing varieties. However, in ever bearing strawberries, runner production is very low because the flower bud development continues throughout the growing seasons. In plains, runners are not produce frequently and if at all produced, it is difficult to protect them from high temperature in May-June. Runner production is invariably a varietal character; however, it can be enhanced with the treatment of different growth regulators. A single spray of GA3 (50-100 ppm) to the new and one-year old planting of ever bearing strawberry inhibits flowering, promotes runner development and increases the yield of marketable runners in the following years. In contrast, runner production is inhibited by soil application of paclobutrazol. The runners are mostly lifted during September-October.

Strawberry is commercially propagated by runner plants. Generally one plant produces 7-10 runners but under proper management, it can go up to15 runners/ plant. It can also be propagated through crowns (3- 5 plants/crown), but division of crowns of older plants is too tedious and expensive for cultivars producing runner plants readily. Runner formation can be stimulated with the application of IBA (100ppm) 10 days before flowering, and also with Morphactin (50 ppm).

For large-scale propagation of virus-free plants, tissue culture is widely used. Under favourable conditions, one strawberry meristem can be multiplied to yield more than one million plants in a year. Plants can be regenerated from meristematic callus, anthers and immature embryos.

PHALSA

Phalsa is commercially propagated by seeds in India and they germinate in 15 days. The seeds should be sown in well prepared nursery beds at 4-5 cm apart in rows spaced at 25-30 cm in July- August with the onset of monsoon. But the seed lose viability greatly within three months, when stored in ambient condition and hence, fresh seed from fully ripe fruits should only be sown. The Phalsa plant is readily propagated by rooting of hardwood cuttings and semi-hardwood cuttings from spring flush, treated with 1,000 ppm NAA, and planted in July. Ground-layers, treated with hormones, have been 50% successful and air-layers, with 85% of success in phalsa. Hardwood and semi-hardwood stem cuttings of Phalsa should be collected from 4 to 5 year old plants and 15 cm long stem cuttings with basal portion should be prepared. For rooting media, sandy soil and farmyard manure (FYM)in ratio of 2:1 by v/v should be mixed thoroughly, cleaned for stones and grasses, then the mixture should be filled in root trainers. The basal ends of the cuttings should be dipped in dilute solutions of 1000 ppm, 1500 ppm, 2000 ppm and Indole-3-Butyric acid by quick dip method for 10 seconds before planting in the rooting medium. Rooting of phalsa cutting depends upon various factors such as pre treatment of cutting, growing condition, environmental factors, etc. which influence the regeneration ability of cuttings. Although, phalsa can strike roots but rooting is not appreciable. So, growth regulators are to be used to improve its high rooting ability. Type of cutting and planting date influence the rooting of phalsa, the stimulation of adventitious root formation in stem cuttings treated with auxins is well known. In addition, the combination with other compounds has been shown to enhance the root formation. Hence, it is possible that optimum use of growth regulators and suitable season would help for rapid multiplication in propagating phalsa cuttings. Rooting efficiency would be better when, it is done in controlled conditions such as mist chamber.

FIG

Figs are one of the easiest fruit crops to propagate. Hardwood cuttings taken when the plants are fully dormant readily root and are most commonly used for propagation of figs. Cuttings should be six to ten inches in length and approximately one-half to one inch in diameter. Place the cuttings in a warm, humid environment such as wrapping them in a moist paper towel and placed in a polyethylene bag for 10-14 days to encourage callus formation. Cuttings can then be planted in pots in a commercial potting media to encourage rooting and shoot formation. Softwood cuttings can also be used for propagation, but a mist system is usually needed for successful plant production.

Cuttings may be taken at the time of pruning or whenever the tree is fully dormant. They should be 3/8 to 1inch in diameter and 8 to 10 inches long. Small, late-season growth will produce very few roots on hardwood cuttings. Cuttings should contain several nodes or "joints," since roots develop in these areas. The base of the cutting should be just below a node. Place the cuttings upright in a garden furrow or cutting bed and space 8 to 10 inches apart. Make sure at least one node is above the soil line, since leaf development must occur at this point. Also make sure the basal end of the cutting is placed underground. The scar where a leaf was previously attached is clearly seen below the bud at each node when the cutting is oriented correctly. Mist propagation of leafy cuttings has become increasingly important. With this technique, even the hardest to root varieties can be grown successfully. Mist increases the atmospheric humidity surrounding the cutting, thus reducing water loss through transpiration. The cooling effect of mist also reduces the rate of respiration, and photosynthesis continues in the intact leaves.

Constant or intermittent mist may be used. With constant mist, a fine spray is applied continuously during the rooting period. With intermittent mist, water is sprayed on the cuttings periodically. Both types have advantages as well as disadvantages. Factors to consider when choosing the proper technique are air temperature and propagation bed drainage.

Figs root readily under mist, and, where it is feasible, consider this technique. Summer months are best adapted to mist propagation. Cuttings about 6 inches long should be placed in a suitable rooting medium (sand, peat or mixture of both) with about 3 inches of the cutting above the top of the medium. A rooting hormone will encourage root development even though most figs root readily without added hormones.

Cuttings should develop a strong, extensive root system within three to four weeks when mist propagation is done properly. After cuttings have developed roots, a gradual adjustment period is desirable, and, in most instances, is necessary to increase survival when they are transferred from the propagation bed. A gradual adjustment can be accomplished by reducing the frequency of misting after root development begins or by moving plants to shaded and partially shaded areas for a couple of weeks before placing in full sun. If possible, locate the cutting bed in partial shade. Cuttings will root fairly well in full sun but progress more rapidly under half-shaded conditions. Cuttings may be taken in late fall or early winter when the trees are dormant and stored in moist sand or sawdust under cool conditions until the weather is suitable for setting them out in the late winter or early spring. The cuttings must not be allowed to dry out during storage and must not be taken until the tree is dormant. Deep, well-drained, well aerated soil is necessary for good results. Cuttings should be watered periodically to prevent drying out. Storing cuttings in inverted position appears to retard shoot development during storage. Rooted cuttings are grown in the nursery bed for one year, before being transplanted to the orchard. Transplant them during the dormant season from November to February.

B) EVERGREEN FRUITS

BANANA

Conventionally bananas are propagated through sword suckers of 1.5-2 kg weight selected from healthy gardens. Tissue cultured plants are becoming popular among farmers as they are not only virus free but also results in uniform and synchronized harvest, gives 35-45% higher yields over conventional planting material.

Healthy sword suckers of 3 months age having strong base, gradually tapering to slender point with one or two narrow sword like leaves at the tip are preferred for planting, since it grows faster and comes to bearing early compared to water suckers. Suckers from high yielding and healthy plant crop are selected. If the suckers are small the crop duration is longer but with more number of fruit. If the suckers are bigger the crop duration is shorter with less number of fruit but of better size. For traditional varieties corms weighing 1.5-2.0 kg and for Cavendish clone corms weighing 1.25-1.5 kg may be preferred.

Selection of suckers and planting: If any damage is noticed to the corm of the suckers at the time of separation of the sucker from the mother plant, the damaged portion of the corm may be clearly sliced off. The top portion of the pseudostem of the sucker may be given a slant cut leaving six inches pseudostem over the corm. This will facilitate easy draining of plant sap, rain water etc., leading to successful establishment of the sucker. But in coastal region the top portion of sucker is retained while planting and deheaded to 2/3 size in about 20 days when growth is visible to maintain uniform height of the plantation. All the old roots of the rhizome should also be trimmed.

Sucker treatment: The suckers thus prepared should be dipped in 0.25% Monocrotophos (2.5ml in 1 litre) and 0.5%Copper Oxychloride (5g in 1 litre) solution for 15-20min and then planted. The selected field must be ploughed with mould board plough to a depth of 25-30 cm and allowed to dry for two weeks. Before planting the field should have fine tilth. The dug up pits may be allowed to dry for 10 days before planting. Just before planting of the suckers, the dug up soil or fresh top earth mixed with (compost) farmyard manure 10kg, 50g folidol may be filled to three quarters of the pit. Banana suckers should not be planted either too deep or too shallow. The suckers must be planted in the centre of the pit in such a way that the corm and another two inches of pseudostem over it should submerge in the soil. Press the soil around the sucker to avoid air spaces, to give firm stand to the stem and for better sprouting of the suckers.

Gap filling: Generally the banana suckers strike roots within 10 to 15 days after planting. Maintaining optimum population for unit area always leads to higher yields. So gap filling should be done within 20 days to ensure uniformity in the crop stand.

PAPAYA

Papaya is normally propagated by seeds, being across-pollinated plant, the plants raised from seeds may have a mixed heritance with wide variability. It is wise to collect seeds from a reliable source. Freshly extracted seeds show low germination due to presence of sarcotesta which acts as an inhibitor. The seeds should be cleaned with sawdust and dried in shade before sowing. The seeds may remain viable for 9 months, if stored in air tight container and kept at room temperature. Seed stored at 4^0–6^0C in polyethylene bags remain viable till 29 months.

Seedlings may be raised at any time of the year except during the hottest, coldest and heavy raining period of the year. Seeds are sown in raised nursery beds at a depth of 1cm in rows spaced at 15cm with 2.5cm spacing in the row. Seedling may be ready within 45-60 days for transplanting. Seedlings can also be raised in polyethylene bags filled with equal proportion of soil, FYM and sand.

***Preparation of stock seedlings*:** Seedlings are raised by sowing papaya seeds before 60 days of budding and transferred to pots/polybags/field when they attain a height of 8 cm. The seedlings of 1-1.5 cm diameter are ready for budding.

Selection of scion: For scion material, vigorously grown female plant is headed back well in advance of budding (about 45 days), to induce axilliary growth.

Side shoots emerging from below the cut point, having a length of 24 cm and 1.2 cm diameter, are taken for bud wood. In this regard juvenility of the plant has to be given due consideration. It has been observed that female plant cut at a height of 30-60 cm gives rise to shoots which have vegetative buds. At a higher level, emerging shoots have reproductive buds only. Using the above rootstock and scion material, patch and shield budding are done during July, August, September and October (Table 4). The top of seedling stock is removed after a week of budding. The buds sprouted after 15 days of budding attain sufficient length after a month. The highest success of 90 per cent is obtained in patch budding if done in the first fortnight of September closely followed by 80 per cent in the second fortnight of August, whereas in shield budding good success is obtained (80 per cent) if done in the first fortnight of September. There is earliness of flowering and increased yield in budded plants as compared to the control plants.

Table 4: Seasonal effect on success in different methods of budding in papaya

Time of budding	Success (per cent)	
	Patch budding	Shield budding
July	60.0	37.5
August	80.0	56.0
September	90.0	80.0
October	50.0	40.0

Source: R.N. Singh, Gorakh Singh and O.P. Rao (1986)

Cutting: Papaya can also be propagated by cutting, but the method is more expensive and laborious, and requires a large number of mother plants; hence, the method is not usually followed. In exceptional cases, where superior types of trees bearing parthenocarpic fruits are met with, it may be of practical use. However, recent findings indicate that rapid propagation of superior papaya clones by cuttings is now possible for individual growers, to ensure the supply of uniform, high quality fruits on market. The rooting of cutting was significantly improved and disease problems in the mist beds were minimized when small cutting were immersed in solution of fungicides (captan, mancozeb), and growth retardants or IBA. Smaller cuttings (80 to 150mm long x 8 to12mm diameter) gave more rapid and better rooting than the larger cuttings and up to 95% rooting was obtained in 4 weeks with 100 cutting/m^2 during the summer months.

***in vitro* propagation:** Shoot cultures were obtained by culturing shoot tips on Murashige and Skoog (MS)-medium supplemented with 0.5 g/l (NAA) naphthalene acetic acid + 0.5 mg/l (BA) benzyle adenine. Among different combinations of growth regulators added with and without adenine sulfate to culture medium, 2 mg/l NAA + 0.6 mg/l (IBA) indole butyric acid + 60 mg/l adenine sulfate was the best for shoot multiplication. The number of proliferated shoots increased with increasing sucrose concentration in culture medium. The best results of root formation (75%) as well as root length (3.15 cm) were obtained using 2 mg/l IBA. Successful transplanting was observed when the plantlets were transferred into pots contained peat moss and sand (1:1).

PINEAPPLE

Pineapple is very easy to propagate vegetatively. Suckers arising in the axil of the leaves on the mainstem form roots and can be used for propagation. Even the crown of leaves above the fruit and parts of the stem itself can be used. Another method of propagation is by slips, which are the suckers, arising immediately below the fruit. Suckers and slips should be preferred for planting as they come to bearing earlier than the crown and produce larger fruits. Before planting, suckers are sorted out into larger, medium and small to avoid competition between plants of different sizes. Too large suckers or slips should not be used for planting. Suckers weighing 400-500g or slips of 350-450g are considered suitable as planting material. Prior to planting curing of slips and suckers for 8-10 days in shade is necessary as fresh suckers planted in moist soil begin to decay. Before planting some of the lower leaves are removed from the sucker to facilitate the formation and entry of roots into the soil. After removing scaly leaves, planting material should be treated with Chlorpyriphos (0.25%) and Carbendazim (0.1%) solution to protect against mealy bugs and heart rot, respectively. Harvested stems can effectively be utilized to produce planting materials in pineapple. The resulting suckers can be utilized for cyclic multiplication continuously. Hence the fast multiplication technique can easily be practiced by the pineapple growers to fulfil their planting material requirement throughout the year.

DATE PALM

There are three techniques to propagate date palm: Seed propagation, offshoot propagation (traditional methods), and the recently developed tissue culture techniques.

1. **Seed propagation:** Seed propagation, also called sexual propagation, although useful for breeding purposes, is not a proper method of date palm vegetative propagation, and should be discouraged. Reasons in favour of discouraging seed propagation, are the following:

- Date palm is a dioecious species and consequently half of the progeny will be males and half will be females, with no certain way to determine at an early stage the sex of the progeny, nor fruit or pollen quality prior to flowering (often only seven years later);
- Female plants originating from seedlings usually produce late maturing fruits of variable and generally inferior quality compared to established clonal palms. In a seedling plantation it is rare that more than 10 per cent of the palms produce fruit of satisfactory quality;
- Date palms are heterozygous, and thus there will be much variation within the progeny, and desirable characteristics of the parent palm may be lost. In other words, it is not true to type propagation and no two seedling palms are alike;
- Seedlings differ considerably with regard to production potential, fruit quality and harvesting time, making them very difficult to market as one harvest;

The above reasons result in waste of time, space and money.

Thus, seed propagation is by far the easiest and quickest method of propagation. However, it is not a true to type propagation technique and no two seedlings will be alike.

There is always an increased demand for date palm off-shoots of superior cultivars. Using off-shoots for propagation of date palm is up till now the main method in order to insure true to type trees. It is usually recommended that the off-shoots should not weigh less than 10 kgs, and can weigh up to 25 kg. Unlike most other fruits which can be propagated true to type from cuttings or by means of grafting or budding, dates can be propagated true to type only from offshoots. These offshoots develop from buds produced in the axils of the leaves, usually at or near the base of the palm. Occasionally offshoots grow on the trunk of the palm several feet above the surface of the ground. For example, the Deglet Noor produces on an average 10 to 12 offshoots, while the Hayany bears 20 to 50 or more offshoots per tree. The best and common means of propagation is by transplanting the suckers, or offshoots when they are 3 to 5 years old and weigh 40 to 75 lbs (18-34 kg). They are usually separated from the parent palm as needed, but in southern Algeria suckers are often put on sale standing in tubs of water. Some offshoots are maintained in nurseries until roots are formed, though most are set directly in the field after a seasoning period of 10 to 15 days just lying on the ground, in order to lose 12 to 15% of their moisture. In parts of Egypt subject to annual flooding, very large offshoots—up to 500lbs (226 kg) are planted to avoid water damage. In general, it is said that at least 2 offshoots can be taken from each palm annually for a period of 10 to 15 years. The potential of tissue culture for multiplication of date palms is being explored in Iraq, Saudi Arabia, India and California (USA). The success of date palm cloning by tissue culture methods, based on organogenesis and somatic embryogenesis, has been investigated by many workers. Organogenesis in date palms has a low efficiency due to the low number of explants that form shootlets in vitro, the long time required for the initiation phase, the low multiplication rate and the strong influence of the variety. Somatic embryogenesis has been obtained from shoot tips which were excised from off shoots.

LITCHI

Litchi can be propagated successfully by cuttings and grafting (splice and inarching). Budding is not commonly practiced. The most common and the easiest method adopted all over the world is air layering.

Air layering: It is also called "marcotting" in china and "gootee" in India. About 2 cm wide ring of bark is removed just below a bud from a healthy and vigorous twig about one year old and 2.5-4.0 cm across. The cut is then surrounded with a mud ball containing sphagnum moss (2 parts of damp moss and 1 part of soil from the foot of old litchi tree is best suited) and wrapped with a polythene sheet. Both ends are tied with fine rope or rubber bands to make it practically air tight. When sufficient roots are formed in about 2 months, the branch is cut below the soil or sphagnum moss and potted in a nursery. July – October is the most appropriate time. After removing the air layers (marcotts) from the mother plant, it is desirable to provide some moist or humid atmosphere by sprinkling water mist for further 2 or 3 weeks. It is necessary to cut back the top of the branch, so as to secure a proper proportion of leaves to root. At least 6 months old marcotted (air layered) plants should be planted in the permanent field preferably in monsoon (rainy season). I BA (2-10 g/litre of water) is the most effective in root promotion in air layering of litchi.

Litchi do not reproduce faithfully from seed, and the choicest have abortive, not viable, seed. Furthermore, lychee seeds remain viable only 4 to 5 days, and seedling trees will not bear until they are 5 to 12, or even 25, years old. For these reasons, seeds are planted mostly for selection and breeding purposes or for rootstock. Attempts to grow the litchi from cuttings have been generally discouraging, though 80% success has been claimed with spring cuttings in full sun, under constant mist and given weekly liquid nutrients. Ground-layering has been practiced to some extent.

In India, a more recent development is propagation by stooling, which has been found "simpler, quicker and more economical" there than air-layering. First, air-layers from superior trees are planted 4 ft (1.2 m) apart in "stool beds" where enriched holes have been prepared and left open for 2 weeks. Fertilizer is applied when planting (at the beginning of September) and the air-layers are well established by mid-October and putting out new flushes of growth in November. Fertilizer is applied again in February-March and June-July. Shallow cultivation is performed to keep the plot weed-free. At the end of 2 1/2 years, in mid-February, the plants are cut back to 10 in (25 cm) from the ground. New shoots from the trunk are allowed to grow for 4 months. In mid-June, a ring of bark is removed from all shoots except one on each plant and lanolin paste containing IBA (2,500 ppm) is applied to the upper portion of the ringed area. Ten days later, earth is heaped up to cover 4 to 6 in (10-15 cm) of the stem above the ring. This causes the shoots to root profusely in 2 months. The rooted shoots are separated from the plant and are immediately planted in nursery beds or pots. Those which do not wilt in 3 weeks are judged suitable for setting out in the field. The earth around the parent plants is levelled and the process of fertilization, cultivation, ringing and earthing-up and harvesting of stools is repeated over and over for years until the parent plants have lost their vitality. It is reported that the transplanted shoots have a survival rate of 81-82% as compared with 40% to 50% in air-layers.

Important Definitions

Scion: Short piece of detached shoot containing several dormant buds, which, when united with the stock, comprises the upper portion of the graft and from which the stem or branches or both grow. It should be of desired cultivar and free from diseases.

Rootstock: is the lower portion of the graft, which develops into the root system of the grafted plants. It may be seedling, rooted cutting, or a layered plant. Primary function of rootstock is to provide anchorage by growing deep into the soil. To regulate the uptake of moisture and nutrients.

Interstock: a piece of stem inserted by means of two graft unions between the scion and the rootstock. It is used: to avoid incompatibility between stock and scion. To make use of winter hardy trunk. To take advantage of its growth controlling properties.

An interstock is sometimes grafted in between a stock and scion to impart certain important characteristics to the unified plant. Interstocks are useful for dwarfing, to overcome stock-scion incompatibility, impart winter hardiness, and reduce disease problems. (The interstock may be only a thin section of wood, a short section of trunk in a fruit tree, or the trunk and lower portions of scaffold branches.) This is often referred to as "double working."

Bud-wood selection: Bud wood selection should be made from a mother plant which should be healthy, true to type and free from diseases and pests. The trees should be certified for the desirable characters by the Certifying agency.

- Choose bud wood from healthy shoots that have seen lots of sun and are moderately vigorous.
- Use the buds in the middle of the shoot: buds at the base are too old and the ones at the tip are not mature enough.
- Scion budsticks should consist of tree branches less than one year old and less than ¼ inch diameter.
- Budsticks should have dormant buds; buds that have sprouted and are actively growing are not suitable.
- Budwood should also be healthy; buds from sick trees may not be successful.
- Budsticks may be round or angular, and should be stiff enough so that when buds are cut the tissue is not smashed.
- Buds should be plump and healthy, and not too closely spaced.
- Do not choose budwood from vigorous or extremely thorny suckers or water sprouts that arise from the tree trunk or limbs.
- Clip the leaves from the budwood, leaving the petioles (leaf stems) intact. The petioles provide convenient handles that can be used to hold a bud when it is cut from the budstick.

Stock-Scion Relationship

Stock-scion relationships for successful asexual propagation

A grafted or budded plant can produce unusual growth patterns which may be different from what would have occurred if each component part of a graftage viz., rootstock and scion was grown separately or when it is grafted or budded in other types of rootstocks. Some of these have major horticultural value. This varying aspect of rootstocks will influence the performance of a scion cultivar or *vice versa* is known as stock-scion relationship.

A. Effect of Stocks on Scion Cultivars

1. Size and growth habit

In apple, rootstocks can be classified as dwarf, semi-dwarf, vigorous and very vigorous rootstocks based on their effect on a scion cultivar. If a scion is grafted on dwarf rootstocks (e.g. M.9), the scion grows less vigorously and remain dwarf only. On the other hand if the same scion is grafted on a very' vigorous rootstock (e.g. M.2) the scion grows very vigorously. In citrus, trifoliate orange is considered to be the most dwarfing rootstock for grapefruit and sweet oranges. On the other hand, in mango, all plants of a given variety are known to have the same characteristic canopy shape of variety despite the rootstocks being of seedling origin. But mango rootstocks like Kalaypad, Olour have been found to impart dwarfness in the scion cultivars. Guava cultivars grafted on *Psidium pumilum* are found to be dwarf in stature. Pusa Srijan guava rootstock also imparts dwarfness in Allahabad Safeda, a commercial cultivar of guava.

2. Precocity in flowering and fruiting

The time taken from planting to fruiting i.e., precocity is influenced by rootstocks. Generally fruit precocity is associated with dwarfing rootstocks and slowness to fruiting with vigorous rootstocks. Mandarin, when grafted on Jambhiri rootstock is precocious than those grafted on sweet orange or orange or acid lime rootstocks.

3. Fruit set and yield

The rootstocks directly influence on the production of flower and setting fruits in oriental Persimmon (*Diospyrous kaki cv.* Hachiya). When it is grafted on *D. lotus*, it produces more flowers but only few mature into fruits. However, when *D. kaki* is used as the rootstock, the fruit set is very high. The influence of rootstock on the yield performance of cultivar has been well documented in many fruit crops. Acid limes budded on rough lemon register nearly 70 per cent increased yield than those budded on troyer citrange, Rangpur lime or its own rootstock. Sweet orange var. Sathugudi budded on Kichili rootstock gave higher yield than on Jambhiri or on its own seedling.

4. Fruit size and quality

Sathgudi sweet oranges grafted on Gajanimma rootstocks produced large but poor quality fruits while on its own roots they produced fruits with high juice content and quality. The physiological disorder 'granulation' in sweet orange is very low if on Cleopatra mandarin seedlings. On the other hand, rough lemon seedling stocks induced maximum granulation. The physiological disorder black end in Bartlett Pear did not appear if *Pyrus communis* was used as the rootstock. When *P. pyrifolia* was used as the rootstock, this disorder appeared, affecting fruit quality.

5. Nutrient status of scion

Rootstocks do influence the nutrient status of scion also. Sathgudi orange trees have a better nutrient status of all nutrients in the leaves when it is budded on *C. volkarimariana* root stock than on its own rootstock or Cleopatra mandarin stocks.

6. Winter hardiness

Young grapefruit trees on Rangpur lime withstand winter injury better than on rough lemon or sour orange. Sweet oranges and mandarins on trifoliate stocks were more cold hardy.

7. Disease resistance

In citrus, considerable variability exists among the rootstocks in their response to diseases and nematodes. For instance, rough lemon rootstock is tolerant to *tristeza, xyloporosis and exocortis* but is susceptible to gummosis and nematode. On the other hand, troyer citrange is tolerant to gummosis but susceptible to *exocortis* virus disease. Similarly, guava varieties grafted on Chinese guava. (*Psidium friedrichsthalianum*) resist wilt diseases and nematodes.

8. Ability to resist soil adverse conditions

Among the citrus rootstocks, foliate orange exhibits poor ability, while sweet oranges, sour orange, rangpur lime rootstocks exhibit moderate ability to resist excess salts in the soil. Similarly, in pome fruits, variation exists among rootstocks to resist excess soil moisture or excess boron in the soil. Myrobalan plum rootstocks generally tolerate excess boron and moisture than Marianna plum root or other rootstocks' viz., peach, apricot or almond.

B. Effect of Scion on Rootstock

1. Vigour of the rootstocks

In apple, it has been found that if apple seedlings were budded with the 'Red Astrachan' apple, the rootstock produced a very fibrous root system with few tap roots. On the other hand if scion 'Goldenburg' was budded on the seedlings, they produced two or three pronged deep roots without fibrous root system. In citrus, if the scion cultivar is less vigorous than the rootstock, the rate of growth and the ultimate size of the tree is more determined by the scion rather than the rootstocks.

2. Cold hardiness of the rootstock

Cold hardiness of citrus roots is affected by the scion cultivar. Sour orange seedlings budded to 'Eureka' suffered much more from winter injury than the unbudded seedlings.

3. Precocity in flowering

Young mango rootstock seedlings (6 months to one year old) were found to put forth inflorescence when the branches from old trees are inarched which can be attributed to the influence of scion on the rootstock.

LITERATURE CONSULTED

Abellanosa, A.L. and Pava, H. M. 1987. Introduction to Crop Science. CMU, Musuan, Bukidnon: Publications Office. 245 p.

Al-Bakr, A. J. 1972. The Date palm: Past, Present and Future, 2nd ed., Al-Watten Press. Baghdad, Iraq.

Ananda, S. A. 1993. Propagation and rootstocks of apple. In: Advances in Horticulture, Vol. 2- Fruit crops (Eds., Chadha, K.L and Pareek, O.P). Malhotra publishing House, New Delhi, India, pp. 587-598.

Ananda, S. A. 1993. Propagation and rootstocks of stone fruits. In: Advances in Horticulture, Vol. 2- Fruit crops (Eds., Chadha, K.L and Pareek, O.P). Malhotra Publishing House, New Delhi, India, pp. 598-606.

Ananda, S. S. 1993. Rootstocks of Stone Fruits. In K. L. Chadha and O. P. Pareek (eds). Advances in Horticulture: Fruit Crops. Malhotra Publishing House, New Delhi, India.

Bal, J.S. 2014. Fruit Growing.3rd Edition. Kalyani Publishers,New Delhi.pp:434.

Bal,J.S.,Gill,S.S. and Sandhu,A.S.2016. Raising Fruit Nursery.2nd Edition. Kalyani Publishers,New Delhi.

Bergh, B.O. 1987. Avocado breeding in California. South African Avocado Growers Association Yearbook, 10: 22-24.

Bose, T. K., Mitra, S.K., Sadhu, M. K. and Das, P. 1997. Propagation of Tropical and Subtropical Horticultural Crops. Naya Prokash, Kolkata, India

Bryant, G. 1995. Propagation Handbook. Stackpole Books: Mechanicsburg, Pennsylvania.

Cepoiu, N. and Marval, M. 1988. Effect of Rootstock and Planting Density on the Growth and Fruiting of some Apple Cultivars. *Horicultura* 31(1): 61-68.

Chandra, A.,Chandra, A.and Gupta, I.C.1994. Arid Fruit Research. Scientific Publishers, Jodhpur, India.

de Lima, L.A., Naves, R.V., Yamanishi, O.K. and Pancoti, H.L. 2010. Behavior of three papaya genotypes propagated by grafting in brazil. *Acta Horticulturae* , 851:343-348

Gunjate, R.T and Tawde, A.B. 1993. Propagation and rootstocks of mango. In: Advances in Horticulture, Vol. 2-Fruit crops (Eds., Chadha, K.L and Pareek, O.P). Malhotra Publishing House, New Delhi, India, pp. 515-522.

Hannah, M. 2002. Nursery stock production using the pot-in-pot technique. Department of Horticulture, Ohio State University.

Hartmann, H. T., D. E. Kester, F. T. Davies and R. L. Geneve. 1996. Plant Propagation, Principles and Practices. 6th ed. Prentice Hall: Upper Saddle River, New Jersey.

Kamprath, H.. 2003. Proposal for a fruit tree orchard. GTZ DED, Blantayre, Malawi.

Khatiwada P. P. and Ghale, M. S. (eds.).1998. Nepal Agricultural Research Council, Khumaltar, Second National Horticulture Research Workshop 13-15 May, 1998.

Khatiwadwa, P. P. 1998. Potential of Summer Grafting on Pear. Proceedings of the national horticulture research workshop13-15 may,1998. Nepal Agriculture Research Council. Khumaltar.

Moll, J.N. and Wood, R. 1980. An efficient method for producing rooted avocado. Citrus and Subtropical Research Institute, Bulletin 99, Nelspruit, South Africa

Peter, K.V., Rajan, S. and Baby L. Markose. 2006. Propagation of Horticultural Crops. (Eds.) New India Publishing Agency, New Delhi.

Rahudkar W.B., Bhujbal B.G., Sonawane, M. and Rajput, H. 2010. YCMOU, Textbook Publication No. AGR 227 Horticulture Nursery Management.

Rajan, S. 2007. Propagation of Horticultural Crops, NIPA. Pp. 206

Resource Book on Horticulture Nursery Management Developed Under National Agricultural Innovation Project, Indian Council of Agricultural Research, New Delhi - 110012.

Salunkhe, D.K. and Kadam, S.S. 1995. Handbook of Fruit Science and Technology: Production, Composition, Storage & Processing. CRS Press, USA, pp. 632.

Samaddar, H. N. and Yadav, P.S. 1970. A note on vegetative propagation of cashewnut, avocado, jackfruit, and custard apple. *South Indian Horticulture*, 18(1/2): 47-49.

Saroj P.L., Singh U.V. 2010. Refinement of aonla budding technique under hot arid ecosystem. *Indian Journal of Horticulture*, 67: 90-95.

Saroj, P.L., Nath, V. and B.B. Vasishta, B. B. 2000. Effect of polycontainers on germination, seedling vigour, root characters, and budding success in aonla. *Indian Journal of Horticulture*, 57(4): 3000-304.

Sharma, R.R. 2002. Propagation of Horticultural crops. Kalyani Publishers, Ludhiana, India.

Sharma, R.R. and Srivastava, M. 2004. Plant propagation and nursery management. International Book Distributing Co., Lucknow, India.

Singh, R. N., Rao, O. P. and Singh, Gorakh 1985. A new approach to papaya propagation. *Current Sci*ence 54 : 1189-90.

Singh, R. N., Singh, G. and Rao, O. P. 1986. Vegetative propagation of papaya through budding. *Indian Journal of Horticulture,*43: 1-8.

Singh, S., Singh, A.K., Joshi, H.K., Bagle, B.G. and Dhandar, D.G. 2007. Jamun- A Fruit for Future. Technical Bulletin published by CHES, Vejalpur (Godhra), Gujarat, CIAH (ICAR), pp. 24.

Westwood, M. N. 1970. Rootstock-Scion Relationship in Hardiness of Deciduous Fruit Trees. *HortScience*, 5(5): 418-421.

Westwood, M. N. 1988. Temperate Zone Pomology. Timber Press Portland, USA.

12

Tissue Culture and Biotechnology

P. Mohanpuria

INTRODUCTION

Plant tissue culture is basically *in vitro* culturing of any type of explants such as leaf, root, bud and stem or even single cell except the seed or embryo on a suitable culture medium under controlled environmental conditions. Plant tissue culture employs the totipotency of plant cells to produce whole plant. The term totipotency was coined by German plant physiologist, Haberlandt (1902) who is known as father of tissue culture techniques.

Definition of Micropropagation

Micropropagation is one of the best and most successful examples of commercial application of tissue culture techniques. The concept of micropropagation was first presented to the scientific community in 1960 by Morel, who produced the virus-free cymbidium through meristem culture. It entails *in vitro* propagation of plants starting with very small plant parts (0.2-10mm) grown aseptically, followed by their establishment in soil under green house conditions. It is the art and science of plant multiplication *in vitro*. This technique has been standardized and commercialized for various fruits and vegetable crops.

Advantages of Micropropagation

Micropropagation as a tool for crop improvement is gaining popularity day-by-day because of the following advantages:

i. Large-scale multiplication of plants in a limited time and space.

ii. Production of disease free plants.

iii. It is independent of seasonal constraints and ensures year round, true-to-type bulking of elite materials.

iv. A small tissue requirement as an explants micropropagation always save the material.

v. Micropropagated field grown plants exhibits vigorous growth and higher yield.

vi. It helps in reducing the breeding cycle of plants.

vii. It is highly beneficial in plants in which vegetative propagation is not possible or is very slow (eg. date palm, papaya, coconut etc.).

viii. Homozygous plants can be possible under *in vitro* conditions.

ix. Production of synthetic or artificial seeds.

x. Creating genetic variability for desirable traits is possible.

xi. Long term storage of the material in lesser space is possible under *in vitro* conditions.

xii. Production of secondary metabolites in tissue culturing.

Limitations of Micropropagation

The followings are the main limitations in micropropagation of plants:

i. Microbial (bacterial and fungal) contaminations in the tissue culture hinder the growth of our propagules because these grow very fast in comparison.

ii. Different procedures carried out in laboratory are technically skilled.

iii. The basic facilities required for micropropagation such as laminar, autoclave, incubation room and green house/hardening room are usually costly.

iv. Protocol of making complete plants from small explants may vary from species to species. Further, hardening of micropropagated plants is also difficult because these plants generally have abnormal leaf morphology, poor photosynthetic efficiency, malfunctioning of stomata (open), reduced epicuticular waxes.

v. The cultured explants of certain plant species like citrus, strawberry, apple etc. secrete phenolic substances into the medium, which cause browning due to oxidation of phenols and/or formation of quinines which mostly affect the growth of cultured explants. Sometimes, the plants having high level of phenols do not respond in tissue culture.

STAGES IN MICROPROPAGATION

Micropropagation of any commercial crop is divided into at least five stages. A brief description of these stages has been here under:

Stage 1: Selection and maintenance of stock plants or mother plants for explants isolation to initiate the aseptic cultures

The success of micropropagation depends upon this preparative stage. The mother plants from which explants have to be excised should be:

a. A certified, true-to-type representative of the desired species or a cultivar/genotype.

b. Healthy and free from insect-pests and diseases, and be maintained on regular pesticide and fertilizer programme. The mother plants should be indexed for the pathogens.

c. Quite vigorous.

Desired explants should be excised from the selected mother plants during the active growth phase (spring season) of the year.

Stage 2: Initiation and establishment of aseptic culture

The establishment of explants depends on several factors such as the source of explants, type of explants such as leaf, root and even it can range in size from 0.1mm to large corms and stems and *in vitro* culture conditions. This stage lasts from one week to 2-3 months or even larger. This stage involves collection of explants, its surface sterilization and culturing on suitable media conditions. Following points should be taken care under this:

a. Surface sterilization: The explants are sterilized by treating with any one of the surface disinfectants such as sodium hypochlorite (0.5-5% for 5-30min), mercuric chloride (0.1-1% for 2-10min), silver nitrate (1% for 5min), bromine water (1-2% for 2-10min), calcium hypochlorite (9-10% for 5-30min), hydrogen peroxide (3-12% for 5-15min) and ethanol (70-95% for 0.5min) etc. depending upon the species/explant used. Sometimes, few drops of detergents like tween-20 or teepole is also added at this stage for effective penetration of surface disinfectant as it reduces the surface tension between explants and disinfectant.

b. Medium composition: The medium should provide the explants with all nutrients that are necessary to nourish and to make the explant grow as desired by adjusting the concentration of PGRs. The basal medium formulation is often standard eg. MS (1962) but more complex media may be necessary for smaller explants. The selection of basal media and PGRs at this stage is of special significance, and may vary according to plant and tissue type.

c. **Controlled conditions**: The explants inoculated in the media should be maintained under optimum light (cool white fluorescent light of 70±5μmol (photons) $m^{-2}S^{-1}$ with a 16-18 hours light photoperiod on a 24 hours light/dark cycle), temperature (25±2 °C) and relative humidity (55± 5%) for its establishment.

Stage 3: Proliferation and rapid multiplication of shoots

Main aim of this stage is to increase the number of propagules or plantlets. For this, repeated sub-culturing is done after every 2-3 weeks to encourage more proliferation of explants. Repeated culture of axillary and adventitious shoots, nodal cuttings, somatic embryos and other organs from stage 1 lead to multiplication of propagules. Generally, higher concentrations of cytokinins stimulate multiplication of axillary or adventitious shoots but higher concentration of auxins are necessary for callus proliferation and somatic embryogenesis. The duration of this stage is usually several months to 1-2 years and largely depends on the choice of propagator.

Stage 4: Rooting of shoots

The shoots obtained from stage 2 are transferred to rooting medium (containing auxins like IBA) to induce root primordial or root formation. Several factors such as PGRs, micro and macronutrients, growth supplements, light and temperature affect the root formation at this stage. For this, media modifications such as reduced or completely eliminated cytokinins, reduced sugars, increased auxins and gelling agents is desirable. The rooting can also be induced when *in vitro* shoots are exposed to stress conditions. For rooting, regenerated shoots are treated with auxins, usually either by putting the shoots on treated with auxin, usually either by putting the shoots on auxin containing medium or by dipping them in auxin solution and planting on hormone free medium (basal medium). In some plant species, transfer of shoots to a cytokinin free media is enough to produce rooting. Many plants such as banana, pineapple, roses, potato, chrysanthemum, strawberry, mint and several grasses etc. are rooted on the half-strength MS basal medium. In addition, the environmental conditions (especially light intensity is increased) in the laboratory are also modified. Good sturdy well-rooted plants are essential for high survival during next stages.

Stage 5: Acclimatization or hardening and transfer to soil/field

The *in vitro* obtained plants are hardened/acclimatized before transfer to field. Because plantlets produced under laboratory conditions are very delicate and are usually incapable of being established in natural field conditions. Hence, these should be hardened for 4-8 weeks by modifying the environmental conditions in the laboratory itself. It is a process by which physiologically and anatomically plantlets adjust from in vitro to ex vitro conditions. The plantlets are maintained under partial shade and high humidity, by covering them with clear plastic bags or by keeping them in mist chambers. Thus, these plantlets are acclimatized or hardened by gradually reducing humidity or by providing diffused light for few days or by removing plastic bags for intermittent period in a day. Sometimes, plantlets are transferred onto moist cotton/papers in trays and are kept in the incubation room with daily change of tray water for few days. Once the stomata start functioning and new and functional roots develop, photosynthetic activity is increased; as a result, plants become autotrophic. This is the final stage of the tissue culture operation after which the micropropagated plantlets are ready for transfer to the greenhouse. These plantlets are then transferred to an appropriate substrate (peat, vermiculite, sand: soil mixture etc.) and are gradually hardened to get established in the field easily.

PATHWAYS OF REGENERATION OR ROUTS OF MICROPROPAGATION

A complete plant can be regenerated from the excised plant parts by number of pathways or routes. Routes of micropropagation is based on the type of explants used for *in vitro* culture. Explants for micropropagation can be (a) having preformed meristem such as axillary buds; (b) lacking preformed meristem such as organogenesis and somatic embryogenesis. Thus, routes of micropropagation can be categorized into 3 types which are as follows:

(A) Axillary bud formation

In axillary bud formation, as the meristematic cells are already present in the explants, meristem tip culture and shoot tip culture are included. It is considered as the convenient route for micropropagation, since it does not include callus stage. The excised meristem tip is typically small (often <1mm in length) and is removed by sterile dissection under the microscope, as in the case of potato. This explant comprises the apical dome and a limited number of the youngest leaf primordia and excludes any differentiated provascular or vascular tissues. The major advantages of using small-sized explants are production of pathogen free, especially virus free plants, minimizing genetic instability and somaclonal variations in the culture. While in shoot tip culture, comparatively a larger sized (often 0.2-10 mm) shoot apex that contains a relatively large number of developing leaf primordia are preferred. In this route, the existing meristems such as shoot tip or nodal bud is cultured in the medium containing more concentration of cytokinin than auxin and the shoot proliferation depends on the cytokinin used.

The most commonly used cytokinins are Benzyl aminopurine (BAP) and kinetin. Both the excision of the explant and the cytokinin rich medium activate the bud, leading to a large number of side shoot proliferation which can be separated later for further culture. These regenerants are considered genetically stable as compared to regenerants from adventitious meristems. Axillary bud formation is considered as the ideal method of micropropagation but it is limited by slow multiplication rate. Although, in micropropagation technique, chances of survival of explants increases with the increase of size of explants but the chances of contamination also increase along with.

(B) Organogenesis and Somatic embryogenesis

In case of organogenesis and somatic embryogenesis routes of micropropagation, preformed meristem is lacking and thus, each of these is of two types: a) *de novo* method/indirect method: Micropropagation proceeds through callus phase. b) adventitious method/direct method: Adventitious shoots are produced from unusual point of origin, other than shoot tip and axillary buds and includes no callus phase. In case of *de novo* somatic embryogenesis, induced embryogenic determined cells (IEDC) in the explants are to be induced for producing embryogenic pathway while in adventitious somatic embryogenesis, somatic embryos are induced from pre-embryogenic determined cells present in intraovular tissues like integument, egg, nucellus and scutellum etc. The somatic embryogenesis is the best technique for rapid and true to type multiplication of plants. The somatic embryos originate either from somatic or vegetative cells and are bipolar in nature i.e the shoot and root meristems in the explants are formed simultaneously. Embryo induction requires high auxin initially followed by low auxin and cytokinin medium. In general, somatic embryogenesis is more often favoured by culture in an agitated liquid medium. Embryogenesis starts with globular stage, then to ear-shaped stage where roots and shoots can be distinguished and finally to torpedo stage in which shoots and roots have vascular connections. Furthermore, the somatic embryos can be encapsulated, thus producing artificial synthetic seeds, which can be used in situations where complete viable seeds are not formed.

In case of organogenesis route, the initiation of adventitious meristem depends on the proper balance between cytokinin and auxin. As a general rule, higher ratio of cytokinin to auxin leads to shoot formation and reverse is true for root formation. The plants produced via this method are not always genetically stable, due to the accumulation of aneuploid and polyploid cells with continuous subculturing of callus causing mixoploids. Further, repeated subculture of callus also reduces its regenerative capacity.

MICROPROPAGATION SYSTEMS

Based on the part of the plant to be cultured, different micropropagation technique for different plant species are followed. Some of them are

1. Meristem Tip Culture

The meristem tip consisting of one or two pairs of leaf primordial are cultured in the culture medium. The plants regenerated after few weeks are hardened and released into the natural environmental conditions. The advantage of using this system is for producing virus free planting material. Rapid multiplication of planting material is also possible by this system. Horticultural crops like potato, dahlia, carnation and orchids are widely produced using this method.

2. Embryo Culture

Embryo rescue and culture is important where post-zygotic incompatibility occurs. In this system, aseptic culture of embryo is to be standardized. The immature embryo is rescued before degradation occurs. The excised embryo is cultured in the semi-solid or solid medium. Apart from normal nutrients in the medium, some additional plant extracts like coconut milk or yeast extract is added for enhanced growth and development of embryos. This system is used to overcome the seed dormancy when distant crosses are made in hybrid production. In horticultural crops, this technique is widely used in grapes where seedless varieties are produced by embryo culture of crosses between seedless X seedless varieties.

3. Callus Culture

A callus is an unorganised mass of cells which are parenchymatous in nature. Callus formation is a natural healing process by plants during wounds, grafts, bud union and rooting of cutting processes. Callus culture requires auxins initially then followed by cytokinins at later stages. It also requires stimulation of cell division. Callus culture involves two stages of organogenesis. The first stage involves the formation of meristems followed by the second stage where the growth of stem buds and roots takes place. The transition between two stages requires different concentration of hormones.

4. Anther Culture

Anther culture technique was developed by Guha and Maheshwari (1964) in Datura plant. This technique is used to produce plants which are useful for producing homozygous lines. This is used to produce haploid plants. In this technique, excised anthers are cultured on culture medium. The anthers are usually excised during microscopic mitosis stage (the stage when the pollen passes from uninucleate to the binucleate condition). These haploids so produced are treated with colchicine for doubling of chromosome to produce the diploid plants.

5. Cell Culture

This technique is very useful for plant breeders as it enables them to induce desirable variability in a single cell and the complete plant can be regenerated from that single cell. In this technique, cell suspension cultures are obtained by homogenising a piece of callus into the liquid medium and shaking with shaker until medium becomes cloudy with suspended cells.

6. Protoplast Culture

Protoplast isolation was demonstrated first by Cocking in 1960. In this technique, the cell wall is removed first by mechanically or enzymatically to get protoplast. This protoplast is then cultured in the culture medium. The advantage with protoplast is that to search protoplast can be brought together by a chemical or physical agent termed fusogen. With this method wide crosses like intergeneric or even interkingdom hybridizations are possible.

7. Ovule Culture

In this technique, unfertilized excised ovules are cultured into the medium. This technique can be used to produce hybrid seeds in wide genetic crosses, which is rather difficult through conventional methods. It is also useful in obtaining the plants in self-incompatible species.

Selection of Explant

Theoretically, all living plant parts are capable of regeneration *in vitro* but depends upon the quality of the source or the mother plant. This mother plant should have the following characters:

a. True to type plant of the desired species.

b. Healthy plants which are pest and disease free.

c. Quite vigorous.

Further, the juvenile tissues respond more readily to *in vitro* culture. Apart from the above characters the type of explants such as root, leaf, stem and *in vitro* culture conditions such as media composition, temperature, humidity and light etc. affect the explants growth and establishment.

FACTORS FOR SUCCESSFUL MICROPROPAGATION

There are certain factors that must be considered for successful micropropagation which are as follows:

i. A complete aseptic conditions are to be maintained while autoclaving the culture media and during work in laminar. The Inoculation room should be fumigated periodically using dilute formaldehyde.

ii. Activated charcoal (1-2%), citric acid (50-100mg/l) and polyvinyl pyrrolidone (PVP) are to be added in the culture medium to avoid from oxidation of phenol is secreted by certain plants results in browning which produces toxic effects.

iii. A well standardized tissue culture medium should be used.

iv. A true-to-type representative of the desired healthy, cultivar/genotype with active growth should be used for micropropgation.

ESSENTIAL REQUIREMENTS FOR MICROPROPAGATION

Basic and essential requirements for micropropagation are as follows:

1. A well-equipped laboratory

The laboratory should have sufficient working area for culturing and other handling processes. The laboratory should have proper electricity and water facilities throughout the day. For controlling temperature, an automatic thermostat with a timer should be installed. All other equipment for micropropagation for example autoclave, laminar etc. should be established.

2. Aseptic condition

The aseptic condition is the most basic need in a micropropagation laboratory. Asepsis starts with surface disinfectants like mercury chloride. The culture medium and the instruments should be sterilized by autoclaving at pressure 15 lbs/inch^2 and at a temperature of 121ÚC for 20 minutes. The thermolabile substances like vitamins and PGRs are to be filter sterilized by 0.22 μm filter.

3. Culture medium

The nutrients requirement of the explant for growth and development is provided by the culture medium. It contains both organic and inorganic salts and their concentration varies from one medium to another. The most commonly used media in micropropagation are MS (Murashige and Skoog, 1962), B5, WPM (Woody Plant Medium), Whites (1963) and Nitsch and Nitsch (1956) Medium.

4. Culture environment

Temperature and light play a very important role in growth and development of explant. The light intensity of 300-400 klux is required in tissue culture laboratory. Duration of light is another factor influencing the regeneration of plants. Generally, 16 hours of light and 8 hours of darkness are required. Similarly blue light favours the fruit initiation and red light stimulates the shoot initiation. The temperature of 27±2ÚC is required for micropropagation of most of the fruit crops.

Media Used for Hardening of Tissue Cultured Plants

Based upon the need, different media in individual or in combinations are used for hardening of the tissue culture raised plants which are as follows:

Garden soil: Garden soil provides nutrients to the plants efficiently. It is also used to enhance the water-holding capacity of soil.

Sand: Sand is usually used in the media where maximum aeration is required. But sand has low water and nutrient holding capacity so water and nutrients should be supplied frequently.

Peat: Peat is a good source of nitrogen and has high water holding capacity.

Perlite: Perlite is added in the media to enhance the aeration but does not contain any nutrients.

Vermiculite: Vermiculite is a good source of magnesium and potassium.

Soilrite mix: Soilrite mix is a mixture of peat, perlite and vermiculite. It contains good water holding capacity and enhanced aeration.

PROBLEMS IN MICROPROPAGATION PROCESS

The following are the main problems encountered during micropropagation:

i. The facilities required for micropropagation are costly.

ii. Most of the procedures in laboratory require labour with technical ability.

iii. Microbial contaminations once appear, have a tendency to quickly spread across all the systems in the laboratory.

iv. Some plants do not respond well to the micropropagation protocol followed like high phenol containing plants.

v. Hardening of plants at field level is a very difficult procedure and varies from one plant to another some plants show very good success in laboratory conditions but fail to establish in the field.

How to Prepare Tissue Culture Media

Most of the frequently used tissue culture media follow the following methods of preparation:

Initially all the stock solutions of major and minor salts, vitamins and PGRs are prepared separately for future use and are stored at 4°C. The required quantity of agar is weighed and dissolved in required double distilled water while heating at 80-100°C. The required sucrose is added. The pH is adjusted to 5.7 or 5.8 after making the final volume with the help of dilute HCl or NaOH. The medium is then poured into test tubes or flasks after boiling. The test tubes are plugged with non-absorbent cotton and covered with Aluminium foil. The medium autoclaved is kept in the incubation room for 4-5 days and then it can be used.

Composition of Murashige and Skoog 1962 Media

The scheme for preparing the stock solutions for MS media is given in the Table 1.

Table 1: Composition of MS (1962) medium

Stock no.	Strength of stock	Constituents	Quantity (g/l)	Use of stock (ml/l)	Actual amount in the culture medium (mg/l)
			2 litre		
1	20X	$NH_4 NO_3$	66.0	50	1650
		KNO_3	76.0		1900
		$KH_2 PO_4$	6.800		170
		H_3BO_3	0.248		6.2
		$MnSO_4.7H_2O$	0.892		22.3
		$ZnSO_4. 7H_2O$	0.344		8.6
		KI	0.033		0.825
		a) $CuSO_4. 5H_2O$	1.0 ml		0.025
		b) $Na_2MoO_4. 2H_2O$	1.0 ml		0.250
		c) $CoCl_2. 6H_2O$	1.0 ml		0.025
			½ litre		
2	50X	$CaCl_2. 2H_2O$	11.0	20	440
			½ litre		
3	50X	$MgSO_4. 7H_2O$	9.250	20	370
			1 litre		
4	100X	$FeSO_4. 7H_2O$	2.780	10	27.8
		Na_2 EDTA	3.728		37.28
			1 litre		
5	100X	Thiamine HCl	0.010	10	0.1

Contd.

Stock no.	Strength of stock	Constituents	Quantity (g/l)	Use of stock (ml/l)	Actual amount in the culture medium (mg/l)
		Nicotinic acid	0.050		0.5
		Pyridoxine HCl	0.050		0.5
		Glycine	0.200		2.0
		*Myoinositol			100
		*Sucrose			30g/
		*Agar			18 g/l

a)Dissolve 100mg $CuSO_4$. $5H_2O$ in 100ml distilled H_2O.
b)Dissolve 1g Na_2MoO_4. $2H_2O$ in 100ml distilled H_2O.
c)Dissolve 100mg $CoCl_2$. $6H_2O$ in 100ml distilled H_2O.

Note: 1. Add 1 ml concentrated HCL in stock solution no. 4 to avoid precipitation.
2. Heat slightly stock solution no. 4 to avoid precipitation.

Plant Growth Regulators (PGRs) Used in Tissue Culture Media

In order to achieve an optimal rate of micropropagation, the basal medium is modified with the addition of PGRs (Table 2) such as auxins, cytokinin and gibberellins etc. These PGRs regulate the growth and development of cultured plant tissues. Auxins promote the cell growth, elongation, and rooting while cytokinins promote cell division and shoot proliferation. Auxins and cytokinins are the most widely used PGRs in plant tissue culture and are usually used together. The ratio of the auxin to cytokinin determines the type of culture established or regenerated. A high auxin to cytokinin ratio generally favours root formation while its inverse favours shoot formation. The gibberellins help in elongation of shoots. Abscisic acid (ABA) inhibits the cell division and promotes distinct developmental pathways like somatic embryogenesis.

Table 2: Plant growth regulators used in tissue culture

Plant growth regulator	Solvent to dissolve	Working concentration (mg/l)
Auxins		
(i) 2,4-Dichlorophenoxy acetic acid (2, 4-D)	Ethanol/1N	0.1-5.0
(ii) Napthalene acetic acid (NAA)	NaOHIN NaOH	0.01-10.0
(iii) Indole-3-butyric acid (IBA)	Ethanol/1N NaOH	0.1-10.0
(iv) Indole-3-acetic acid (IAA)	Ethanol/1N NaOH	0.01-3.0
Cytokinins		
(i) Adenine	Water1N	0.1-5.0
(ii) 6-Benzyl amino purine (BAP)	NaOH1N	0.1-10.0
(iii) 6-Dimethyl allyl amino purine (2-ip)	NaOH1N	1-30
(iv) Kinetin	NaOH	0.1-5.0
(v) 1-Phenyl1-3-(1,2,3- thiadiazol-5 yl) urea (Thidiazuron)	DMSO/Ethanol	0.001-0.05
(vi) Zeatin	1N NaOH	0.01-5.0
Others		
(i) Gibberellic acid (GA_3)	Ethanol	0.01-5.0
(ii) Maleic acid hydrazide (MH)	1N NaOH	0.01-10.0
(iii) Phloroglycinol	Water	50-150

Precautions: Always store the stock solutions in refrigerator. Never use the precipitated stock solutions for media preparation. IAA stock solution should be kept in dark coloured bottles to avoid oxidation. Heat labile PGRs like GA_3 and zeatin are prepared and filter sterilized through 0.22 µm filter to a sterile container. The filtered solution is then aseptically added to the culture medium, which has been autoclaved and allowed to cool to approximately 35-45 °C.

How to Prepare Cytokinin Stock Solution

Take 50mg of cytokinin in a clean dry test tube. Dissolve it in few drops of 1N NaoH. To this, add double-distillled water to make the volume to 100ml.

Protocols for Fruit Crops

Micropropagation technique has been successfully adopted in many horticultural crops in India which are listed below in Table 3.

Table 3: Micropropagation technology developed for horticultural crops in India

Fruit Crop	Institution, which standardized technology
Annona	IIHR, Bangalore; BARC, Mumbai; NCL, Pune
Banana	IIHR, Bangalore; TNAU, Coimbatore; NCL, Pune
Citrus	NBRI, Lucknow; NRC Citrus, Nagpur; University of Jodhpur, Jodhpur
Grape	IARI, New Delhi
Guava	GBPUAT, Pantnagar
Papaya	IARI, New Delhi
Pineapple	BARC, Mumbai
Strawberry	TERI, New Delhi
Date palm	AAU, Gujarat

Plate 1-4: Micro-propagation in fruit plants at various stages in Laboratory (See colour version on page 422)

LITERATURE CONSULTED

Bal, J.S. 2014. Propagation and nursery management. Micropropagation. In: Fruit Growing. Pp: 41-48. Kalyani Publishers, New Delhi.

Bal, J.S. S.S. Gill and Sandhu A.S. Sandhu, 2016. Propagation of different fruit plants. Micro propagation. In: Raising Fruit Nursery. Pp: 56-57. Kalyani Publishers, New Delhi.

Bhojwani S.S and Dantu, P.K. 2013. Plant Tissue Culture: An Introductory Text. Springer India.

Chanana, Y. R. and M.I.S. Gill. 2008. Revised Propagation and Nursery Management. Pdf. Nsdl. niscais. res. In.

Chawla, H. S. 2002. Introduction to Plant Biotechnology. 2nd Edition, Science Publishers, Inc., USA.

Cocking, E.C.1960. Some effects of chelating agents on roots of tomato seedlings. *Biochem Jour*. 76:51-52.

DoreSwamy, R. and Sahiyram, L. 1993. Tissue Culture and Biotechnology. In: Advances in Horticulture Vol. 2. Ed. by K. L. Chadha and O.P. Pareek. Malhotra Publishing House,New Delhi.

Guha, S. and Maheshwari,S.C. 1964. *In vitro* production of embryo from anthers of Datura. *Nature,*Lond., 204:497.

Haberlandt,G.1902.Culurversuche mit isolierten Pflanzenzellen. Sitz-Ber. Mat. Nat. Kl. Kais. Akad. Wiss.111:69-92.

Kaur A., Mohanpuria P., Sarao N.K., Cheema G.S., Sidhu G. Singh and Sandhu J.S. 2013. Sugarcane Micropropagtion: A laboratory Manual (No. COA/2013/Manual 23). School of Agricultural Biotechnology, Punjab Agricultural University, Ludhiana, India, Pages 59.

Lambardi M., Ozudogru, E.A.and Jain, S.M. 2013. Protocols for micropropagation of selected economically-important horticultural plants. Methods in Molecular Biology, vol.994. Springer Science+Business Media,Humana Press.New York.

Mishra S. P. 2009. Plant Tissue Culture. Ane Books Pvt. Ltd, New Delhi.

Morel,G.M.1960. Producing virus free cymbidium.*Bull.Am.Orchid.Soc.*,15:214.

Murashgi,T. and Skoog,F.1962. A review for rapid growth and bioassay with tobacco tissue culture. *Physiol.Plant.*,15:472-97.

Nitsch, J.P. and Nitsch, C.1956. Auxin-dependent growth of excised *Helianthus tuberose* tissue, *Am.J.Bot.*,43:839-51.

Sharma, R.R. 2013. Propagation of Horticultural Crops: Principles and Practices. Kalyani Publishers, New Delhi.

White, P.R.1963. The cultivation of animal and plant cells. Handbook of Tissue Culture. Jaques Cartell, Pennsylvania.

13

Seed Germination and Dormancy

D.P. Sharma and Rimpika

SEED GERMINATION

Germination is the growth of a plant contained within a seed; it results in the formation of the seedling, it is also the process of reactivation of metabolic machinery of the seed resulting in the emergence of radicle and plumule. The seed of a vascular plant is a small package produced in a fruit or cone after the union of male and female reproductive cells. All fully developed seeds contain an embryo and, in most plant species some store of food reserves, wrapped in a seed coat. Some plants produce varying numbers of seeds that lack embryos; these are called empty seeds and never germinate. Dormant seeds are ripe seeds that do not germinate because they are subject to external environmental conditions that prevent the initiation of metabolic processes and cell growth. Under proper conditions, the seed begins to germinate and the embryonic tissues resume growth, developing towards a seedling.

The most common example of germination is the sprouting of a seedling from a seed of an angiosperm or gymnosperm. In addition, the growth of a sporeling from a spore, such as the spores of hyphae from fungal spores, is also germination. Thus, in a general sense, germination can be thought of as anything expanding into greater being from a small existence or germ.

Seed germination is a complex physiological process that it is response to environmental signals such as water potential, light, nitrate and other factors . Poor seed germination is the major limiting factor of apple (*Malus x domestica* Borkh.) for large scale production and cultivation under cold arid conditions. Seed germination in general can be controlled by many factors like natural germination (growth) inhibitors. These are the derivatives of benzoic acid, cinnamic acid, coumarin, naringenin, jasmonic and abscisic acid (ABA). It has been postulated that seed coat (testa) of many plant species contains considerable amount of germination inhibitor, which prevent their seed germination. The first stage of germination consists of ingesting water and an awakening or activation of the germplasm. Protein components of the cells that were formed as the seed developed became inactive as it matured.

Control is exercised by four classes of plant hormones: inhibitors such as abscisic acid which block germination; auxins which control root formation and growth; the gibberellins which regulate protein synthesis and stem elongation; and cytokinins that control organ differentiation. Gibberellic acid (actually a group of related substances called gibberellins) was discovered as a metabolic byproduct of the fungus *Gibberella fujikuroi*. Gibberellic acid (GA_3) is known to be concerned in the regulation of plant responses to the external environment. Recent studies have revealed that cold stratification has a direct effect on production of gibberellins (GAs) in seeds of *Arabidopsis thaliana*.Exogenously applied GA_3 overcomes seed dormancy in several species and promotes germination in some species that normally require cold stratification, light, or after-ripening. GA promotes the production of enzymes such as endo-â-mannanase, which loosen cell walls in the endosperm, thereby reducing resistance to radicle emergence. In some temperate species, dormancy is broken by a period of warm temperatures followed by cold stratification. This response is most often associated with morphophysiological dormancy ; however, seeds with morpho-physiological dormancy have underdeveloped embryos .In order to accelerate this method, it can be combined with some treatments such as chemical applications or mechanical seed coat removal. Many investigators have studied the effects of exogenous growth regulators on seed germination. Gibberellins eliminated the chilling requirements of peach and apple seeds and increased their germination.

Temperate fruit production in India is based mainly on apple. In India apple ranks 5th in production and roughly occupies 4.5% of fruit area in the country. The apple is mostly cultivated in Jammu and Kashmir followed by Himachal Pradesh and Uttarakhand and Jammu Kashmir alone accounts for 60% of its production in India. Since more than 60% area of Jammu and Kashmir state falls in Ladakh region hence this study was undertaken to extend and popularize its cultivation in this vast area with prime objective to increase the seedling rootstocks of apple so that it can establish easily and survival rate is improved for increasing the apple fruit production in this nontraditional area. Sexually raised plants are long lived, have extensive and powerful root system with wide and deep distribution in the soil which accounts for persistent and adequate annual growth of absorbing roots thus helping them acclimatize easily in the environment. Also rootstocks exhibit a great effect on production efficiency yield quality, adaptability, tree vigor, impart resistance to biotic (diseases and insects/pests) and aboitic stresses (salt tolerance and water logging) of scion cultivar with the main function of seedling rootstocks is to provide anchorage by growing deep in to the soil and also regulate uptake of moisture and nutrients. Hence this study was under taken to standardize apple seed germination with regard to different concentrations of gibberellic acid so this large number of apple fruit trees may be established in this nontraditional area of Ladakh India for commercial orcharding and get huge returns of apple fruit production in this remote area of India under cold arid conditions for the benefit of country in general and farmers in particular.

Chemically induced germination of unstratified apple (*Malus x domestica* Borkh.) has two possible uses. First, by eliminating stratification one might obtain seedlings immediately after seed maturation. Second, wide familial variation in germination time might be reduced by chemically treating either partially stratified or unstratified seed. Many investigators have studied the effects of exogenous growth regulators on seed germination. The application of gibberellins increases the seed germination percentage by increase of the amino acid content in embryo and they cause release of hydrolytic enzyme required for digestion of endospermic starch when seeds renew growth at germination. Gibberellic acid acts synergistically with auxins, cytokinins and probably with the other plant hormones; what might be called a system approach, or synergism. The overall development of plant is regulated by the growth hormones, nutrient and environmental factors. They also vary in their germination requirement. It is not known in which concentrations these hormones will cause a response in the cell. This investigation with growth hormones will help in determining which of the hormonal concentration is suitable for seed germination and proper seedling growth. Hence the main objective of this study was carried out to find out the effect of GA_3 on germination of apple seed, their subsequent growth and survival of seedlings. In view of the above back ground, the present investigation was undertaken to study the influence of growth substances like Gibberellic acid (GA_3) on seed germination, radical and plumule elongation; to draw the information of timing and control of seed germination and seedling growth of a species in their natural habitat. Experiment, seeds of all treatments were placed on filter paper moistened with 3% fungicide solution (Captan) in Petri dishes were placed in an incubator at 21±1°C and 70– 80% humidity with darkness. Germination was measured in 3-day intervals during 30 days. All seeds with at least a 5 mm long radicle were considered as germinated. Germination percentage per treatment was calculated as the average of three replicates with 50 seeds. Percentage data of germination were subjected to the angle transformation and the analysis of variance was performed. After long dipping, seeds were sown in the pots in the month of March. Observations on germination of seeds, length of seedlings, diameter of seedling, number of leaves per seedling, length and width of leaves, length and diameter of root and survival of seedling were recorded and compared with control statistically. These were soaked for 30, 35 and 40 h in the above concentrations as per treatment combination and only double distilled water for the control set.

From the above discussion, it is concluded that different concentrations of Gibberellic acid (GA_3) endorsed germination percentage and growth of seedlings in apple seed i.e. radical and plumule elongation. In this study, in order to remove dormancy in apple seeds, exogenous GA_3 application has been most successful in breaking dormancy at 500 ppm concentration. This it is recommended from this research that that the concentration of growth regulators GA_3 500 ppm for 40 hrs favours the increased enzymatic activity that leads to the favourable environment for the seed germination as well as the growth of the radicle and plumule leading to better growth and survival of seedlings. Hence it is inferred that an adaptation and practical application of these findings in nurseries might also have an economic effect.

Many species of seed do not germinate when placed under conditions which are regarded as favourable for germination, namely an adequate water supply , a suitable temperature and the normal composition of the atmosphere. These seeds are viable , and can be induced to germinate after various treatments. Such seeds are said to be dormant or in a state of dormancy. Seed dormancy is a well known phenomenon in deciduous trees. Peach seeds are dormant at the time of fruit harvest and normally require stratification at 2 °C to 5 °C under moist conditions for 10 to 12 weeks to bring about the resumption of growth. This phenomenon has also been observed in some vegetable crops and ornamental flowers. There are several methods for inducing or promoting the germination of dormant seed. The after-ripening process as described above, and the application of growth regulators, like gibberellic acid and kinetin are two ways of breaking dormancy of both fruit buds and seed. Light and temperature also exhibit a great influence on the germination of dormant seed. The nature of the mechanism involved in the breaking of dormancy by gibberellic acid and by the chilling process is an intriguing problem. Biochemical and physiological changes in peach seed as affected by gibberellic acid or chilling has been investigated but our knowledge of the dormancy mechanism is still incomplete. Nucleic acid metabolism may be involved in gibberellic acid- induced germination and chilling may also involve a stimulation of nucleotide synthesis.

Seed germination depends on both internal and external conditions. The most important external factors include right temperature, water, oxygen or air and sometimes light or darkness. Various plants require different variables for successful seed germination. Often this depends on the individual seed variety and is closely linked to the ecological conditions of a plant's natural habitat.

Germination response is affected by environmental conditions during seed germination.

Water : Water is required for germination. Mature seeds are often extremely dry and need to take in significant amounts of water, relative to the dry weight of the seed, before cellular metabolism and growth can resume. Most seeds need enough water to moisten the seeds but not enough to soak them. The uptake of water by seeds is called imbibition, which leads to the swelling and the breaking of the seed coat. When seeds are formed, most plants store a food reserve with the seed, such as starch, proteins, or oils. This food reserve provides nourishment to the growing embryo. When the seed imbibes water, hydrolytic enzymes are activated which break down these stored food resources into metabolically useful chemicals. After the seedling emerges from the seed coat and starts growing roots and leaves, the seedling's food reserves are typically exhausted; at this point photosynthesis provides the energy needed for continued growth and the seedling now requires a continuous supply of water, nutrients, and light.

Oxygen: Oxygen is required by the germinating seed for metabolism. Oxygen is used in aerobic respiration, the main source of the seedling's energy until it grows leaves. Oxygen is an atmospheric gas that is found in soil pore spaces; if a seed is buried too deeply within the soil or the soil is waterlogged, the seed can be oxygen starved. Some seeds have impermeable seed coats that prevent oxygen from entering the seed, causing a type of physical dormancy which is broken when the seed coat is worn away enough to allow gas exchange and water uptake from the environment.

Temperature : Affects cellular metabolic and growth rates. Seeds from different species and even seeds from the same plant germinate over a wide range of temperatures. Seeds often have a temperature range within which they will germinate, and they will not do so above or below this range. Many seeds germinate at temperatures slightly above 60-75° F (16-24° C) [room-temperature if you live in a centrally heated house], while others germinate just above freezing and others germinate only in response to alternations in temperature between warm and cool. Some seeds germinate when the soil is cool 28-40° F (-2- 4° C), and some when the soil is warm 76-90° F (24-32° C). Some seeds require exposure to cold temperatures (vernalization) to break dormancy. Some seeds in a dormant state will not germinate even if conditions are favourable. Seeds that are dependent on temperature to end dormancy have a type of physiological dormancy. For example, seeds requiring the cold of winter are inhibited from germinating until they take in water in the fall and experience cooler temperatures. Four degrees Celsius is cool enough to end dormancy for most cool dormant seeds, but some groups, especially within the family Ranunculaceae and others, need conditions cooler than -5° C. Some seeds will only germinate after hot temperatures during a forest fire which cracks their seed coats; this is a type of physical dormancy.Most common annual vegetables have optimal germination temperatures between 75-90° F (24-32° C), though many species (e.g. radishes or spinach) can germinate at significantly lower temperatures, as low as 40° F (4° C), thus allowing them to be grown from seeds in cooler climates. Suboptimal temperatures lead to lower success rates and longer germination periods.

Light or darkness : Can be an environmental trigger for germination and is a type of physiological dormancy. Most seeds are not affected by light or darkness, but many seeds, including species found in forest settings, will not germinate until an opening in the canopy allows sufficient light for growth of the seedling. Scarification mimics natural processes that weaken the seed coat before germination. In nature, some seeds require particular conditions to germinate, such as the heat of a fire (e.g., many Australian native plants), or soaking in a body of water for a long period of time. Others need to be passed through an animal's digestive tract to weaken the seed coat enough to allow the seedling to emerge.

TYPES OF GERMINATION

(i) Epigeal germination

In this, the cotyledons are raised out of the soil and generally become green and photosynthetic. In dicots, they are pushed up by rapid extension of hypocotyl before growth of the epicotyl. Epigeal germination occurs in bean, caster, mustard, tamarind, sunflower etc.

(ii) Hypogeal germination

In this type of germination, the cotyledons remain underground. Hypocotyl growth is restricted. The epicotyl grows to raise the first leaves out of the soil. Hypogeal germination occurs in dicotyledenous seeds of gram, pea, mango, ground nut etc and in monocotyledons like rice, maize, wheat etc.

(iii) Viviparous germination

This is a special type of germination occurring in mangrove plants. These plants generally grow in salty lakes, sea coasts and deltas. Here, the seed geminates while still attached to the parent plant.The embryo emerges out of the fruit with a massive radicle pointing downwards. Due to increased weight, the seedling separates from the parent plant and establishes itself in the muddy soil below.

PROCESS OF SEED GERMINATION

The process of seed germination includes the following five changes or steps.Such five changes or steps occurring during seed germination are: (1) Imbibition (2) Respiration (3) Effect of Light on Seed Germination(4) Mobilization of Reserves during Seed Germination and Role of Growth Regulators and (5) Development of Embryo Axis into Seedling.

(i) Imbibition

The first step in the seed germination is imbibition i.e. absorption of water by the dry seed. Imbibition results in swelling of the seed as the cellular constituents get rehydrated. The swelling takes place with a great force. It ruptures the seed coats and enables the radicle to come out in the form of primary root.

Imbibition is accomplished due to the rehydration of structural and storage macromolecules, chiefly the cell wall and storage polysaccharides and proteins. Many seeds contain additional polysaccharides, not commonly found in vegetative tissues. Seeds packed dry in a bottle can crack it as they imbibe water and become swollen.

(ii) Respiration

Imbibition of water causes the resumption of metabolic activity in the rehydrated seed. Initially their respiration may be anaerobic (due to the energy provided by glycolysis) but it soon becomes aerobic as oxygen begins entering the seed. The seeds of water plants, as also rice, can germinate under water by utilizing dissolved oxygen.

The seeds of plants adapted to life on land cannot germinate under water as they require more oxygen. Such seeds obtain the oxygen from the air contained in the soil. It is for this reason that most seeds are sown in the loose soil near the surface. Ploughing and hoeing aerate the soil and facilitate seed germination. Thus the seeds planted deeper in the soil in water-logged soils often fail to germinate due to insufficient oxygen.

(iii) Effect of Light on Seed Germination

Plants vary greatly in response to light with respect to seed germination. The seeds which respond to light for their germination are named as photoblastic. Three categories of photoblastic seeds are recognized: Positive photoblastic, negative photoblastic and non-photoblastic. Positive photoblastic seeds (lettuce, tobacco, mistletoe, etc.) do not germinate in darkness but require exposure to sunlight (may be for a brief period) for germination.

Negative photoblastic seeds (onion, lily, Amaranthus, Nigella, etc.) do not germinate if exposed to sunlight. Non-photoblastic seeds germinate irrespective of the presence (exposure) or absence (non-exposure) of light.

In these light sensitive seeds, the red region of the visible spectrum is most effective for germination. The far-red region (the region immediately after the visible red region) reverses the effect of red light and makes the seed dormant. The red and far-red sensitivity of the seeds is due to the presence of a blue-coloured photoreceptor pigment, the phytochrome. It is a phycobiloprotein and is widely distributed in plants.

Phytochrome is a regulatory pigment which controls many light-dependent development processes in plants besides germination in light- sensitive seeds. These include photo-morphogenesis (light-regulated developmental process) and flowering in a variety of plants.

Phytochrome and Reversible Red-Far-red Control of Germination

The pigment phytochrome that absorbs light occurs in two inter-convertible forms Pr and Pfr. Pr is metabolically inactive. It absorbs red light (660 nm.) and gets transformed into metabolically active Pfr. The latter promotes germination and other phytochrome-controlled processes in plants. Pfr reverts back to Pr after absorbing far-red (730 nm.).

In darkness too, Pfr slowly changes to Pr. Owing to this oscillation of phytochrome between Pr and Pfr status, the system has been named as "reversible red—far-red pigment system" or in brief phytochrome system. Treatment with Red light (R) stimulates seed germination, whereas far-red light (FR) treatment, on the contrary, has an inhibitory effect.

Let us examine seed germination in positive photoblastic seeds e.g. lettuce (*Lactuca sativa*). When brief exposure of red (R, 660 nm.) and far-red (FR, 730, nm.) wave lengths of light are given to soaked seeds in close succession, the nature of the light provided in the last exposure determines the response of seeds. Exposure to red light (R) stimulates seed germination. If exposure to Red light (R) is followed by exposure to far-red light (FR), the stimulatory effect of Red light (R) is annulled.

This trick can be repeated a number of times. What is crucial for seed germination is the quality of light to which the seeds are exposed last. This also indicates that responses induced by red light (R) are reversed by far-red light (FR).

Whole of this can be shown as given ahead

R..	Germination
R + Fr	No germination
R + FR+ R	Germination
R+FR+R+FR	No germination
R+FR+R+FR+R............	Germinarion

Light requirement for seed germination may be replaced by hormones such as gibberellins or cytokinins. Several development processes of plants controlled by phytochrome may be mimicked by appropriate hormones given singly or in combination with other hormones at the correct time.

(iv) Mobilization of Reserves during Seed Germination and Role of Growth Regulators

During germination the cells of the embryo resume metabolic activity and undergo division and expansion. Stored starch, protein or fats need to be digested. These cellular conversions take place by making use of energy provided by aerobic respiration.

Depending upon the nature of the seed, the food reserves may be stored chiefly in the endosperm (many monocotyledons, cereal grains and castor) or in the cotyledons (many dicotyledons such as peas and beans). Thorough investigations in the mobilisation of reserves from the endosperm to the embryo via a shield-like cotyledon (scutellum) has been done in several cereal grains.

The outer layer of special cells (aleurone layer) of endosperm produces and secretes hydrolyzing enzymes (such as amylases, proteases). These enzymes cause digestion i.e. breakdown of the stored food such as starch and proteins in the inner endosperm cells.

The insoluble food is rendered soluble and complex food is made simple. These simpler food solutions, comprising of sugars and amino acids thus formed, are diluted by water and passed towards the growing epicotyl, hypocotyl, radicle and plumule through the cotyledon.

Gibberellic acid plays an important role in initiating the synthesis of hydrolyzing enzymes. Gibberellin, therefore, promotes seed germination and early seedling growth. Assimilation of this food by the growing organ induces growth and the seedling soon assumes its ultimate shape.

It is very significant to note that the dormancy inducing hormone, abscisic acid (ABA), prevents the germination. The concentration of ABA has been shown to increase during the onset of dormancy of the embryo during seed development in several kinds of seeds.

When young embryos of cotton are removed and grown in culture, they continue to grow without the development of any dormancy. Dormancy in such cases can be induced by the addition of ABA at a crucial stage of growth.

(v) Development of Embryo Axis into Seedling

After the translocation of food and its subsequent assimilation, the cells of the embryo in the growing regions become metabolically very active. The cells grow in size and begin divisions to form the seedling.

DORMANCY

Dormancy is a condition where seeds will not germinate even when the environmental condition (water, temperature and air) are permissive for germination. This is an important survival mechanism for some species because these species do not germinate unless adverse climatic conditions end. For certain species chilling temperature for certain length of time help in the termination of dormancy.

Exogenous dormancy: Physical dormancy and Chemical dormancy

Endogenous dormancy: Morphological dormancy, Physiological dormancy,Non-deep physiological dormancy, Photo dormancy, Intermediate physiological dormancy, Deep physiology dormancy and Epicotyl dormancy

Double dormancy: In some species e.g. *Cerci's occidentalis* seed coat is impervious and at the same time the embryo is also dormant. Such seeds require two years for breaking of dormancy in nature. In the first spring the micro-organisms act upon the seed making it weak and soft and then embryo dormancy is broken by receiving chilling temperature in the winter. Combinations of two or more types of dormancy are known as double dormancy.

Secondary dormancy: Secondary dormancy is due to germination conditions. Secondary dormancy is a further adaptation to prevent germination of an imbibed seed if other environmental conditions are not favourable. These conditions can include unfavourably high or low temperature, prolonged darkness, and water stress. It is two types:

Thermo dormancy: High temperature induced domancy.

Conditional dormancy: Change in ability to germinate related to time of the year.

Two basic kinds of dormancy have been recognized. One is the influence of external factors such as light, temperature, water, etc. The other inherent dormancy, is a condition brought about by or accompanying the ripening of the ovule and/ or the maturation of the embryo. Environmental conditions and genotype are factors which influence seed dormancy, and these two influences may be mutually dependent and sometimes cannot be separated. The causes of seed dormancy are varied and may be quantitative, but in general fall into the following major classes:

1. Rudimentary embryos
2. Physiologically immature embryos (inactive enzyme system)
3. Mechanically resistant seed coats
4. Impermeable seed coats
5. Presence of germination inhibitors.

In the case of a rudimentary embryo, an after-ripening requirement must be satisfied to allow time for the postharvest maturation of the embryo. In other instances, a low temperature may be necessary to bring about physical changes in order that the seed can germinate. The seed coat can prevent germination either by limiting the permeability of water and gases as in the case of legumes or mechanically limiting the enlargement of the embryo. Removal of these mechanical barriers by 4 chemical or mechanical treatments may allow germination to proceed in some seeds. In some cases, seed coats alter the growth substance relationships of the enclosed issues. Hence they are closely interrelated with seed dormancy. Dormant seed subjected to an optimal dormancy breaking treatment, may lose their ability to germinate even under favourable conditions. This phenomenon is called secondary dormancy. Secondary dormancy may develop spontaneously in seed due to changes occurring in them, as in some species of Taxus and Fraxinus. Sometimes secondary dormancy is induced if the seed are given all the conditions required for germination except one. Among the factors which have been shown to induce secondary dormancy are restriction of gaseous exchange, high or low temperature, prolonged exposure of light requiring seed to darkness and of dark-requiring seed to light.

Function of Seed Dormancy

Seed dormancy is considered as an aspect of growth cessation. A dormant system has only two possible immediate fates: resumption of growth or death. The occurrence of seed dormancy has a significant advantage to plants by the potential for growth during unfavourable conditions. The adaptation significance of seed dormancy might be considered an ecological mechanism in which a more favourable time for germination may result in greater survival of seedlings.

Hormonal regulation of four phases of dormancy

1. Inductive. 2. Maintenance , 3. Trigger and 4. Germination.

Inductive phase: It is characterized by a marked decline in the hormone level, and is present during the development of the seed.

Maintenance phase: Little is known about the development of dormancy in the seed, but certain events during the maturation of seed inevitably lead to the onset of dormancy. These events may be environmentally triggered e.g. photoinduction, thermoinduction , or chemoinduction. Onset of dormancy may be controlled by the critical balance of an inhibitor-promotor complex(es). During seed maturation, the balance between an inhibitor and promotor may be shifted in favor of the inhibitor component to bud imposing dormancy. This shift may be accomplished by a decrease in the synthesis of the promotor, a build up of inhibitory intermediate metabolites, or by a direct.

Trigger phase: This phase of seed dormancy represents a period of sensitivity to a specific environmental condttion. A triggering agent may be responsible for inducing germination, but need not be present continually. The triggering agents are varied depending on the different types of seed dormancy. It may be a photochemical on as in photoblastic seeds, a thermo-chemical reaction as in after - ripening (stratification) or an inhibitor removal by scarification, leaching or seed coat removal.

Germination phase: The subsequent process after dormancy is seed germination. This phase is marked by an increase in hormone and enzyme activity. The early stage of germination seems to involve enzyme activation and degradative reactions, while the later stages are associated with the translocation , mobilization and similation of organic nutrients. A germination agent, presumably a naturally occurring hormone (auxin, gibberellin and /or cytokinin) is believed to be required. It appears to function via an inhibitor-promotor complex. Several such complexes may be involved in the germination response of any one species, with each complex being responsible for some specific process, such as degradation of seed coat or mobilization of nutrients. From the above observations, the over-all control of seed dormancy seems to involve a reduction in the growth-promoting hormone content during maturation, i.e., dormancy onset. Under suitable environmental conditions a trigger factor is activated which increases the hormone content. The hormones (germination agents) then perform some functions, probably activate preexistent hydrolytic enzymes, and/or stimulate the synthesis of additional enzymes via DNA depression. These degradative reactions supply appropriate monomers for the respiratory activity of the embryo, resulting in germination.

Gibberellin and Seed Dormancy

Endogenous gibberellins as functional hormones Although gibberellins were discovered as products of a 9 fungus which parasitizes a higher plant, it is now recognized that gibberellins are a constituent of normal green plants. This finding comes from the bioassays of extracts of seeds of various species and pea shoots, which all show similar physiological effects from gibberellin or gibberellin-like substances. Circumstantial evidence has been accumulated showing that there is a positive correlation between endogenous gibberellin levels and certain developmental trends. Treatments of gibberellin can induce flowering of some photo-periodically sensitive and some cold-requiring plants. It has also been shown that the induction of flowering may bring about a natural rise in endogenous gibberellin content pointed out that Kato and Ito have reported that the gibberellin levels are higher in expanding leaves of apple than in those that have completed expansion , and that levels are higher in terminal buds of vigorous shoots than in those of weak ones . In all cases by far the highest concentration of gibberellin was in the seed. Even in the seed, gibberellin levels remained very low until fruit growth had nearly been completed, maximum levels always being reached after fruit growth had ceased. The marked rise in gibberellin level which occurred at that time was very strongly correlated with the period of maximum growth rate of the seed; after seed growth was complete the endogenous gibberellin levels declined rapidly. In regard to the seed dormancy and gibberellin levels, gibberellin treatments can overcome some types of dormancy, and as seeds emerge from the dormant condition there may be a natural rise in endogenous gibberellin contents. Gibberellin content increased during the chilling process of hazel seed, and suggested that the gibberellin was possibly responsible for overcoming dormancy. The fact that gibberellins are present in dormant seed in many cases, and that gibberellin breaks seed dormancy lead one to assume that an effective or critical gibberellin level is involved in breaking dormancy.

Possible roles of gibberellic acid in the breaking of seed dormancy: Promotive effects of exogenous gibberellin in the germination of nondormant seeds have been reported in wheat, barley, and rice. Similar responses have also been reported in many plant species. Yet, one of the most dramatic effects of gibberellin is the breaking of seed dormancy. Gibberellic acid will break dormancy in light-dependent seed. They found lettuce seed germinates spontaneously when pretreated with gibberellic acid solution. Subsequent studies show that the application of exogenous gibberellin induces the germination in many dormant seeds. As to the role of gibberellins in promoting seed germination, many hypotheses have been proposed. Gibberellin may influence the elongation of the embryonic axis; it can promote the release of readily soluble food material from food reserves or it can induce quantitative and qualitative changes of protein, amino acids and auxin in the embryo. Gibberellins also play an important role in the regulation of nucleic acid synthesis. In reviewing the problem of dormancy within the framework of molecular biology. GA3 acted as a chemical signal, that activates the cells of the aleurone layer into creating a hydrolytic enzyme (α-amylase). The activity of α-amylase in isolated barley endosperm increased markedly in response to an application of GA 3. In addition, they noted that the synthesis of amylase was inhibited by actinomycin.

Advantages of Seed Dormancy

- Permitting germination only when environmental conditions favour seedling survival e.g. temperate species.
- Creation of a "seed bank"
- Dormancy can also synchronize germination to a particular time of the year.
- Seed disposal can be facilitated by specialized dormancy conditions e.g. modification of seed covering through digestive tract of a bird or other animals.

Methods of Breaking Dormancy

Scarification

1. Mechanical Scarification
2. Acid scarification
3. Hot water Scarification

Stratification

1. Open Field Stratification (Plate 1.)
2. Refrigerated stratification(Plate 1.)

Germination in Stratified Seeds of Temperate Fruits

The seeds of temperate fruits like apple, pear, peach, apricot, almond, walnut etc. need moist chilling for breaking dormancy. Stratification is done by placing the seeds in alternate layers of moist sand at 7.2°C. The period required for stratification and per cent germination achieved varies in different fruits (Table 1). Apple seed is stratified in sand at 4-7°C for 70-75 days results 70-75 per cent germination. Stratified seeds of apple become plump and swollen (P1&P2)

P. 1: Open field stratification

P. 2: Refrigerated Stratification

P. 3: Stratified seed of apple (Plump and swollen)

P. 4: Properly Stratified seed of apple (inside view)

(See colour version on page 423)

Table 1: Stratification period and percentage of germination in seeds of temperate fruits

Kind of Fruit	Stratification period (days)	% germination
Apple	70-75	70-75
Kainth (*Pyrus pashia*)	30-35	90-95
Peach	60-70	55-60
Apricot	45-50	75-80
Almond	45-50	85-90
Walnut	95-100	80-85
Pecan	70-75	75-80

Hormone changes during stratification

A tri-phasic change in endogenous hormones is typical for many seeds (Fig. 1); generally there is:

- A reduction in ABA level (major hormone responsible for seed dormancy)
- Increased synthesis of cytokinin and gibberellin.

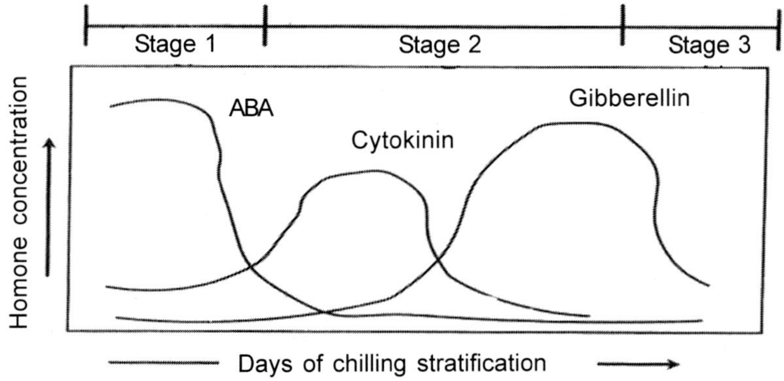

Fig. 1: Triphasic changes in hormone production during chilling stratification

Role of Hormones in Seed Dormancy

Plant hormones affect seed germinations and dormancy by affecting different parts of the seed. Embryo dormancy is characterized by a high ABA/GA ratio, whereas the seed has a high ABA sensitivity and low GA sensitivity. To release the seed from this type of dormancy and initiate seed germination, an alteration in hormone biosynthesis and degradation towards a low ABA/GA ratio, along with a decrease in ABA sensitivity and an increase in GA sensitivity needs to occur. Different growth regulators like kinetin, GA, ethrel and BA and Chemical Thiourea are recommended for breaking dormancy in temperate fruits (Table 2).

Table 2: The recommended concentrations of growth hormones in temperate fruits

Crops	Concentration
Apple	Thiourea - 5000 ppm
	Kinetin - 25 ppm
	GA - 50ppm
	Ethrel -100-200 ppm
Pear	GA - 150 ppm
	Thiourea - 5000 ppm
Peach	Thiourea - 5000 ppm
	GA - 400 ppm
	BA - 400 ppm
Walnut	GA - 250 ppm
	Ethrel - 1000 ppm

LITERATURE CONSULTED

Chakrabarti, N. and Mukherji, S.2003. Effect of Phytohormone pretreatment on nitrogen metabolism in Vigna radiate under salt stress. *Biol Planta*, 46: 63-66.

Chauhan, J.S., Tomar, Y.K., Singh, I.N., Ali, S. and Debarati, A.2009. Effect of growth hormones on seed germination and seedling growth of black gram and horse gram. *J American Sci.* 5(5):79-84.

El-Barghatgi, M.F. and El-Bakkosh, A. 2005. Effect of some mechanical and chemical pre-treatments on seed germination and seedling growth of Quercus coccifera (Kemes Oaks), *J Jerash Private University.*

Martinez-Gomez, P. and Dicenta,F.2001 Mechanisms of dormancy in seeds of peach (*Prunus persica* (L.) Batsch) cv. GF 305, " *Scientia Horticulturae*, 91: 51–58.

Rayees,A. Wani, Tanveer,H., Malik, A.R., Jahangir,A.B. and Dar, N.A.2014. Studies On Apple Seed Germination And Survival of Seedlings As Affected By Gibberellic Acid Under Cold Arid Conditions. *International Journal of Scientific & Technology Research*, 210-216.

Yamaguchi,S. and Kamiya,Y.2002. Gibberellins and light-stimulated germination," *J. Plant Growth Regul,* 20:369–376.

Yamaguchi,S.and Kamiya,Y.2000. Gibberellin biosynthesis: Its regulation by endogenous and environmental signals. *Plant Cell Physiol*, 41: 251–257.

14

Propagation Media, Mixtures and Containers

S.S. Gill and J.S. Bal

INTRODUCTION

For survival of young seedlings and rooting of cuttings it is essential that they maintain a healthy water and air balance especially to make up the water lost in transpiration. Water stress within the cuttings hinders metabolic activities and cell regeneration and consequently initiation of roots. So medium should be such that it provides adequate water required for rooting and growth. Also these substrates must provide initial nutrients necessary for plant growth. These should also contain micro-organisms that seedlings may need.

PROPAGATION MEDIA

Characteristics of Good Media

It is utmost important to know the characteristics of media and mixtures used for propagation of plants in containers, known as containerized seedlings or multiplied under field conditions. The quality of these media has great influence in plant propagation as these substrates provide the seedlings with nutrients, water and air for good development. Also these media contain microorganisms which are essential for plant growth.

There are several materials of different types are known to be used for seed germination as well as rooting of cuttings. It has also been seen that properties of these media and mixtures influence a lot, the germination of seed and growth of young seedlings. The following are some of the characteristics of the good media used for multiplication of plants to be kept in kind before their judicious use.

1. Media used for rooting of cuttings must be firm and dense, so that they can hold the cuttings firmly during rooting. If it is not so, it is possible that in loose media, cuttings may get disturbed after irrigation which will lead to decrease in rooting percentage.
2. Media used should be such that it has sufficient water holding capacity. If the water drains easily, more irrigation are required thus increasing the cost of production of plants. Also if a single irrigation misses, it will adversely affect rooting of cuttings.
3. The media should be porous, so that the excess water drains away and plants get good aeration. Less porous media causes dampness which leads to appearance of fungal disease a cause of death of germinating seedlings and tender roots of cuttings. In less porous media hard crust formation on the upper layer of soil due to rain or irrigation water will certainly affect seed germination.
4. The media should not have high salinity level and not too acidic. It is better if it has neutral pH. The salinity has many harmful effects on plant production. Thus the media used for propagation should avoid this negative characteristic. Saline soils are often avoided for even growing higher plants. However, some species from the forest plantation can tolerate high salinity level like Acacia. In fruits, pomegranate can be counted for such soils for plantations.
5. Initial fertility level of media is very important. The fertility level should be such that it is capable of supplying of nutrients to germinating seedlings and growth of young plants. Otherwise plant will remain stunted and it may not become saleable in the coming planting season.
6. There are many micro-organisms which are beneficial for plants and most important are mycorrhiza, Frankia, non-symbiotic nitrogen fixers-azotobacter etc. So the media used for propagation must contain these micro-organisms especially mycorrhiza strains and rhizobia. These not only help better plant growth, also ectomycorrhiza protect plants against pathogens such as Phytophthora, Pythium, Rhizoctonia due to formation of mantle.
7. It should be free from weed seeds, nematodes and various noxious organisms. Media containing weed seeds will itself create problem as these weeds after germination compete with the plants for nutrients, moisture and even sunlight. Since in the initial stage of rooted cuttings, hoeings are avoided for non-disturbance of tender roots, weeds under such situation become nuisance for nursery growers. Presence of nematodes cause nodes on the roots which hinder plant growth.

8. Media used should be such that it does not disintegrate while sterilization, otherwise beneficial microorganisms will get destroyed.

Media Used for Propagation and Growth of Seedlings

Different types of media used for propagation and growth of newly produced plants are as follow.

1. Soil

Soil, an important medium used for propagation is composed of materials in the solid, liquid and gaseous states and they must be present in soil in the proper proportion for the satisfactory growth of the plants. Solid part of the soil is composed of inorganic and organic forms. The inorganic fraction is basically composed of mineral matter derived from weathered parent material. The organic portion is composed of both living and dead organisms, which included bacteria, fungi, earthworms, plant roots and animal remains in various stages of decay.

The field soils are inappropriate for seedlings production and often lack the necessary plant nutrients. Thus, the field soils are not often used for container raised seedlings unless they are amended with other ingredients. For such soils, organic matter in the form of decomposed manure, compost, rice husk or inorganic materials such as vermiculite, perlite are used.

The commonly used artificial soil medium is composed of one part of nursery soil, one part of sand and one part of well rotten farmyard manure. Along with nutrient components, soil texture is also very important character to be considered before using it for seedlings growth.

Soil must have granulated texture and it should also be such that no crust is formed on the upper layer of soil, otherwise it will inhibit seed germination. Some of good soil used for pot seedlings such as sand loam has sand 75%, silt 14% and clay 11%. Clay loam soils have sand 34%, silt 39%, and clay 27%.

2. Sand

Though sand is one of the most artificial medium used for seed germination, it does not contain any nutrients or buffer capacity but it does facilitate air and water movement. It is well aerated and useful for extensive root growth. Since sand has low water holding capacity thus cutting planted for rooting need frequent irrigations.

Sand is generally composed of small rock grains with size ranging from 0.05-2.00 mm diameter. Quartz sand is generally used for propagation. It is usually recommended that this material should be mixed with dried well rotten farmyard manure and some soil to increase its water holding capacity. Sand should be preferably fumigated before use as it contains weed seeds and some harmful fungi.

3. Peat or Moss Peat (Bhal)

Peat is of various types and is mainly composed of remains of aquatic, marsh, bog or swap vegetation which is kept in water under partially decomposed conditions. The different types of peat are classified as under.

a) **Moss peat**: A decomposed product of the moss grass derived from genus *Sphagnum*. It is highly acidic with high moisture holding capacity. Also contains some nitrogen usually one per cent. The high pH (3.5-4.5) of this material needs to be reduced before use with other medium used for propagation.

b) **Reed-sedge:** It consists of remains of grasses, sedges, reeds and other swamp plants. It is acidic to neutral with pH value between 4.5 to 7.0. It has low moisture holding capacity as compared to moss peat.

c) **Peat humus:**It is an advanced state decomposed material of *Hypnum* moss or reed sedge and also has low moisture holding capacity and its nitrogen content is between 2 to 3.5 per cent.

d) **Hypnum peat:** It is composed of decomposed plants from the genus *Hypnum*. It has neutral pH (7.0) to slightly alkaline (8.0). Hypnum peat has usually light brown to greenish brown colour. When it is used in mixture, it should be broken apart and moistened before mixing it with other medium.

4. Sphagnum moss

It is the young residue or living portion of acid-bog plants of the genus *Sphagnum*. It is relatively sterile as compared to other media used for propagation. Sphagnum is light in weight and has high water holding capacity i.e. 10 to 20 times more than its own weight. It has pH between 3.5 to 4.5, so it is acidic in nature. Before use this material is shredded. Generally it is used as germination medium and also in layering method of propagation. It has fungistatic properties due to which damping off disease occurs in seedlings.

5. Vermiculite

It is a mineral which expands too much when heated and such mineral is called micaceous mineral. Chemically it is a hydrated magnesium aluminium iron silicate. It is very light in weight and has neutral reaction i.e. pH is 7.0 and is insoluble in water. The particle size is 2-3mm in diameter and widely used for propagation. It has also high cation exchange capacity and can hold nutrients in reserve.

6. Perlite

It is silicaceous material of white-grey colour and this material is of volcanic origin. It is neutral in reaction (pH7.0) with very good water holding capacity. The aeration increases when it is mixed with other medium. It is commonly used for propagation of *in vivo* and *in vitro* produced plants.

7. Compost

It is good humus material with fine quality of moisture holding when mixed with soil medium. It is also used as mixture with soil medium to increase organic matter content. Before use it is essential to sterilize the compost to avoid disease incidences.

8. Leaf mould

It is also known as decomposed leaves and often used for soil amendment. Sterilization of leaf mould is necessary to avoid disease incidences, nematodes presence and to kill insects etc.

9. Shredded bark,Saw dust and Wood savings

Such materials have lower rate of decomposition, thus should be broken apart and moistened before use. The fresh material may be toxic in nature, thus mixed with other medium when it is well decomposed.

PROPAGATION MIXTURES

Mixtures are defined as the mixing of different media keeping in view the type of seeds to be germinated and kind of propagating material.

Nursery soils are sometimes inappropriate for seedlings production. Also, they lack necessary plant nutrients required for plant growth. These soils do not have required porosity which is necessary for aeration of plants. Thus, organic matter or other inorganic materials such as vermiculite, perlite are mixed with the soil to increase its plant production capacity.

Better Potting Mixtures

The potting mixture of better texture is prepared by adding sand and some organic matter such as peat moss or saw dust or shredded bark.During preparation of these mixtures,soil should be screened and make it uniform and large particles are eliminated. The materials added to mixture are normally be dry and moistened slightly later on. The ingredients are placed in a pile arranged in layers and turned with shovel till get well mixed in the mixture prepared for propagation. The preparation of soil mixture should preferably be taken place at least a day prior to use. The soil mixture should be just slightly moist at the time of use so that it does not crumble.

While preparing these mixtures, it is advantageous to add some inorganic fertilizers like superphosphate, potash and even some nitrogenous fertilizers can also be added to the mixture to meet the initial requirement of propagating materials. Loam clay soil alone is unsatisfactory for pot mixture. They are often heavy and poorly aerated or tend to become sticky after watering. Upon drying they may shrink forming a hard and cracked surface.

Different Potting Mixtures for Propagation and Seedling Growth

Due to continuous raising of seedlings, the fertility level often decreases and regain that potting mixtures are prepared to provide appropriate nutrition to the plants. Also, these mixtures provide necessary aeration and moisture to the growing seedlings.

The following are some of the potting mixtures which are used for containerized seedlings,

a) Mixture for common use

Mixture of sand and well rottened farmyard manure (1:1) is commonly used for propagation of plants through cuttings. Such mixtures are also rich in nutrients and before use they are treated with chlorpyriphos 20EC to avoid the attack of termites. This mixture is well drained and hold sufficient moisture for root development.

b) Mixtures for young seedlings and pot rooted cuttings

In such mixtures, sand; one part; loam soil; two parts, peat moss or leaf mould; one part are mixed.

c) Mixture for grown up nursery stock and containerized seedlings

It contains sand: one part; loam soil: two parts; peat moss: one part. These mixtures may be compensated with inorganic fertilizers like, superphosphate, potash and nitrogenous fertilizers

d) Mixtures for short period

This mixture can be retained for a week period. For one cubic yard mixture, it contains Sand: one part; peat: one part; Hoff and blood meal: one kg; KNO_3 150gm; K_2SO_4 150gm; superphosphate: one kg; dolomite: one kg; $Ca(CO_3)_2$: one kg.

These mixtures can be safely sterilized by steam or chemicals.

e) Mixtures for long period

This mixture can be stored for longer period. The ingredients included in this mixture are Sand (0.5-0.05mm size): one part; peat moss (0.5-0.05 mm size): one part. It has moisture holding capacity 48%.To each cubic yard of mixture addKNO_3 150gm: K_2SO_4 150gm; single superphosphate: one kg; dolomite lime: 3 kg; $Ca(CO_3)_2$: 3 kg.

Sand used in the mixture does not allow the mixture to become compact and peat increases the water holding capacity.

PROPAGATION CONTAINERS

Benefits of containers for propagation

The evergreen fruit plants are mainly planted in the field with earth ball and this process of transplanting reduces mortality rate after transplanting. But digging of earth ball and packing the same is costly affair as it involves lot of labour and it is time consuming too. But when such species are raised in containers, it becomes easy to transport and no labour is required like lifting the plants with earth ball. Thus, in view of the above advantage the use of plant containers has increased enormously for the last couple of years. Also, containerized plants can be shifted to safer place during natural hazards like heavy winds, or rains or thunder storms.

Different Types of Containers used in Nursery

The plants produced in containers are called containerized seedlings. These containers used for propagation are available in different forms, size and in different materials. Polystyrene, Polyethylene, Fibre and Paper are the common types of containers available in the market. Some new forms and materials are constantly being developed and displayed in the markets. These are tested to be used as containers for raising young plants. In India, even leaf cups made from easily available materials like moss grass are also used for raising plant seedlings. Fibre cubes, open tubes of bamboo, baskets or another light and cheap material like earthern pots, paper, polythene bags are commonly used for plant propagation. The following are some of the containers commonly used for nursery plant production:

a) **Polythene bags:** This is most commonly used containers in the tropics and subtropics area of the country. These containers are made up of transparent white or black opaque sheet of polythene. They are made locally and are easily available material and also cheaper. To have good aeration and drainage, these bags are perforated. The polythene used for preparing these bags is of gauge 150, 200 or 250.

b) **Earthen pots:** These are made from soil and with or without bottom and are broken to take out the plant with earth ball when planted in the field. These earthen pots are relatively costly and also more difficult to transport as compared to polythene containers.

c) **Moss cylinders:** These are cylindrical tubes of moss about 15 cm in length and 5 to 8 cm in diameter. They are often used by nurserymen especially raising Eucalyptus plants. Also fruit nursery production people use these moss cylinders. These are filled with nursery soils and after planting seedlings in these cylinders, they are kept under shade.

d) **Fibre cubes:** Mostly used in foreign countries, but now in India nurserymen have started using them for plant propagation. They are light in weight with better strength and are easy to transport them.

e) **Flats:** They are usually shallow, wood or plastic or metal trays with draining holes in the bottom. The flats are mobile. They are useful for germinating seeds or rooting of cuttings. Durable wood of cypress, cedar or red wood should be used for flats.

f) **Clay pots:** The familiar red clay flower pots are long used for growing young plant. They are porous, lose moisture readily and are heavy. They are easily broken and their round shape is not economical due to occupying more space. After continued use, toxic salt accumulation build up and require soaking in water before reuse. However, they can be steam sterilized.

g) **Plastic pots:** They are round or square in shape and are reusable. Plastic pots are non-porous and light in weight. Some types are fragile and require careful handling. Plastic pots cannot be steam sterilize. However, some of the common plant pathogen can be eliminated by giving three minute dip in hot water (70^0C).

h) **Metal containers:** They are used containers from canneries, restaurant and bakeries. Large number of nursery plants are grown and marketed each year in 4-20 litre sized Cans. Such Cans are also available especially prepared for growing plants with drainage holes at the bottom. They are enamelled to retard rusting. In areas having high summer temperature, use of light coloured containers (white or aluminium) may improve growth by avoiding heat damage to the roots. During winter season, dark coloured Cans are good for growing seeds and cuttings.

i) **Baskets:** Baskets made of bamboo strips or *Tamrix* are also used for transporting nursery plants to plantation sites. They are cheap and even can be buried with earth ball in the soil.

j) **Milk cartons:** These are suitable for on-farm production of plants.

Modern Alternative for Plant Domestication

A modern alternative for plant domestication, so called 'root trainers' are used now-a-days. These are rigid containers with internal vertical ribs that directs the roots to grow downward, thus avoiding the spiralling of roots. These have also a large drainage hole at the bottom allowing air and root pruning.

Mixture Properties and Growth of Seedlings

The substrate is an essential input for seedlings propagation and its importance for raising container plants cannot be under estimated. These substrates provide the seedlings with nutrients, water and air for good plant development. They also contain the micro-organism that seedlings may need for their proper growth.

Substrate Properties Influencing Seedlings Growth

a) Physical Properties: The following points are considered under physical properties.

Water holding capacity: It is essential to avoid frequent watering to growing plants.

Porosity: It ensures aeration to the plants. Less porous soil may have dampness problem which may be the cause of appearance of some diseases as well hindrance of seed germination due to crust formed on the upper layer of soil.

Plasticity: It is an essential property of the mixture for better growth of the plants.

Bulk density: It also require property of mixture and media.

b) Chemical Properties: Chemical properties include the following aspects of media and mixtures.

Fertility: Initial fertility of mixture is essential for better start of young seedlings.

Acidity (pH): Too acidic or media with high pH may not be helpful for plant production.

Buffer capacity or cation exchange capacity (CEC): A property of mixtures related to release of nutrients to the growing plants.

c) Biological Properties: This relates to availability of appropriate rhizobia and mycorrhiza strains in the mixtures.

LITERATURE CONSULTED

Bal, J.S., Gill, S.S. and Sandhu, A.S. 2016. Raising Fruit Nursery 2nd Edition. Kalyani Publisher, New Delhi.

Chauahan, S.K., Gill, S.S., Khajuria, H.N. and Chauhan R. 2006. Production Technology and Management of Agroforestry Models (Eds.). Agrotech Publishing Academy, Udaipur.

Dwivedi, A.P.2006. A Text Book of Silviculture. International Book Distributors, 9/3 Rajpur Road, Dehradun 248001, India.

Kanwar, J.S. and Bal, J.S. 2004. Practical Manual of Propagation and Nursery Management. PAU, Ludhiana. Pp:15-16.

Sharma, R.R. 2013. Propagation of Horticultural Crops. Kalyani Publishers, New Delhi. Pp: 20-24.

15

Care of Propagated Plants

J.S. Bal

INTRODUCTION

The initial care of young seedlings and propagated plants is very necessary and important. Efforts should be made in such a way that the young plants can be propagated and grown to a saleable size with no insect-pests and diseases attack and nutrient deficiency. The supply of healthy plant material to the growers helps in good reputation of a nursery as well as to make fruit crops more profitable.

Sometimes young nursery plants suffers from various problems like attack of insect-pests and diseases, nutrient deficiency, scanty or excessive moisture and stagnation of excess or rain water in the field. The young seedlings/plants require light and frequent irrigation and need special care to make them grow well and to save them from adverse weather conditions. They must be protected from intense heat and cold. The young nursery plants growing on light or sandy soils can exhibit deficiency symptoms of macro and micro elements. Such nutrient deficiencies can easily be corrected by foliar or soil application of desired nutrient.

1. PROTECTION FROM ADVERSE CLIMATE

a) Protection from High Temperature

The atmospheric temperature during hot summer months of May-June goes very high in Northern Indian plains. High mean temp sometimes goes very higher i.e. 40^{0}C in the month of April. It leads to sunburn on leaves or plants. The young nursery plants and seedbeds need attention and adequate protection against hot sun. Mango, citrus and litchi plants are more liable to damage from high temperature and dry weather than other fruit plants. The suitable measures should be adopted to save the young nursery plants from sunburn injury and adverse effect of high temperature.

i) **Irrigation:** The water need of nursery plants increased manifold during hot summer months of May-June in north Indian conditions. The frequent application of irrigation minimise the damage occurred due to sunburning and from hot desiccating winds during these months. By giving repeated need based irrigation raises the humidity around the nursery area and can prevent the adverse effect of heat to some extent. The frequency of watering can be reduced with the advent of rains in the months of July.

ii) **Erection of Thatches:** The thatches can be easily prepared by using farm waste materials viz. Rice straw (Parali), Sarkanda or Date palm leaves. By erecting thatches in the nursery rows or near the seed beds, protection against high temperature to young seedlings and raised seed beds can be provided effectively. The *Jantar* should be sown during February-March around nursery row and seed beds provide good protection from high temperature and hot velocity winds. *Arhar* can also be raised in the nursery rows to provide shade and minimise sun injury to young seedlings in nursery rows and on the seedbeds.

iii) **Planting Wind Breaks:** Wind breaks trees should be raised around the periphery of fruit nursery at least two years before the planting of mother trees for taking bud wood and rootstocks. Mango seedlings, shisham eucalyptus, arjan, jamun and mulberry are the tall growing most effective windbreaks trees. Avoid planting wind break trees if plants of same species are raised in the nursery to save them from the inoculum of different diseases and insect-pests attack.

b) Protection from Cold

The young nursery plants should be protected from the occurrence of frost during winter months i.e. December-January in northern Indian states and upto February-March in hilly areas. Sometimes both late frost and early frost also cause huge damage to young plants. Weak and sickly plants and those suffering from malnutrition are more liable to damage from cold and frost injury. Such plants and nursery seedlings become buddable after longer times. Mango, papaya, litchi, citrus, guava and aonla plants are more susceptible to cold and frost injury. Thus, the suitable protection measures against frost and severe cold should be adopted timely.

i) **Raising Frost Tolerant Fruit Plants:** It is advisable to raise only frost tolerant fruit plants in the nursery where occurrence of frost is common during winter months. The temperate fruit plants viz. pear, peach, plum, almond and other plants like ber, guava, loquat and grape should be multiplied in such localities. Among citrus plants, mandarin, sour orange and sweet orange can tolerate cold better than kagzi lime and sour lime. The plants of kagzi lime need more protection from cold.

ii) **Erection of Thatches:** The young seedlings and seedbeds should be provided thatches or kullies for protection from cold. These thatches can be prepared from sugarcane trash, rice straw, *sarkanda*, maize stalk or cotton sticks and erected around the seedbeds and in the nursery rows for protecting young seedling. The south-west side should be kept open to allow sufficient sunlight. Mango, papaya and kagzi lime and rootstock seedling of Jatti Khatti and Cleopatra should be protected from cold by erecting suitable thatches. Nursery beds can also be covered with transparent polythene sheets for protection from cold. These sheets allow the sunlight to enter to the beds and raise the temperature of the bed which enhances the growth of seedlings. During day time when sun is high, these polythene sheets can be removed to avoid excessive shading on the seedbeds and helps in minimising the attack of damping off. Jantar or arhar plants can also raised between the nursery rows to provide protection to the tender plants from cold.

iii) **Creating Smoke in Field:** The atmospheric temperature of the nursery area can be raised few degree by creating smoke when there is chance of occurrence of frost. The heaps of grass, trash, dry leaves etc should be kept ready at several places in the fruit nursery. These heaps should be burnt slowly during the period of frost occurrence. In this way, the young nursery plants can be effectively saved from cold and frost warning.

iv) **Light and Frequent Irrigation:** The young nursery plants should be given light and frequent watering especially during the period when chance of frost damage are imminent. This is the simple and most practical approach to ward off the damage from cold. By giving light irrigation during frost threatening period, temperature of nursery area can be raised by 1-2^{0}C, thus save the plants from immediate adverse effect of frost occurrence.

v) **Maintaining Healthy Plants:** The nursery plants should be maintained healthy by giving timely nutrition, irrigation and protection from insect-pests and diseases. Such healthy plants can bear sudden low temperature and frost damage during severe winter. However, weak and sickly plants suffer heavily from the adverse effect of low temperature.

vi) **Planting Wind Breaks:** Before planning to start fruit nursery, provision for planting wind break plants should be made around the periphery of the marked area. The cold waves from north and south damage nursery plants more. Therefore, wind break plants should be planted well ahead of nursery raising, both on north and south side. Tall growing plants like eucalypts and arjun should be inter planted with low-headed plants like ber, mango seedlings, jamun, karonda etc. This combination of windbreak plants provides better protection from high velocity cold winds.

2. MANURING AND FERTILIZATION

The nursery plants should be given manure and inorganic fertilizers from time to time to make them grow healthy. The organic manures like farmyard manure, compost, poultry manure, oil cakes, vermicompost and leaf mould are applied in the soil. The physical condition of the soil is improved considerably by adding inorganic manures. The manures should be added much before the preparation of nursery beds or planting of rootstocks or mother trees in the field. Green manuring of the nursery area must be given well in time for better mixing in the soil. Inorganic fertilizer like urea, diammonium phosphate and superphosphate are applied to nursery plants on the basis of their need. These fertilizers are applied for better growth and development of nursery plants. Inorganic fertilizers can also be applied through foliar spray on the nursery plants which is very effective than the soil application.

Deficiency of Nutrients and its Correction

Nitrogen deficiency occur in many nursery plants. The young leaves turn light green to light yellow (chlorosis) with short and slender stalks. The plants showed stunted growth. The deficiency of nitrogen can be easily corrected by applying urea or ammonium sulphate in the soil or with foliar application of urea (1%) during active growth period of nursery plants. Some of the fruit nurseries also show deficiency of phosphorus and potassium. Foliar spray of DAP at 2 per cent and single super phosphate at 1 per cent and potassium sulphate at 1 per cent can be applied for correction of phosphorus and potassium deficiencies in nursery plants.

The deficiency of some micro nutrients like zinc and iron occur commonly in nursery plants. Zinc deficiency in nursery plants is characterised by interveinal chlorosis of young leaves and shortening of internodes. Leaves remain small and erect (rosette). In severe cases, it results die-back of twigs. Soil application of zinc sulphate @20-25kg/ha or foliar application @0.5 per cent correct the zinc deficiency in plants.

Iron deficiency in nursery plants show chlorosis in young leaves with major vein remaining green. The stalk of leaves are short and tender. The deficiency can be corrected either by applying 40-50kg/ha ferrous sulphate into the soil or by foliar application of 0.5 per cent ferrous sulphate.

3. WATER NEED OF YOUNG NURSERY PLANTS

The young nursery plants require more frequent and light irrigation for their better growth and development. Excess application of water to grafted plants and nursery buds is harmful as the young plants have shallow root system. Ultimately, it leads to rotting of roots. However, if water is available in deficient state to the young nursery plants, they will desiccate particularly in light soils during summer months.

Flood System of Irrigation: In general, flood method of irrigation is followed in fruit nursery. Irrigation water used in flooding in excessive as the entire field is to be wetted to meet the need of the root system. The nursery field must be properly levelled to avoid stagnation of water in low lying areas. Water stagnation for few days may damage the nursery plants due to root rot diseases.

Sprinkler System of Irrigation: This system of irrigation is also used to irrigate nursery plants under water shortage conditions. The water is supplied with pressure through the sprinkler attached to pipes. The sprinklers are adjusted in such away to overlap one-fourth area covered by other sprinklers. These are further moved to another area for watering. It is suitable for uneven or sloppy lands and in the areas where subsurface water is unfit for irrigation.

Drip System of Irrigation: Drip irrigation is useful and economical method of irrigation and give maximum water use efficiency. It supply water near the root zone of the plants through drippers/emitters. The availability of nutrients to the plants is very high with drip system. Evaporation losses are negligible with this system. It is costly system of irrigation. However, in limited availability of water or in scanty rainfall areas, drip system is most efficient and useful.

Irrigation Frequency to Nursery Plants

The water requirement of nursery plants is usually high during summer months. The evergreen fruit plants require more irrigations than the deciduous fruit plants. Irrigation to apple, pear, peach, plum and grapes is done at longer interval during winter. But, the evergreen plants are irrigated even during winter to escape danger of frost. The nursery raised on light sandy loam soils require water supply more frequently than on heavy clay soils.

4. WEED CONTROL IN NURSERY

Why Essential to Eradicate Weeds

In the nursery weeds grow frequently everywhere especially in plant rows, water channels, paths and roads. The weeds compete with the nursery plants for water, nutrients and light and harbouring dangerous pests. This results in poor growth of the seedlings. Therefore, for normal growth and better success of nursery plants it is very important to manage the weeds in the fruit nursery. Management of weeds is very costly input and add maximum to cost of their propagation. The weeds like motha, baru grass and khabbal grass supress the growth of nursery plants. During nursery operation like hoeing, manuring and watering, weeds create hindrance in smooth working. Eradication of weeds should be undertaken at priority to keep the nursery area free from weeds.

Types of Weeds Grown in Fruit Nursery

The climate of tropical and sub-tropical regions is very favourable for seed germination and rapid growth of weeds. All types of weeds like annual, biennial and perennials are found growing in the fruit nurseries. The following are common weeds grown in different types of fruit nurseries.

Common Name	Scientific Name	Common Name	Scientific Name
Motha/Dila	*Cyperus rotundus*	Guli danda	*Phalaris minor*
Khabbal or Barmuda Grass	*Cynodon dactylon*	Maina	*Medicago denticulata*
Bathu	*Chenopodium album*	Wild Senji	*Melilotus parviflora*
Baru Grass	*Sorghum halepense*	Chulai	*Amaranthus viridis*
Congress Grass	*Parthenium hysterophorus*	Dab or dip	*Typha catifolia*
Jangli Palak	*Rumex dentatus*	Itsit	*Trianthema portulacastrum*
Bhang (Indian hemp)	*Cannabis sativa*	Jangli Gajjar	*Dacus carota*
Blue Pimpernel or Billibutti	*Anagalis arvensis*		

Among annual weeds, motha and itsit are commonly grown kharif season weeds during rainy season from June to October. The rabi season weeds like chulai and maina start growing during second fortnight of October and set seeds before summer i.e. April-May. The biennial weeds like Jangli gajjar takes two season to complete their life cycle. The perennial weeds like baru grass, motha and congress grass grown for two or more years in the nurseries and also on water channels, paths and roads.

Weed Control at Nursery Stage

Before starting fruit nursery, deep ploughing in the field should be done to remove the weeds at first stage. Then, while preparing nursery beds and doing shallow hoeing's in nursery plots and seed beds, weeds can be controlled by manual and chemical methods.

i) **Manual Weed Control:** The weeds can be controlled by various methods like manual, mechanical, biological, use of mulch and chemical. The conventional method of controlling weeds through manual means is very expensive and labour intensive. Mechanical control of weeds in nursery is not possible due to dense planting that will damage the root system of seedlings. The manual weeding is usually carried out at 15-20 days interval depending upon the growth of weed. This method is time consuming and back breaking. In perennial weeds, during hand weeding parts of the roots left in the soil that will require repeated hand weeding. Mulching with paddy straw and black polythene is also helpful in checking weeds in the nursery as these are easy, cheaper and less time consuming method.

ii) **Chemical Weed Control:** The use of herbicides is easy and less labour intensive method of weed management. Chemical control of weeds have also been found to be economical in many other crops. Pendimethalin @1.5*l*/acre as (past-emergence) post planting of seedlings resulted minimum weeds up to 90 days in guava nursery. In mango nursery, to check broad spectrum of weeds, post-emergence (4th week of October) spray of glyphosate @3*l*/ha is found effective. The pre-emergence (1st week of October) spray of diuron @4kg/ha is also proved useful for the control of monocot and dicot weeds. In pear nursery, pre-emergence spray (1st week of March) of hexuron @2.5kg/ha or post-emergence (4th week of March) of glycel @1.5*l*/ha was beneficial in checking the different types of weeds.

5. REMOVAL OF STOCK SPROUT

The stock sprout below the bud union of budded/grafted plant should be removed occasionally till the complete union of stock and scion. The sprout emerged very fast on the rootstock portion which directly competes with the new growth of the scion for water and nutrient. Removal of such unwanted sprouts on the rootstock is very important to improve the health of the grafted plants for getting ready for sale in the shortest possible period.

Sprout Delayed Union: The union between stock and scion gets delayed if unwanted sprout is not removed timely. Even in many cases the union between stock and scion do not prove successful and scion shoots remain stunted in growth and ultimately dries up. The removal of stock sprouts should be continuous feature even after the graft union has taken place successfully. By doing so, the food material will be properly utilized by the scion shoots to grow normally.

Sprout Removal from Ground Level: The sprouts growing from the ground level on the rootstock portion should be removed regularly till the grafted plants are not lifted from the nursery. In evergreens plants like guava and citrus emergence of sprout on the rootstock portion near the soil level is very common. This happens due to continuous occurrence of growth in evergreen fruit plants throughout the year. In case of deciduous fruit plants they goes dormant during winter and unwanted growth on stock portion also stopped.

Single Stem in Deciduous Plants: The plants propagated from cuttings like grapes, pomegranate, fig and plum should be trained to single stem only after rooting has taken place. The plants from cutting trained to single stem grow fast than those trained with retention of more than one single shoot.

Sprout Removal in Evergreen Plants: The plants propagated during August-September should be watched regularly and all stock sprouts should be removed timely. In citrus plants like sweet orange and mandarin stock sprouts should be removed at weekly interval in the month of October. During next month in November, the stock sprout from these citrus plants especially in kinnow should be removed after 10-15 days. Due to onset of winter, such sprouts from citrus plants should be pinched off at least once in the month of December. The sprouts from the stock portion of budded plants of different fruits should be removed once in a week during the growth period before the onset of winter. In ber T-budding is normally done in June with higher success. The side shoots emerged after budding should be removed and the scion shoots should be trained to single stem.

Removal of Bandage

After successful union between stock and scion, tying material in the form of bandage over the grafting joints should be removed for normal growth of the plants. The retention of bandage at the grafting joints for a longer period may result girdling of the area. Such plants show stunted growth at later stage and do not prove in healthy state. Therefore, all the tying material of any kind should be removed from the grafting joints after proper union and before the sale of plants.

6. TIMELY PLANT PROTECTION MEASURES

The fruit nursery should be kept free from the attack of insects-pests and diseases by give regular and timely spray of insecticides and fungicides. The young nursery plants attract many insect pests and diseases at early stage and cause heavy damage. The growth of young plants is restricted and they look sickly and unhealthy.

Measures Against Insect-Pests

Insect-pests cause damage in various ways by making holes in leaves, eating the leaf margins, spotting on leaves, chewing leaves and damage different parts of the plant. Various measures and techniques are followed to protect the nursery plants from different pests. The important methods of pest control like cultural, physical, mechanical, chemical and biological methods are adopted to control the insect-pests in the fruit nurseries.

In citrus, psylla and leaf minor are the most common pests. Citrus psylla cause serious damage to nursery plants throughout the country. It can be controlled by spraying 125 ml Rogor 30 EC (dimethoate) or 20 ml Imidacloprid 200 SL or 16 g Actara 25 WG (thiamethoxam) in 50 litres of water during the appearance of pests in March and again in early September. Leaf minor makes zig-zag silver white galleries on new leaves which get twisted. To check this spray 50 ml sumicidin 20EC (fenvalerate) or 10 ml Ripcord 10EC (cypermethrin) or 125 ml Hostathion 40EC (triazolphos) or 20 ml corocodile/confidor 17.8 SL (lmidacloprid) in 50 litres of water during April-May and August-September.

Mango hopper is very active during February-March and suck the sap from the tender leaves. For its control spray 100g sevin or hexavin 50 WP (carbaryl) or 80 ml of Malathion 50EC in 50 litres of water. Guava shoot borer is a serious pest in guava nursery and damage tender shoots and nursery plants. The pest can be controlled by spraying 50 ml (0.1%), Durmet 20EC (chlorpyriphos) or 40 ml (0.1%) Ekalux in 10 litres of water.

In temperate fruits leaf hoppers and aphids suck the cell sap from the leaves and cause serious damage to the plants. These can be checked by spraying 100 g hexavin 50 WP (carbaryl) or 80 ml Rogor 30EC (dimethoate) in 50 litres of water. Collect and destroy the chaffer and other defoliating beetles in kerosenised water. Adopt sanitation measures in the nursery.

Measures Against Diseases

In fruit nursery plants diseases are caused by different source viz. fungi, bacteria, viruses, micro-organisms and mycoplasma. In addition to this adverse climate, humid conditions, abiotic stresses, nutrient imbalance, acidic and alkaline soil conditions further help in stimulation of diseases incidence in plants. The diseases can be controlled through spray application or soil and seed treatments.

In citrus nursery for the control of canker, scab and wither tip or dieback and in ber for the control of leaf mould and leaf spot diseases, spray with Bordeaux mixture 2:2 250 or 50% copper oxychloride (0.3%) is very effective. The spray should be given during March, July and September for the control of wither tip and scab and during October, December and January for the control of canker.

Virus and virus like diseases can be effective checked by adopting preventing measures like by using disease free bud wood, tolerant rootstocks and sterilizing the budding tools.

In mango if seedlings over three months old showed swelling in the leaf axil then it is suspected due to mango malformation (bunchy top). Then remove the affected seedlings and destroy them. To check damping off in papaya treat the seeds with cerasan/agrosan/thiram before sowing. The soil of the seed beds should be sterilized with 2.5 per cent formaldehyde solution prior to sowing of seeds. The shot hole disease in temperate fruits can be checked by spraying captan/ziram/thiram (0.2%). Anthracnose and cercospora leaf spot disease in grapevine plants results dark brown circular spots on leaves around the mid rib and main veins. These diseases can be controlled by adopting following spray schedule.

i) Bordeaux mixture 2:2 250 in end March, end May and mid-August.

ii) Spray Bavistin 50WP 1g/litre of water in end April, Mid-July and early September.

iii) Spray Score 25EC 100 ml in mid-July, mid-August and mid-September in 100 litres of water.

LITERATURE CONSULTED

Anon. 2011. Package of Practices for Cultivation of Fruits, PAU, Ludhiana.

Bajwa, G.S., Minhas, P.P.S, Brar, S.S. and Bal, J.S. 1992. Studies on Weed Control in Pear Nursery. Emerging Trends in Temperate Fruit Production in India. NHB Technical Communications No.1: 185-90.

Bajwa, G.S., Bal, J.S., Brar, S.S. and Minhas, P.P.S. 1993. Studies on Chemical Weed Control in Mango Nursery. *Punjab Hort. J.*, 33(1-2):40-43.

Bal, J.S. 1985. Care of young fruit plants. The Horticultural Bulletin, 27(9):1-2.

Bal, J.S. 1986. Raising fruit plants. The Horticultural Bulletin, 28(5):1-5.

Bal, J.S. 1992. Raising a fruit plant nursery. The Tribune, 112(182):6.

Bal, J.S. 2014. Fruit Growing, 3rd Edition. Kalyani Publishers, New Delhi.

Bal, J.S. Gill, S.S. and Sandhu, A.S. 2016. Raising Fruit Nursery, 3rd Edition. Kalyani Publishers, New Delhi.

Boora, R.S., Dhaliwal, H.S., Singh, J., Bons, H.K and Arora, N.K. 2014. Studies on weed management in guava *(P. guajava)* nursery under North-Western Subtropical irrigated zones of India. *Annals of Horticulture*, 7(1):1-5.

Sharma, R.R. 2013. Propagation of Horticultural Crops: Principles and Practices. Kalyani Publishers, New Delhi. Pp. 87-93.

16

Protected Cultivation

K.S. Gill and Rachna Arora

INTRODUCTION

With the change in world scenario from plentiful resources to limited with population increasing geometrically, shrinking land and climate change, the pressure on agriculture is mounting continuously. Meeting the nutritional requirements of population by optimum supply of food is the pre-requisite for sustainability of life under such conditions. Rapid urbanization is causing scarcity of cultivable land. Therefore, the productivity per unit area of the crops needs to be increased manifold. Horticulture, in general, has attained a pride status in the field of modern agriculture. However, radical changes are expected in horticultural practices in near future as the situation in terms of area as well as production is not as comfortable as the food grain production.

Many people around the world want to find quality fruit in the markets in every season. To increase the productivity of horticultural crops, hi-tech practices need to be adopted, such as, growing these under protected or controlled environment so that the optimum yield levels can be achieved. Fruits, vegetables and flowers can be successfully grown under greenhouse environment to achieve the desired objectives. Greenhouse structures are based on the principle of greenhouse cultivation. It is a frame structure covered with transparent material like glass, polyethylene, polycarbonate etc. and the solar radiation is able to pass through the greenhouse cover. The crop, floor and other objects inside the greenhouse absorb the heat admitted inside the greenhouse. These objects in turn emit long wave thermal radiations for which the greenhouse covering material has lower transparency thus the inside air temperature is increased. This is called 'Greenhouse Effect'. Greenhouse cultivation is the most practical way of achieving the objectives of protective horticulture, in which, the natural environment is modified by using advanced engineering technologies so that optimum yield levels are attained with improvement in quality without any adverse effect on environment. In this specialized system, the soil-climate ecosystem i.e. soil temperature, humidity, solar radiation, environmental temperature etc is modified which may alter the conventional cropping cycle and stabilize and improve the production and quality with efficient utilization of water and nutrients, minimized pesticides or chemical residue, increased photosynthetic efficiency and reducing transpirational losses. Earlier, the technology was considered as means of increasing productivity in western countries but has tremendous scope indeveloping countries and is becoming popular now a days. By adopting it, the growers can now look forward to a better and additional remuneration for high quality produce.

STATUS, SCOPE AND IMPORTANCE

A new dimension to produce more in a limited area

Presently, the area under protected cultivation of horticultural crops is only around 40,000 ha out of which large portion is in northern plains which is not being utilized with great success. There is tremendous potential for increasing the area under low cost polyhouses for production of high quality off-season produce. There is a negligible scope to expand cultivation horizontally, hence, to achieve increased production increasing productivity per unit of space and time is the only way. Obviously, protected cultivation has come as a boon to small and marginal farmers who can now think of earning descent livelihood from their small land holdings which otherwise was a distant dream for them. Purpose of protected cultivation in India is to provide early harvest, protection from frost, diseases and pests. Use of dwarfing rootstocks and advanced construction materials has made cultivation of fruits inside the greenhouses possible. In recent years, applications of KNO_3, growth substances, micronutrients, defoliation and use of mineral oils etc are tried in protected structures in addition to open conditions to meet the chilling needs, breaking dormancy and getting high yields. For dwarf varieties and high density planting, the fruit plants are being grown in pots too.The countries like Israel and Holland haveutilized the protected cultivation structures to the maximum. In European countries like Spain, Egypt and Italy, the area under protected cultivation is expanding very fast. China is the leading country in Asia having maximum area, followed by Japan and South Korea. Fruit crops such as orange, grapes, mango etc are being grown in Egypt in nethouses for increasing the yield and avoiding climate change impacts. Strawberries are being grown most popularly under protected cultivation structures in Asian, European and Mediterranean countries. The protected cultivation in India started mainly to extend the marketing period of vegetables and fruits by extending the growing season. The protected cultivation is destined to play an increasingly important role in India as a means of sustainable crop intensification leading to optimum water use efficiency with better control of product quality without any adverse effect on environment. Commercially, it started in India during early nineties in southern states. Faced with the constraints of limited land holdings, rapid urbanization, declining crop production, increasing population and declining bio-diversity, demand for food has increased manifold and protected cultivation has offered fruits. Moreover, the greenhouse cultivation of horticultural crops has a potential to create additional jobs in the rural sector. The design of the polyhouse (as per location) and its construction by reputed companies is the major factor for successful growing of various potential crops in different geographical conditions.

Advantages: The competitiveness in markets and the ever-increasing quality standards for growers have increased the importance of each small agricultural production unit. It is within these units that the advantages of the modern poly house are expressed - the ability to control crop growth almost regardless of external factors, while producing the high quality of produce year-round and ensuring the farmer a stable flow of income.

- Basic protection of crops from extreme temperatures, solar radiations, high wind velocities by the use of protected cultivation structures and modifying the inside environment characterizes the importance of these in modern times, securing the harvests without harming the environment.
- To get uniform and better quality produce, the horticultural produce is raised under protected cultivation structures. Round the year supply of produce to the markets as well as off season production can be achieved through it thus increasing marketable supply and economic returns.
- Promotion of polyhouse cultivation can provide boost to exports of agricultural produce from India. Protected horticulture helps in self employment generation due to year round activities. Thus, protected cultivation technology has given growers a powerful management tool for production of crops. Generally high value commercial crops are grown under protected cultivation as protected cultivation requires high level of technology and adequate economic returns over investment. As such, quality, productivity and remunerative prices are the major considerations for selecting poly house crops. Moreover, precise environment and control over nutrient application push plants to new limits of growth and high productivity from small land holdings.
- Effective utilization of natural resources in intensive agriculture is the main objective in today's scenario which can be achieved through protected cultivation. Moreover, use of different mulching material (organic wastes,

plastic films etc) aids in decreasing evapo-transpirational losses and prevents the weed growth.

- Large substantial demand of fruits in big cities can be adequately met with more and more installation of protected cultivation structures. The nurseries with greenhouse facility can improve the quality of propagating material with reduction in time of preparation of saplings or fruit plants.
- Harvesting time is regulated to reduce the post harvest losses. There is relatively risk free production (controlled temperature, RH, light intensity and/or CO_2) as compared to open environmental conditions, which, in turn, reduces the business risk. Automation of cultivation practices like irrigation, fertilization etc saves time.
- Plants can be provided with proper nutrition through drip fertigation. Plant protection measures are applied with minimum use of pesticides.
- The sterilized growing media and soil less cultivation can be used, thus providing the plants with soil free from diseases and other harmful micro organisms.
- If adopted on a large scale, the yield predictions and planning of crop production schedules for consumer needs can be done with more precision to cope up with the timely availability of growing demands of population. Protected cultivation causes precocity as well as more yield in fruit trees than open field conditions. An economic analysis revealed that the production cost of grapes grown in greenhouse was higher than open field grapes, however, greenhouse grapes were more profitable than open field because of premium quality. Furthermore, greenhouse grape production provided 3-4 months earliness to producers fetching premium prices in the market.

The major constraint in adoption of greenhouse cultivation technology in India is high initial cost of establishment. These include construction material and equipment. Environment control devices increase the operational costs too. Many state governments in India are offering certain benefits or subsidies to encourage the coverage under these structures. So the cost effective low and medium-cost greenhouses, low tunnels and other related facilities have to be developed accordingly. Moreover the information or database on crop management for Indian conditions is lacking in many crops especially fruits which need to be standardized for successful implementation of protected cultivation programs. For this, research and development activities need to be accelerated with changing climate.

TYPES OF GREENHOUSE

Greenhouse cultivation is a highly intensive enterprise requiring substantial labour and capital inputs. Efficient designing and construction of polyhouses (location specific) is essential for success of greenhouse technology. Therefore before starting a polyhouse, a grower must take into consideration various aspects like climate or soil type of the area, crop demand of the region, availability of labour, electricity and water, road facility, nearness to the market etc.Greenhouse production requires constant vigilance in terms of maintenance, crop production and handling. The National Committee on Use of Plastics in Agriculture (NCPA, 1982) recommended specific greenhouse designsfor different regions of India because of varied agro-climatic conditions. For example ,in North-East, greenhouses have to be constructed as rain shelters to permit off-season production. The coastal and southern regions require naturally ventilated polyhouses for efficient and round the year cultivation.As per the requirement of specific area according to climate, the designs roof slope can be increased or decreased and height can also be varied in order to increase or decrease ventilation as per the suitability to local conditions. The varieties suitable for green house cultivation with long production span as well as tolerance to various biotic and a biotic stresses have to be selected, ensuring higher productivity, better quality, lesser pesticides use.

Many NHM and RKVY projects have given a boost for popularization of polyhouse cultivation.There are many different types of greenhouse structures used for crop production. Although there are advantages in each type for a particular application, in general there is no single type greenhouse, which can be constituted as the best. Different types of greenhouses are designed to meet the specific needs. The different types of greenhouses based on shape, utility, material and construction are briefly given below:

1. Structures Based on Shape

For the purpose of classification, the uniqueness of cross section of the greenhouses can be considered as a factor. The commonly followed types of greenhouses based on shape are:

Lean to type greenhouse: The greenhouse is constructed against one side of the building using the side of building as one side of greenhouse. The main advantage is that it is relatively inexpensive and well supported structure. But it has smaller size and is poorly ventilated with poor temperature control.

Even span type greenhouse: This structure is also used for smaller greenhouses. The structure is constructed on leveled ground with two side slopes and is attached to a building at one gable end. The size is bigger and has more cost than lean to type but the design is more flexible. Single and multi even span greenhouses are being used in various regions.

Uneven span type greenhouse: Earlier used in hilly terrains and slopes. Roof width as well as sides are unequal. Wider roof faces south for maximum sunshine. These cannot support automation much.

Lean to type

Even span type

Uneven span type

Ridge and furrow type

Saw tooth type

Quonset type

Ridge and furrow type: Two or more frames of greenhouse built together forms the ridge and furrow type structure. The single large interior structure reduces the cost of installation per unit area, labour as well as automation cost. The furrow or gutter carries the rain water away efficiently. These are widely used in European countries and India.

Saw tooth type: These are similar to ridge and furrow type greenhouse but they have provision of natural ventilation in it giving a saw tooth like appearance.

Quonset greenhouse: In these, the pipe arches or trusses (which support the covering material of a greenhouse) are supported by pipe purling which runs longitudinally throughout the length of the greenhouse. These are either constructed free or interlocking ridge and furrow style and need lesser capital for construction as compared to the gutter connected ones. These are constructed when small isolated area is sufficient for cultivation.

2. Structures Based on Technological Interventions

In general, the higher the level of technology used, the greater the potential for achieving controlled growing conditions. Three types of polyhouses mainly constructed are: fully climate controlled (hi-tech greenhouses), partially climate controlled and modified naturally ventilated.

(i) High technology or high cost greenhouses

These polyhouses are equipped with many automated environment control devices such as temperature sensors and control mechanism, humidity, ventilation and light control units, regulated irrigation and fertigation system, digitally operated door lock – open system etc.The structural frame is made up of galvanized iron (G.I.) pipes. Cladding may be plastic film (single or double), polycarbonate sheeting or glass. Hi-tech greenhouses are constructed to achieve superior crop performance, sustainable environment and enhance the cultivation period of the crops. Evaporative cooling and the heaters are used to maintain the required temperatures inside greenhouse as and when needed. Cost of installation is estimated to be as Rs 2000 per square meter. Although these structures are capital intensive, they offer a relatively productive, environmentally sustainable produce for the agribusiness industry.

(ii) Partial climate controlled or medium cost or medium technology greenhouses

These greenhouses offer a compromise between cost and production with semi-automatic control arrangement owing to minimum investment. The basic cost of installation of these greenhouses is half to that of high tech or fully climate-controlled greenhouses.This type of greenhouse is also constructed using galvanized iron pipes/mild steel pipes. But the problem associated with MS pipes is rusting after some time, which also damages the covering sheet. These greenhouses are also glazed with UV stabilized polyethylene sheet. Exhaust fans with thermostat are provided to control the temperature. Evaporative cooling pads and misting arrangements are also made to maintain a favourable humidity inside the greenhouse but these are semi-automated and need extra care while operating.The cost varies from Rs 800 to 1000 per square meter. These types of greenhouses are suitable for low lands and mid hillareas.

(iii) Naturally ventilated or low cost greenhouses

These are simple greenhouses with low initial investment. The frame may be galvanized iron pipes, bamboos, wooden logs or steel pipes, or any other local material but no environment control systems are provided in the structure. The top of the greenhouses is covered with the plastic and the sidewalls have the insect proof nets from ground or at 2-4' to a height of 5 to 8' with or without manually roll-able plastic cover. Cost of construction varies from Rs 250-300 per square meter. The height of dome is generally less than 3 m. The initial cost of these greenhouses is less than half to that of semi climate controlled greenhouses and are easy to erect. Little or no automation is used in these structures. These polyhouses need ventilation, shading and misting during summer and rainy season (depending upon prevailing weather and location) to bring down the temperature and regulation of humidity whereas vents can be closed during winters. These greenhouses with certain modifications are also suited for growing crops in high hills and cold desserts. In the higher hills, the climate is mild during summer and rainy season and high value crops *viz.*, capsicum, tomato, cucumber, beans, sarda melon *etc.* can be more profitable as compared to open cultivation. Cold regions have been immensely benefited from this technology due to availability of an additional growing season to the inhabitants. In North-East parts, simple bamboo structures have been found useful to raise vegetable crops for domestic consumption and supply to local markets. In general, the smaller construction with stakes of bamboo erected above the nursery beds and covered with plastic film of 200 gauge thickness,are the examples of cheapest low cost structures.

Net House

Net Houses are used for the production of plants in warm climates or during summer months by the nurserymen. Apart from all types of nursery, flowers and foliages which require shade can also be grown in shadehouses. These shade structures make excellent holding areas for field-grown stock while it is being prepared for shipping to retail outlets. Shade net houses are mostly covered with either lath (lathhouses) or polypropylene shade netsupported by poles. Nets are available in many colours. Effect of various colours of shade net on growth and production of fruit trees has been investigated by various workers. Use of white and yellow shade net for growing mango trees in Egyptian conditions resulted in increase in number of leaves, plant height and stem diameter as compared to open field conditions. Green shade nets are popular for flowers and vegetable nursery in Indian conditions.

Factors Affecting Fruits Cultivation under Protected Structures

The most determinate factor in the horticultural production worldwide is climate. Extreme temperatures, humidity, water availability and quality, atmospheric CO_2, wind velocities, variation in sunshine etc have affected fruit production to a great extent. That's why, controlled atmosphere, in contrast to open field conditions offers more CO_2 enriched atmosphere and less extremities in the microclimate inside to enhance the productivity. Fruit trees, because of their size are normally grown under even span greenhouses and the trees in central part have to be trained to a central leader with pyramid shape and the ones close to the sides with an open centre. Under solar greenhouses, cherries, nectarines, strawberries, papaya etc are grown commercially in many Asian countries like China, Japan and Korea. Strawberry can be produced 30-35 days earlier than normal period and period of fruit availability is extended by 45-50 days when planted in low tunnels of 50 per cent shade net.

Instrumentation is a tool for growers to develop near and long term strategies in crop management and resource saving. Technology development for instrumentation has revolutionized the protected agriculture. The multi-channel data loggers with large memory capacity are available. These have provision for connecting various types of sensors which, later on, can be retrieved by connecting a computer. The other methods for getting the data from data loggers are by satellite receiver or by modem. The retrieved data can be viewed in a spreadsheet format. Such technology development simplifies measurement of microenvironments and plant interactions, permitting greater understanding. The various environmental control parameters are: temperature, humidity, light intensity and carbon dioxide concentration.These factors are measured to supplement the suitable environmental condition for maximum biomass production. Soil environment is managed up to some extent in an open field conditions, however, the aerial environment is the most crucial since it not possible to manage it under open field condition.

Temperature: Hence, it becomes the first priority to measure and control in greenhouse systems.Temperature is a primary factor influencing all basic physiological processes such as photosynthesis, respiration, nutrient uptake, translocation, disease incidence and ultimately crop productivity and quality.High temperatures can increase the capacity of air to absorb water vapours and consequently generate a high demand for water. More evapo-transpiration deplete the water reservoir in the soils creating water stress in plants. This water stress not only reduces the crop productivity but also accelerates the fruit ripening. Prolonged low temperature results in slower loss of flowers and fruits. Papaya and strawberry are both extremely sensitive to extreme temperatures and water stress conditions. During extreme cold weather, high temperatures can be maintained in naturally ventilated greenhouses by closing the side curtains and roof vents by trapping the heat generated by solar radiation. In strawberry, flower bud induction, development of runners, fruit size and sugar accumulation is greatly dependent on temperature inside the polyhouse. Photosynthetic rate is highest between 15-27°C temperature.

Soil temperature is important factor for growth of horticultural crops. If soil temperature drop below minimum,root tissue necrosis can occur.Root zone temperature affects root length, branching, nutrient and water uptake and crown formation where leaves and flowers are developed.

Thermostat is an automatic device which senses the temperature inside the polyhouse. But these do not have good precision because of working on mechanism of expansion and physical displacement of liquid filled inside a metal tube. Therefore they need constant calibration. Computerized temperature control devices are more accurate to use. Thermometers also have high precision and are quite economical to use. Digital thermometers display the temperature measured by the thermocouple. An instrument called thermo hydrograph can record the temperature and air humidity on special graph strips. Infrared thermometers are also being used which measure the magnitude of radiated infra red energy from an object and display it into temperature reading after converting it into electrical signal.

Relative humidity: The growth and development of fruits depends upon the air temperature and humidity. Strawberry plants are highly sensitive to dry climate. Both in greenhouses as well as low tunnels, need high humidity. The fruits develop tip burn and calyx burn during hours of low relative humidity. Overhead high pressure foggers are recommended particularly during the day. In Florida, the inverted-seed wedge grafted plants of sapota when planted in intermittent mist in greenhouse produce new roots within 2-4 weeks and scion begins to grow within 3-6 weeks. Extremely high humidity levels provide ideal environment for pathogens.

Hygrometers are instruments used for measuring relative humidity, consisting of two thermometers, one of which includes a dry bulb and one of which includes a bulb that is kept wet to measure *wet-bulb temperature*. The wet bulb is kept moistened and it displays reading with evaporation proportional to drop in humidity of internal air. These instruments and some other relative humidity sensors need to be re-calibrated every year or two to maintain the accuracy. Vapor pressure saturation deficit (VPD) is a better indicator to understand potential transpiration.Thermo hydrographs measure both temperature and relative humidity efficiently when kept in a shade.

Light intensity: Light is a critical component to photosynthesis and proper plant growth. The need for light can differ from one plant species to another and even from one cultivar to another. The successful greenhouse grower will be mindful of the various crops being grown in the greenhouse and their individual lighting needs, and will ensure plants have sufficient spacing to receive the available light. All other plant needs can be met in the greenhouse, but unless light is carefully evaluated and managed, plants will not grow properly and yields will be lower or fail entirely. Fruit thinning patterns of different fruit crops will depend upon their training systems also. For example in peaches and nectarines, the fruit thinning patterns in open centre and Y systems will be quite different from each other and should be performed by keeping more fruit in sun exposed canopy for proper colour development and quality. Many studies demonstrated that papaya has high adaptability for photosynthesis in net house cultivated system. In papaya, broad and horizontal leaves intercept more radiation in the upper canopy combined with more light interception and more photosynthesis. This results in high yield potential of papaya in net house conditions. Intensity of solar radiation also influences the photosynthetic activity and yield of fruit crops. Long photoperiods tend to reduce the losses of photosynthesis through respiration. Moreover, the supplementary lighting results in predicting the production planning more accurately than natural light conditions.

Two types of sensors are often used in greenhouse crop production. Quantum sensor is used for measuring photosynthetic photon flux (PPF), the photons (400-700 nm) usable for photosynthesis received on a horizontal surface. Energy sensor or pyranometer is generally used for quantifying much broader solar radiation (300 - 2500 nm). These both are to be maintained dust free for efficient operation.The data on day length of the entire year for the site of polyhouse construction as well as for the crops to be cultivated must be known. The variation of sunshine is directly related to the differences in day length. The use of artificial light or blinds made of black plastic film can be provided according to the shorter or longer photoperiodicity of the crops, respectively. Moreover, the cloudiness is also a critical factor which needs to be considered. Levels of

UV light also need to be measured as high intensity can be detrimental to plants. The UV meter is used to determine the UV filtering capacity of greenhouse shades and glass barriers.

Carbon dioxide concentration: Carbon dioxide is one of the most limiting factors in photosynthesis. The amount of CO_2 required for optimum plant growth depends on the plant type, temperature, light intensity, leaf area, air, velocity, humidity, water stress etc. But the requirement differs from plant to plant. CO_2 enrichment has shown to increase the growth, development, fruiting and yield of fruit crops. Additionally, it could be used as a way to reduce the time taken by the fruit plants to become buddable by encouraging their vegetative growth. High tunnels enriched with CO_2 concentration are found to have increased the berry size of raspberry up to 12 per cent. Studies have shown that papaya plants grown under high CO_2 concentrations in polyhouse conditions showed more nitrogen use efficiency. A potential for synergism between CO_2 and light is also found. Especially in tomato, the studies have shown that the plants receiving both CO_2 and supplemental light yielded 12-15 per cent more than control with more number of fruits per plant.

Concentration of CO_2 is greatly reduced inside the greenhouse as a result of photosynthesizing plants if vents are tightly closed, especially in cold weather. Regular monitoring of CO_2concentration is important for optimum crop production and plant health. The levels are replenished by air exchange.

Components of Protected Structures

Climate control inside the greenhouse is regulated by ventilating, screening, heating and cooling.

Ventilation : Natural ventilation in greenhouses is done by opening the vents or part of greenhouse cover or rolling up the sheet present on walls, roof or front or back entrance. It needs to be done when both temperature and relative humidity has increased or to regulate CO_2 concentration inside the greenhouse. In warmer climate, this has to be done frequently. The ventilators provided at the topmost part are the most effective as the heat accumulates in the uppermost region of the greenhouse. Vents can be opened and closed manually or mechanically. The openings can be sealed with insect nettings to avoid entry of insects inside.

Screening: Screening is done to protect the plants from sun. The screen material is supported inside the framework with the help of extra wire supports. It is spread during the hot sunny day when temperature is high and avoids the overheating inside the greenhouse also reducing the evapotranspiration to some extent.

Cooling Systems: Simplest method to achieve cooling is evapotranspiration which involves a continuous water supply as well as air exchange causing evaporation and consequently, a depression in temperature. Two main cooling systems commonly used are fan pad system and use of foggers misters. Fan and pad cooling system is the most common and widely used system in which one side wall of greenhouse is equipped with exhaust fans and the other side has porous pads composed of corrugated cardboard material or cellulose. Cellulose material of 100 mm thickness is preferred for use. These pads are kept wet with the continuous supply of water. When the exhaust fans are running, the outside hot air is drawn in through the pads and the moisture absorbs the heat letting it cool down while passing through it. The salt deposits or algal growth in pads can be avoided by drying, cleaning and disinfecting them regularly.Evaporative cooling systems are also used in polyhouses. An evaporative cooler may be installed outside the greenhouse to cool outdoor air which is then discharged inside. The hot air is exhausted through outlet shutters.

Heating Systems: In cold regions, heating systems are advisable for getting better produce. Centralized boilers, having steam or hot water are used for bigger greenhouses. Unit heaters are used for localized heating in smaller spaces. Infra red heaters for indirect radiant heat can be used. Solar heating system is used to heat water during day which is stored and circulated in the pipes to achieve the desired heating.

Watering System: Micro irrigation system is the best for watering plants in a greenhouse. These systems have high water use efficiency, small volumes of water can be distributed over long distances, development of weeds is greatly reduced, evapo-transpirational losses are less and fertigation is possible. Micro sprinklers or drip irrigation equipments can be used. Basically the watering system should ensure that water does not fall on the leaves or flowers of fruit plants as it leads to disease and scorching problems. In micro sprinkler system, water under high pressure is forced through nozzles arranged on a supporting stand at about 1 feet height. This facilitates watering at the base level of the plants. Drip irrigation is more popularly being used in greenhouse cultivation.

Fertigation : The fertigation refers to the simultaneous application of water and fertilizers to the root zone of the plants. In it, an automatic mixing and dispensing unit is installed in which fertilizers are dissolved separately in tanks and are mixed in a given ratio and supplied to the plants through drippers.

SHIFT FROM SOIL TO SOIL LESS CULTURE

Strawberry is a very high value and sensitive crop amongst fruit crops and need protection from wind, rains, frost and extreme high temperatures. Its cultivation was earlier performed in soil using conventional cultivation methods. Soil disinfection is a common practice in its cultivation and the chemicals such as methyl bromide leave residues. Moreover, the excessive use of pesticides in open conditions resulted in highest rate of pesticide contamination in this crop. Greenhouse cultivation provides better controlled environmental conditions for plant development and fruiting with reduced use of pesticides.

Growing Media: The materials other than soils *in situ*, in which the plants are grown, is the growing media. A healthy, disease free and enriched growing media is an important component of greenhouse crops. Contrary to the open conditions, soil alone as growing media is not preferred as it has less loose aeration character and on watering, the O_2 levels decrease resulting in suffocation of roots. That is why, decomposed plant residues or animal manures are mixed with sand to improve the aeration character of the mixed media. More stabilized materials which resist decomposition on exposure to moisture,like peat moss, saw dust etc are used. Perlite and vermiculite are used to substitute sand. In addition, these media have good water holding capacity, cation exchange capacity, proper pH level, light weight and low cost. Disinfection of growing media is essential which can be done by chemical drenching/ fumigation, steaming, pasteurization, solarisation etc.

Soilless Culture or Hydroponics Cultivation without the use of soil using other materials like pure sand, gravels, peat, vermiculite, perlite, rock wool, coconut fiber, sphagnum peat moss or mixture of some of them. Sometimes, hard media is used as compressed straw pallet or rock wool. All the elements required for the plant growth are the same; hydroponics only takes away the soil requirements.The plant roots are immersed in nutritive solution allowing the roots to absorb nutrients. Media used include rockwool and sawdust. Currently, in countries like United States, Australia, Japan, Korea etc, soil less culture is being used in strawberry being a shallow rooted crop as it provides an important advantage by avoiding the need to sterilize the soil, allows greater ability to manage the cultivation with minimizing the use of pesticides and more yields than soil culture. Vertical hydroponic systems are also being used to utilize vertical space in the greenhouse. Hydroponic systems involve high construction cost and more research is needed to develop economically as well as environmentally sustainable production systems.

Aeroponics: Roots are kept in closed air spaces and the nutrient solution is pumped frequently by mist system, without being planted in normal soil. This system is more difficult to set up and manage than hydroponics. However, the chances of soil borne disease incidence is potentially eliminated.

These techniques are quite costly and need to be tried inside the protected structures only. These systems are tried to meet food requirements mainly at places where soil is not suitable for cultivation or normal production is not possible. Moreover, the nutrients can be applied in very small quantities without wastage under completely controlled environment in readily available form. Proper equipments are needed to minimize the errors in these techniques.

NURSERY PRODUCTION IN GREENHOUSE

For healthy, disease free and off season nursery production, some of the improved methods can be used such as polythene bags, plug trays etc. For fruit nursery, polythene bags are filled with soil, FYM and cocopeat and are punched at the bottom for drainage. The growing media such as cocopeat, vermiculite, perlite can be successfully used. The technique is quite economical and the nursery raised is of better quality. But polythenes are hazardous to environment because of problem of their disposal. Plug trays are a suitable option for vegetable nursery raising.

Mulching: Protected cultivation techniques ensure risk reduction, if not complete elimination, in adverse climatic conditions. Mulch application is an effective practice to restrict weed growth, reduce compaction, conserve moisture and reduce the effect of soil borne diseases through soil solarization. Mulches have been found to increase crop growth, yield and early harvest in some cases. As covering materials, mulches help in soil temperature moderation also. Selection of type of mulch material has to be made by taking into account the climate, type of crop, cost of mulch and quality and durability of the mulch. Natural mulches such as dried leaves, crop (wheat, paddy, maize) straw, saw dust, grass clippings, farmyard manure, compost etc have been in use for many years in commercial production and can be applied inside greenhouse too. Organic mulch materials are bulky and costly to transport. Synthetic mulches are more popularly being used. This mainly includes polythenes of different width, thickness and colour. Black low density polyethylene (LDPE) mulch is most commonly used. Infra red transmitting mulches and reflective mulches like aluminium of polyethylene are found to repel certain insects but being costly, these are very scarcely being used commercially. Wavelength selective mulches also known as infrared transmitting (IRT) mulches, selectively absorb photosynthetically active radiation (PAR) and transmit infra red radiation. Disposal of plastic waste mulch is of great concern as these do not get decomposed. The plastic waste has to be graded, cleaned and recycled. Research efforts are already directed in development of cost effective degradable mulch materials with desirable characteristics.

Soil Treatment: Soil in the net house should be sterilized before the nursery sowing and transplanting of crop. Sterilize the soil in net-house with 2% formalin solution. Mix 20 ml of formalin in one litre of water and drench in the soil @ 4-5 litres/m^2. Cover the soil with polythene sheet for 48-72 hrs. Then remove the sheet and stir the soil for 3-4 days for complete elimination of formalin before transplanting of the crop.All cultural practices should be followed to raised the healthy crop.

CROP MANAGEMENT PRACTICES UNDER PROTECTED STRUCTURES

As the area under polyhouse in India is likely to increase manifolds in the near future, so will be the insect- pest problems. The main objective is to produce pest and damage freeplants tolerant to minor insect pest or mite damage. Mild and less toxic pesticides can be used. Integrated pest management (IPM) will be an important tool in the management of these pests. The primary goal of IPM is to optimize pest control in an economically and ecologically sound way.

1. **Sanitation:** The goal of sanitation is to eliminate all possible sources of the pest in the greenhouse. It reduces the need of multiple applications of pesticides and the risk of pesticide resistance.Careful observation in order to prevent the entry of nematodes is the best mean to manage nematodes. If infestation is severe then the polyhouse should be deep ploughed and left fallow for sometimes to reduce the population. Weeds inside and near the outside of the greenhouse can harbour pests, It is best to pull the weeds inside the greenhouse rather than spray them, since insects may survive the spray and migrate on to crops. It is important to keep the area around the exterior of the greenhouse free of weeds. Plant debris from previous crops can also be a source of pests. Removal and disposal of all previous crop debris is important. Ideally, the greenhouse should be thoroughly cleaned and left empty for one week prior to beginning the next crop.

2. **Solarization:** Soil solarization is a non-chemical method of soil disinfection process of covering moist soils with clear polyethylene to trap solar radiation and raise soil temperatures. Most soil inhibiting pests, nematodes, pathogens and weed seeds are killed this way. Solarization is most effective when applied for atleast 30 days in mid summers.

3. **Sterilization of Growing Media:** Harmful microbes of growing media should be eradicated by pasteurization or fumigated with chemical fumigants. Pasteurization with aerated steam is preferred over use of chemicals as these may leave toxic residues in plant products and may have negative effects on environment.

4. **Crop Scheduling:** When two or more monocrops are grown each year, the pathogen populations are perpetuated. Crop rotation has beneficial effect on plant health and yield due to changes in soil microflora.

5. **Double Door Entrance:** The entry of insect pest and disease infestation through unprotected entrance door is quite common and often neglected. The greenhouse glazing if punctured or torn must be repaired immediately. The adding of double door adds extra protection when the second door is opened by the workers only after the first one is closed.

6. **Disease Free Plant Material:** Selection of un-infested plants, cuttings or transplants is critically important. Inspect carefully all the plants brought to the production area and discard or treat those found to be infested. Inside the greenhouse, quarantine and inspection of all plants upon arrival from other greenhouse and regular monitoring of stock plants used for propagation should be done.

7. **Resistant Cultivars:** Using a resistant cultivar is the most successful method in an IPM strategy. One of the challenges in greenhouse vegetable production is the need for multiple resistances.

8. **Insect Screening:** Insect screens need to be installed in all roof vents, doors and cooling pads to avoid the entry of aphids, whiteflies and thrips.

9. **Regular Inspection:** Regular crop monitoring and scouting isimperative in a consistent and uniform pattern at regular intervals. Pheromone traps are installed and yellow sticky cards available commercially should be hanged with the help of strings.

10. **Cultural Practices:** Avoidance should be the strategy to control insect pests and diseases. Optimum and balanced fertilization and irrigation practices to help reduce conditions which predispose plants to infestation should be adopted. Temperature, humidity and light may also be important factors for both the pests and the use of certain pesticides.

11. **Integrated Pest Management:** Integrated pest management (IPM) offers a sustainable approach for dealing with greenhouse insect pests. Greenhouse whitefly and thrips should be controlled biologically as much as possible through the use of their predators and parasites in addition to pheromone traps. Available bio- pesticides can be used for their control too. Non-chemical methods should be high on the list of priorities. Many of the biorational pesticides like soaps, oils, neem products, *Bacillus thuringiensis* (Bt) and others can play an important role in an IPM programme of greenhouse.

FUTURE PROSPECTS AND THRUST AREAS

Horticulturists have to play a significant role in present scenario of climate change and some sustainable agriculture development pathways have to be adopted. With the advancement in agriculture, various types of protected cultivation structures and cultivation practices suitable to specific agro-climatic zone have emerged. There is a great demand of superior quality produce from these structures with a lot of scope for further improvement. As warming of the climate gathers alarming pace posing a danger to agriculture, greenhouse cultivation can be the answer. The globalization of the markets has increased the competitiveness between the regions across the globe emphasizing the need for quality products without affecting the environment adversely. Protected cultivation at the same time is highly capital and knowledge intensive requiring substantial initial investment and scientific skills. Moreover, consistent efforts are needed to develop technologies suitable to agro-climate of our country to match with global standards in terms of productivity and quality. Greater impetus should be provided to efficient mechanization for different cropping operations. There is an utmost need to develop strategic interventions to promote the adoption of high yielding varieties, standardization of crop specific agro-techniques, season based research work for development and supply of quality plant material, promotion of IPM etc.Experts need to be deployed to educate and guide the farmers on regular basis regarding choice of crop suitable to a climatic zone and various crop management practices keeping into consideration various other factors like marketing and income assurance etc.The farmers should be encouraged to form their own marketing co-operative societies in order to reap the benefit of scale economies (low cost of handling, transportation, packaging and storage) and better bargaining and collective strength. Financial assistance such as low interest loans should be made available to poor farmers for adoption on a large scale. Insurance of greenhouse structure and crops grown in it provides a further boost. Adoption of good agricultural practices and integration of cultural, physical, biological and chemical measures to grow crops with minimal use of pesticides are the major considerations for successful commercial models.

LITERATURE CONSULTED

Cruz, J. L., Alves, A. A.C., Lecain, D. R., Ellis, D. D. and Morgan, J. A.2016. Interactive effects between nitrogen fertilization and elevated CO2 on growth and gas exchange of papaya seedlings. *Scientia Horticulturae,* 8: 202: 32-40.

Kumar, A. and Ahad I. 2012. Growth, yield and fruit quality of strawberry under protected cultivation in South Kashmir. *Adv Hort Sci* ,26(2): 88-91.

Medany, M. A, Abrabbo, M. A. A., Farag, A.,Hassanein, M. K. and Abou-Hadid, A. F..2009.Growth and productivity of mango grown under greenhouse conditions. *Egypt J Hort.* 36 (2): 373-82.

Ozkan, B., Fert, C. and Karadeniz, 2007. Energy and cost analysis for greenhouse and open field grape production. *Energy*, 32: 1500-1504.

Polat, A A., Durgac, C. and Caliskan, O. 2005. Effect of protected cultivation on precocity, yield and fruit quality in loquat. *Scientia Hort.,* 104(2): 189-98.

Reddy, P. P. 2016. Sustainable crop production under protected cultivation. ISBN 978-981-287-952-3. Springer Singapore, Heidelberg New York, Dordrecht, London.

Singh, B. 2014 .Protected cultivation of horticultural crops in India: Challenges and opportunities. Keynote address presented in: *3rd International Conference on Agriculture and Horticulture* October 27-29, 2014 Hyderabad International Convention Centre, India.

SECTION - 4

POST-HARVEST FRUIT TECHNOLOGY

17

Fruit Maturity and Ripening

K.S. Thakur and Satish Sharma

INTRODUCTION

Fruits of commercial importance comprise various combinations of tissues that may include an expanded ovary, the seed and other plant parts such as the receptacle (apple, strawberry), bract and peduncle (pineapple). They may develop from a single ovary of a flower as in simple fruits (apple, banana, mango, peach, plum, tomato etc.), number of ovaries of a single flower as in aggregate fruits (strawberry) or from a number of ovaries of different flowers as in compound fruits (pineapple). They are derived from members of a wide range of families of both monocotyledonous and dicotyledonous plants. Inspite the huge diversity in origin, structure, composition and climate adaptabilities that exists among different fruits, there is considerable similarity in the pattern of maturation and ripening that occur in different fruits.

FRUIT GROWTH AND DEVELOPMENT

After maturation fruits undergo a series of physicochemical and physiological changes during various stages of their growth and development through which they acquire the attributes which make them acceptable to the consumers. The most obvious of these changes is an increase in fruit size. Many fruits follow a pattern of growth that follows a single sigmoidal growth curve (Table 1).

Table 1: Growth patterns observed in fruits

Single	Double	Triple
Apple	Apricot	Kiwifruit
Avocado	Black carrot	
Banana	Blue berry	
Date	Cherry	
Mango	Fig	
Melon	Grapes	
Orange	Olive	
Pear	Peach	
Pineapple	Plum	
Strawberry	Raspberry	
Tomato		

In this pattern an initial phase of slow growth, characterized by active cell division, is followed by a period of major increase in size, fresh and dry weight and is dominated by cell expansion and final period of declining growth rate during which ripening is usually initiated. In addition to these fruits other fruits show more complex growth pattern exhibiting double or even triple sigmoidal growth curves. In these patterns there are two or three rapid growth phases which are interspersed by one or two periods of little or no growth. The double sigmoidal growth pattern is observed in many fruits (stone fruits, fig, grapes). In such fruits the two rapid growth phases are designated period I and III, while the interspersing slow growth phase is designated period II. Different periods of fruit growth are associated with the development of different tissues in various fruits. In stone fruits the endocarp and seed reach full size in period I; in period II lignification of endocarp occurs along with the development of ovary and endosperm while in period III expansion of mesocarp tissues occur in which the changes occurring during ripening are indicated.

After fruit set, the major physiological stages in the life of a fruit are growth, maturation, ripening and senescence which occur sequentially with progressive increase in age of fruit (Fig.1). However, a clear distinction between the stages cannot be made as it is difficult to assign specific biochemical or physiological parameters to describe the various stages because the nature and timing of these parameters is not identical in different commodities. Growth involves cell division and subsequent cell enlargement which influences the final size of the fruit. Maturation usually commences before growth ceases and includes different activities in different commodities. Ripening is generally considered to begin during the later stage of maturation and to be the first stage of senescence. Senescence is defined as the period when anabolic (synthetic) biochemical processes, give way to catabolic (degradative) leading to aging and finally death of the tissue. Development and maturation of fruits are completed only when it is still attached to the plant, but ripening and senescence may proceed on or off the plant. Fruits are harvested either when mature or when ripe.

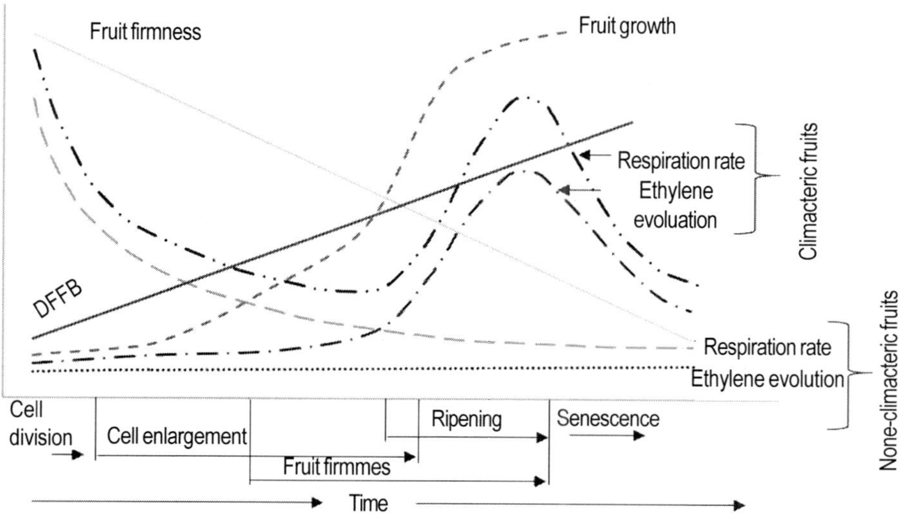

Fig. 1. Diagrammatic representation of the relative changes in important characteristics during growth and maturation of fruits

MATURATION AND MATURITY INDICES

The stage of development of a fruit at harvest has a profound effect on its storage life and postharvest quality. It also influences the way a commodity is handled, transported, stored and marketed. An understanding of the concept of maturity and how it is estimaed, is therefore of utmost importance for successful postharvest handling of the fruits. Fruits that consumers find acceptable to use are often said to be 'mature' or 'ripe'. To most people mature and ripe mean the same thing when describing a fruit and even the dictionary does not explain the difference between the two terms appropriately. However, they are considered to be distinct terms for different stages of fruit development because the stage of development at which the fruit has maximum consumer acceptability does not necessarily coincides with the point at which it is physiologically mature. For this reason the terms 'commercial' and 'horticultural' maturity have been coined to describe 'the stage of development when a plant or plant part possesses the pre-requisites for utilization by the consumer for a specific purpose. It is considered as 'that stage at which a commodity has reached a sufficient stage of development that after harvesting and postharvest handling, its quality will be at least the minimum acceptable to the ultimate consumer.' On the other hand physiological maturity is defined as 'the stage of development when a plant or plant part will continue ontogeny even if detached.' Maturation is the process through which the plant or plant part attains physiological or horticultural maturity.

Judgment of optimum maturity for harvesting a fruit, which after postharvest handling and marketing will yield fruits of the desired quality, is one of the most important decisions to be made by the grower. Most fruits reach their best eating quality when allowed to ripen on the tree. However, it is not possible to allow on-tree ripening because such fruit will be well past their prime eating quality by the time they reach the consumers due to the ongoing physiological changes. Fruits harvested at the immature stage are smaller in size, hard, sour, more subject to shrivelling and they fail to ripen normally and do not attain the desired quality characteristics upon ripening. On the other hand over ripe fruits are likely to become too soft and mealy with insipid flavour which are prone to physical damage and rotting during postharvest handling. Fruits picked either too early or too late in the season are also more susceptible to physiological disorders than those harvested at proper maturity. Fruits are normally picked mature but unripe so that they can withstand the rigours of postharvest handling, especially when shipped over long distances and also to obtain the most acceptable quality characteristics with adequate shelf life. Harvesting of fruits at proper maturity results in fewer storage problems and ensures best quality upon ripening and hence maximum acceptability. Growers must pick fruit when it attains the maturity characteristics as set by the purchaser and failure to do so may result in rejection of fruits by the consumer and hence reduced income.

DETERMINATION OF HARVEST MATURITY

Many subjective and objective methods for determining the maturity of the fruits have been used from time to time with fair to good degree of success and newer methods are being suggested with advancement in available technology. For maturity measurements to be carried out in the orchards, the tests must be simple and should require inexpensive equipments. The test should preferably be objective (a measurement) rather than subjective (an evaluation) and ideally it should be non-destructive. It should also relate consistently to the quality and post-harvest life of a commodity for all growing areas during different seasons. Maturity indices should be standardized separately for different crops/cultivars and orchards in different growing areas over a number of seasons. It is also recommended not to rely on one index/character while determining maturity. A number of tests have been developed to assist growers in making an informed decision for correct harvest times.Maturity of fruits can generally be judged on the basis of computational, physical, biochemical and physiological characteristics.

Computational Indices

Calendar date is the most commonly used maturity index in which the orchardist starts harvesting his crop on a specific date each year. The calendar date is finalized by the grower on the basis of his experiences during earlier years. It is reasonably accurate if there is not much variation in the full bloom date and weather conditions during the growth period over different years. However, wide fluctuations in the date of full bloom may occur in different seasons which may lead to harvesting of immature or over-mature fruits if the full bloom date happens to occur too late or too early in the season. It was therefore suggested that the specific time period,which is required for the completion of biochemical changes that must occur before the fruit attain maturity,must be measured and this led to the concept of counting the number of days between full bloom and harvest (days from full bloom to harvest) which must be allowed to elapse for proper maturation of fruits. However, year to year variations in weather conditions after full bloom affect fruit growth rate and harvest maturity. Moreover it is difficult to define the full bloom date in fruit having a prolonged blossoming period. To overcome these shortening of this method and to include the effect of temperature on flowering and fruit development the use of degree days or summation of heat units is suggested for more accurate prediction of harvest date. Accumulated heat units for optimum maturity are computed by the time in relation to temperature above a certain minimum base temperature (around 50°F for most fruit crops) for each day from the date of full bloom to the expected harvest date. Thus, a day with an average temperature of less than 70°F would provide 20°F degree days and a day with an average temperature of less than 50°F would provide zero degree days of heat units. The degree days or heat units accumulated for each day between the full bloom date and the expected harvest date are standardized over a number of seasons before being recommended.

MATURITY INDICES RECOMMENDED FOR HARVESTING DIFFERENT FRUITS

Maturity	Indices	Fruits
Computational Methods	i. Calendar date	For all fruits
	ii. Days from full bloom to harvest (DFFB)	For all fruits mainly apple, banana, kiwi, mango, olive, papaya, pear, pineapple
	iii. Mean heat unit	Apple, pear, grape, mango, ber, litchi etc.
	iv. T- stage	Apple
Physical Methods	i. Fruit retention strength	Apple, raspberry
	ii. Fruit size and surface morphology	Cherry, almond, grape (cuticle formation), banana (angularity), litchi, nuts, mango, pineapple, sapota, walnut, kiwifruit, glossiness of some fruits (development of wax)
	iii. Weight	Watermelon
	iv. Specific gravity	Cherry, mango, ber
	v. Colour	
	a. Skin	Litchi, olive, pineapple
	b. Flesh	Mango,
	c. Seed	plum, strawberry apple, pear, apricot
	vi. Flesh firmness	Pome and stone fruits, fig
	vii. Juice content	Citrus fruit
	viii. Aroma	Guava
	ix. Abscission	Cantaloupe melon
	x. Acoustic and vibration test	Pineapple, melons, mango, avocado
Biochemical Methods	i. TSS	All fruits
	ii. Titratable acidity	Pomegranate, citrus, papaya, kiwifruit, nectarines and peaches, plum, apricot
	iii. TSS/acid ratio	Grapes, citrus, papaya, kiwifruit
	iv. Sugars (Total and reducing)	Pome and stone fruits, grapes
	v. Starch content	Apple, pear
	vi. Starch-iodine test	Apple, pear, banana
	vii. Tannin content	Persimmon and dates
	viii. Oil content	Avocado
Physiological Methods	i. Respiration rate	Apple and pear, tomato
	ii. Ethylene evolution rate	Pear, apple
	iii. Fruit opening	Nutmeg, chow chow, ackee
Others Methods	i. Bioelectrical conductance	Peach, apple
	ii. Electric properties	Peach, watermelon, apple
	iii. Electromagnetic methods	Avocado, pear, olive, prunes, apple, peach
	iv. Near-infrared Reflectance	Apple peach, pineapple, mango
	v. Radiation	Peach, orange

Physical Indices

Under normal crop loads and a uniform cultural schedule most fruit crops attain specific physical characteristics by the time they become mature enough to be harvested and careful observations of such changes can give an idea about when a fruit should be harvested. However, physical characteristic are influenced by a number of variables such as soil type, fertigation, irrigation, cultivars, crop load and other cultural and climatic conditions which vary among places and seasons. Hence these characters should be used in conjunction with other indices in assessing fruit maturity.

With the approach of maturity fruit retention strength decreases and it becomes easier to separate the fruit from the parent plant because of the development of the abscission layer. An experienced grower can use this character very effectively to judge fruit maturity. Constant measuring of fruit is vital as to determine when the fruit size has reached a marketable size. Size can be measured using a Cranston Fruit Sizing Loopor verniercalipers. When fruit shoulders and sutures are well developed and filled out, most fruits are considered to be mature. To be dependable this maturity measure needs to be coupled with other tests such as skin colour.

Most fruits exhibit changes in colour as they mature and ripen because of destruction of chlorophyll and exposure of underlying pigments or synthesis of new pigments which give characteristic colours to them. Colour development depends on exposure to light, therefore the position of the individual fruit and training of trees will greatly effects the development of pigments in the skin. Skin colour is probably the most common method of assessing fruit readiness to harvest, but it is not considered the most accurate indicator of maturity. However, experienced growers often use background colour as a means of assessing harvest maturity in most fruits. Observing changes in flesh and seed colour may also give useful information about the maturity status of some fruits.

Flesh firmness can be a very good indicator of maturity and it also relates closely to the eating experience for the consumer. With increasing maturity fruits become softer mainly due to the dissolution of pectins comprising the middle lamella of cell walls. Fruits harvested with pressure readings too high will generally offer a dry mealy texture when consumed. Firmness is measured using a penetrometer which measures the resistance required to drive a plunger through the fruit flesh.

Biochemical Indices

Various biochemical constituents also show changes with fruit growth and their concentration may either increase or decrease with advancing maturity and at a particular stage of development fruits generally contain specific amount of these constituents.

During maturation and ripening starch is generally broken down into sugars as a result of which starch content decreases and sugar content increases. The extent of degradation of starch can easily be recorded by the starch-iodine rating test and is a useful method of determining the maturity of starchy fruits such as apple and pear. It is a simple test and can be readily performed in the field. A cross-sectional slice through the middle of the fruit is placed in a shallow pan containing 1.0 per cent potassium iodide and a few crystals of iodine for a couple of minutes and rinsed under tap water. Portions of the slice containing starch turn blue-black while portions where starch has been converted into sugars do not show any change in colour. Fruits start ripening from the core area and starch disappearance progresses outwards with increase in maturity. A starch index chart is prepared by allocating a numerical value to each distinct stage of starch disappearance for later use.

Soluble solids content (SSC), also referred to as total soluble solids (TSS), is a measure of the sugar, organic acids and other soluble components in the juice of the fruit. SSC is measured using a refractometer to determine the percentage of pure sucrose, which is the major constituent in fruit juice, and is expressed as °Brix. It can be used alongwith other indices, such as firmness to judge fruit maturity. TSS values are most commonly used as maturity indices for harvesting different varieties of grapes and citrus fruits.

As fruits mature the level of acidity decreases gradually. Measuring this on farm, in most cases, would not be possible due to the complicated nature of the test. The ratio of SSC: acid content has been found to be more closely related to the quality of the fruit than just acid or SSC content alone and are most commonly used for assessing maturity of grapes. Sugar: acid or TSS: acid ratio is also better related to palatability of the fruit than to either sugar or acid concentrations alone. Oil content increases with the onset of maturity in some fruits like avocados where it is considered to be a reliable index of maturity. It should not be less than 8.0 per cent by weight excluding the weight of seed and skin. In citrus fruits maturity is accompanied by an increase in juice content and minimum juice contents have been recommended for different fruits.

Physiological Indices

Respiration and ethylene production are the physiological processes which show dramatic changes, especially in climacteric fruits, as they mature and begin to ripen. Climacteric fruits show a sharp increase in respiration and ethylene production rates as ripening is initiated. These fruits can be harvested once the increase in respiration and ethylene production has been initiated. A delay in harvesting of fruits after the initiation of climacteric rise in respiration is said to reduce the storage potential of fruits. Though it is a reliable method of judging fruit maturity it is difficult to use, as measurements of carbon dioxide and ethylene require sophisticated instruments and trained personnel to use them and therefore have limitations for use under field conditions. The increase in ethylene production may precede, coincide or follow the increase in respiration in different fruits.

Other Indices

A number of other methods of assessing maturity have been developed and used successfully in different fruits. As fruits mature and ripen, they show changes in the internal optical properties due to the conversion of bound water into free water which changes the light scattering properties of the fruit. Using this property, the light transmittance technique, which was initially developed for tomatoes, is being used for assessing maturity of other fruits after incorporating certain modifications. Similarly, acoustic and vibration tests, electro-magnetic resonance (NMR), near-infrared reflectance (NIR), X- and gamma-radiations have been used successfully to evaluate fruit quality at harvest on experimental basis but have limited application in the field. Efforts are also on to develop biosensors which may be applied to 'sniff' fruits so as to detect the presence of volatiles which may be produced as ripening is initiated.

FRUIT RIPENING

Ripening: Man has long been familiar with the overt aspects of fruit ripening but it is only during recent times that an understanding of the physiological processes associated with ripening has been achieved, although the molecular regulatory signals are still unknown. Ripening is defined as 'the composite of the processes that occur from the latter stages of growth and development through the early stages of senescence and that result in characteristic aesthetic and/or food quality, as evidenced by changes in composition, colour, texture, other sensory attributes'. Ripening is a complex physiological process which transforms a physiologically mature but inedible plant organ into an edible one. It marks the completion of development and the onset of senescence. A ripe fruit is the result of biochemical and physiological changes that occur at the terminal stage of its development. It renders the fruit edible and hence imparts value to fruits as agricultural commodities. Ripening changes normally takes place when the fruits are fully developed and mature and many of them generally occur independently of one other.

Respiration

Ripening requires the synthesis of new pigments, flavour compounds, novel proteins and mRNAs which require energy as well as carbon skeleton building blocks. These are supplied in fruit tissue through the process of respiration. Though all fruits carry out respiration there are marked differences in their rates and patterns of respiration. On the basis of the respiration pattern followed during ripening fruits are classified as climacteric or non-climacteric.

1. **Climacteric fruits:** Climacteric fruits show a sharp rise in respiration rate with the onset of ripening. They exhibit a characteristic peak of respiratory activity during ripening which is termed as respiratory climacteric. Climacteric marks the beginning of ripening after harvest and fruits ripen to their optimal quality off the tree. The respiratory peak generally corresponds to the optimum eating ripeness in most fruits and magnitude of the respiratory peak varies enormously amongst fruits (Fig. 2). In general fruits with highest respiratory rates are more perishable as they tend to ripen more rapidly. In this category fruits can be harvested mature and ripened off the plant either on their own or by exposure to ethylene, although they are capable of ripening on the tree if not harvested at the mature stage. Fruits like banana, mango, guava, papaya, avocado, persimmon, custard apples, peach, plum, apricot, etc. show this type of ripening. The desired quality of fruit is achieved only after it is harvested after proper maturation or on tree ripened fruits whereas fruits harvested at the immature stage fail to ripen normally.

2. **Non-climacteric fruits:** These fruits do not show a rapid increase in their respiration rate during ripening. Their respiratory rates remain unchanged or they simply exhibit a gradual decline in their respiration as senescence is approached. However there are still marked differences in the magnitude of respiration rates between different fruits of this category (Fig. 2.). The general correlation between high respiratory rate and short shelf life also holds true for non-climacteric fruits. These fruits are not able to continue the ripening process once they are harvested from the plant. In general, these fruits do not exhibit dramatic changes in their composition after harvest and should be allowed to ripen to their optimal quality on the tree. In fruits like citrus, grapes, litchi, loquat, pineapple, phalsa, ber, cherry, strawberry etc. there is no specific lag period between maturation and ripening. Some citrus fruits are degreened after harvest with ethylene but this does not alter their edibility.

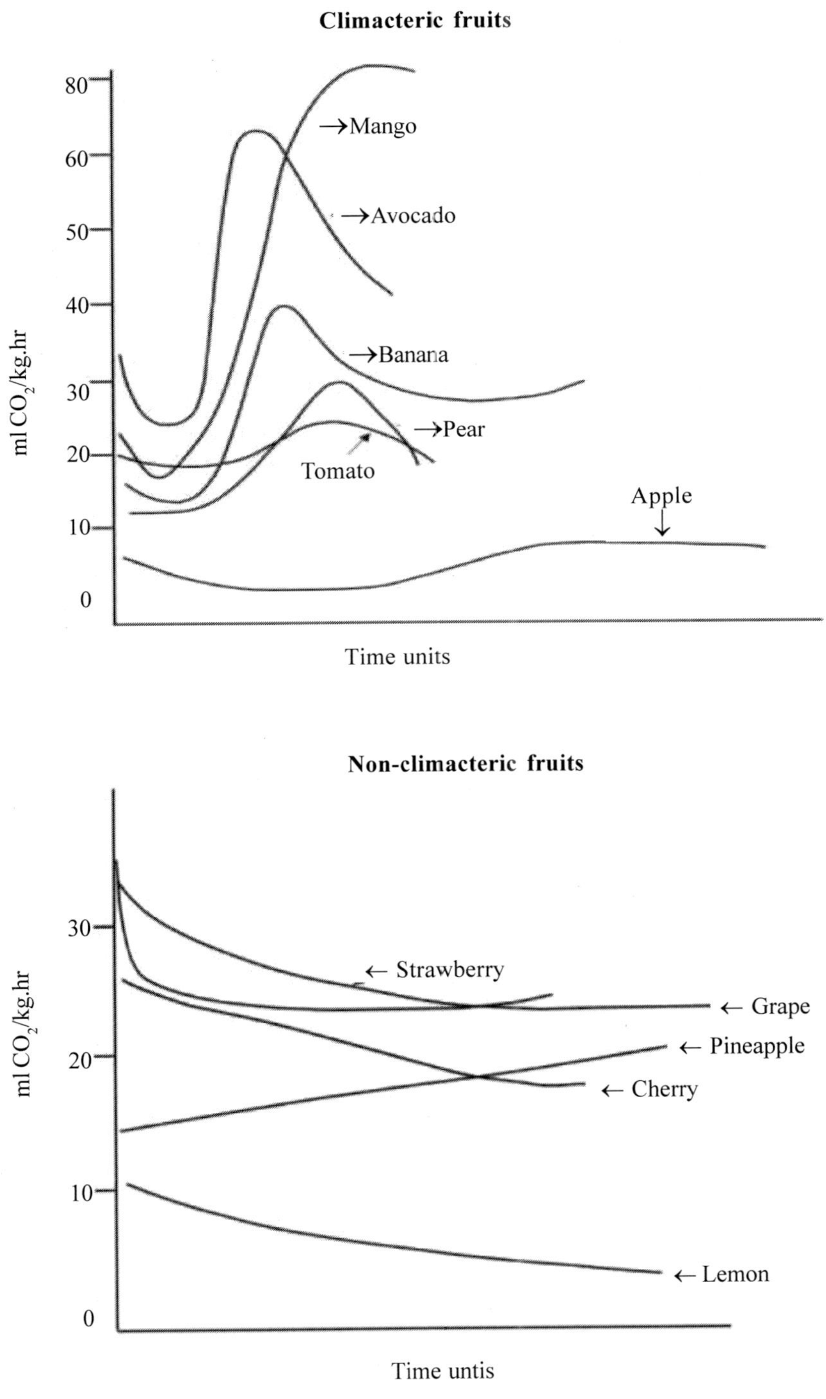

Fig. 2: Respiratory patterns during ripening of climacteric and non-climacteric fruits

Ethylene and Fruit Ripening

All plant developmental processes, including ripening, are under the control of plant growth regulators. Exogenous application of any of the major growth regulators can influence fruit ripening in one way or another. While other growth regulators generally act to delay ripening ethylene and abscisic acid tend to enhance the ripening process. Ethylene is considered as a fruit ripening hormone as it is routinely found to be associated with ripening of fruits.

All plant tissues are capable of producing ethylene, albeit at very low levels. Climacteric fruits show a burst in ethylene production during ripening and it generally precedes the respiratory climacteric in most fruits, though in some fruits it may coincide with or follow the respiratory climacteric. Like the respiratory climacteric different fruits have different peak levels of ethylene production. The patterns of ethylene production in non-climacteric fruits resembles their respiratory pattern and exhibit a decline in the ethylene production rates with advancing maturity and during ripening. However, the rates of ethylene production vary among different fruits. Two control systems are postulated to explain the differences in the ethylene production between climacteric and non-climacteric fruits. System 1 is common to both climacteric and non-climacteric fruits as both types of fruits are capable of basal ethylene production. However, System 2 is unique only to climacteric fruits and is responsible for the autocatalytic production of ethylene during ripening, whereas non-climacteric fruits lack this system.

Climacteric and non-climacteric fruits also differ in their responses to application of exogenous ethylene. Mature green climacteric fruits exhibit an increase in their respiration rates when treated with different concentrations of ethylene. With a gradual increase in the concentration of ethylene application the respiratory climacteric is initiated earlier, although the magnitude of the peak remains constant for all the concentrations. The fruits do not show any further rise in respiration with repeated ethylene treatments once the respiratory climacteric has been obtained by earlier treatments. However, non-climacteric fruits exhibit an increase in the respiration rates with the magnitude of the increase being proportional to the concentration of ethylene treatment (Fig. 3.) Repeated treatments can bring about such responses repeatedly in non climacteric fruits.

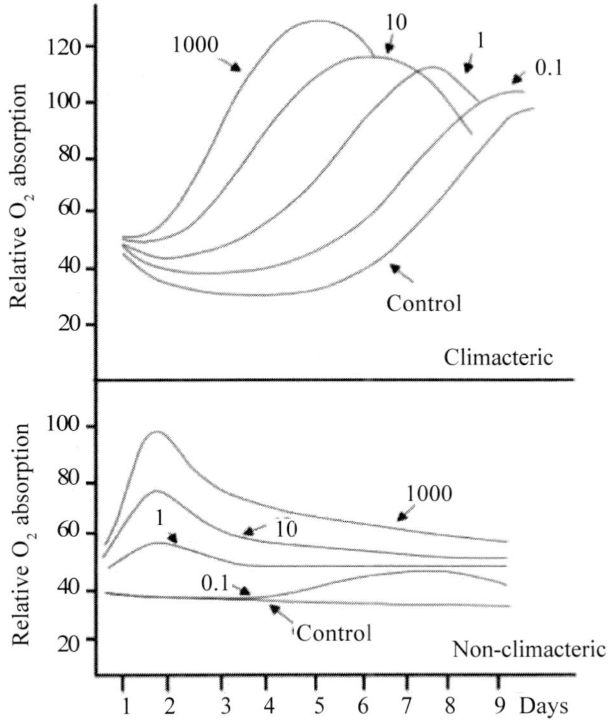

Fig. 3: Effect of different concentrations of exogenous ethylene treatments on respiration rates of climacteric and non-climacteric fruits

Changes During Fruit Ripening

A number of changes take place during fruit ripening which are believed to be under genetic control. Though variable among species, these changes generally include modification of cell wall ultra structure and texture, conversion of starch to sugars, decrease in organic acid content, changes in pigment biosynthesis and accumulation, increase in production of flavours and aromatic volatiles and increased susceptibility to postharvest pathogens. During ripening different unrelated pathways and processes are coordinated to act efficiently and in a synchronous manner to impart desirable, flavour, colour and texture to fruit. These processes collectively confer edibility on the fruit and are, therefore, of economic importance.

Texture: The softening process is an integral part of the ripening of almost all fruits. Changes in the texture of the fruit during ripening result from changes in the structure and composition of their cell walls. As fruits mature and ripen, they exhibit a decrease in texture mainly due to the dissolution of pectins comprising the middle lamella of cell walls. Apart from this, loss of turgor due to postharvest dehydration of fruits and degradation of starch in fruits like banana also contribute to fruit softening.

The softening process associated with ripening of climacteric fruits is under enzymatic control. The loss of intercellular cohesion resulting from dissolution of the middle lamella in ripening fruits is brought about mostly by the action of pectolytic enzymes and is normally accompanied by an increase in the concentration of soluble pectic polysaccharides. The enzymes believed to be involved in fruit softening are D-galacturonases (polygalacturonase), pectinmethyl esterases (PME) and cellulases. The activity of these enzymes is low or absent in unripe fruit and increases during the ripening process. The enzymes involved in the hydrolysis of cellulose, leading ultimately to the production of glucose, are grouped under the generic name cellulases. Perhaps softening is initiated by the action of the cellulases on the interwoven cellulose fibrils of the cell wall thereby enabling the pectic enzymes to penetrate into the middle lamella. Celullase activity is always found in fruit, and the level of activity often increases during ripening.

Color: Change in fruit surface color is one of the most familiar and readily perceived changes during ripening of many fruits. In developing fruit green background color is derived mainly from chlorophyll which may also be found in the internal tissues of some fruits. In most fruits ripening is accompanied by the degradation of chlorophyll which in turn unmasks previously present pigments, particularly B-carotene, and such a process occurs during degreening of lemons. In other fruits the loss of chlorophyll is accompanied by the biosynthesis of anthocyanins or carotenoids. Anthocyanins are water soluble pigments and are responsible for the typical blue, violet, pink and red colours in fruits like grapes, jamun, strawberry and some varieties of mango and banana. Carotenoids are a group of fat soluble pigments which are responsible for red and yellow colours in fruits like mango, ber, banana and papaya. β-carotene and lycopene are the typical carotenoid pigments present in fruits.

Carbohydrates: Carbohydrates are present as simple sugars or complex polymers, comprising hundreds of sugar monomeric units, and are the most abundant group of constituents present in fruits after water. The main sugars present in fruits are sucrose, glucose and fructose with the predominant sugar varying in different fruits. Sugars originate from photosynthetic assimilates, provided by other plant parts and the mechanism of assimilation during maturation and ripening differs amongst fruits. Many fruits accumulate the bulk of their carbohydrates prior to the onset of ripening. Ripening commonly leads to an increase in the sweetness of such fruits as a result of release of simple sugars from starch or other reserve carbohydrates and the interconversion of the released sugars. Starch is composed amylose and amylopectin, both of which ultimately yield glucose units upon hydrolysis by the action of <-Amylase, B-Amylase and glucosidase. Such fruits can be harvested at the mature green stage and still develop acceptable flavour on ripening. Other fruits like strawberry, cherry and grapes continue to accumulate sugars from the plant during ripening and may be commercially unacceptable if they are harvested at the mature green stage as they fail to develop full flavour.

Organic acids: The acidity of the fruit is derived from organic acids which are stored principally in the vacuole. The main organic acids in the fruits are malic acid, citric acid, quinic acid, tartaric acid, isocitric acid, oxalic acid etc. besides a variety of other acids present in minute quantities. The organic acid levels in fruits generally decline during ripening, possibly due to their utilization as respiratory substrates due to the greater metabolic activity during ripening. Many fruits exhibit respiratory quotients of >1 during ripening which suggests the utilization of organic acids as respiratory substrates. In some fruits a small fraction of organic acids may also be converted into sugars through the process of gluconeogenesisin. However, in banana one or more acids may increase during ripening resulting in an increase in the total acid content of the fruit.

Amino acids and Proteins: A decrease in the free amino acid content of fruits occurs during the onset of ripening and it is attributed to their incorporation into proteins which are necessary for the synthesis of various enzymes required for completion of the ripening process. In many fruits increases in protein synthesis occurs early during the climacteric period and reach peak values by the time climacteric maximum is reached. The increase in amino acid incorporation into proteins at the onset of ripening, and similarly the incorporation of RNA precursors into RNAs, including ribosomal RNA, and moreover the production of new mRNA species in ripening fruits suggest that transcriptional and translational processes are associated with the onset of ripening. Apparently the production of new proteins represents an effective enzyme system involved in their production.

Phenolic compounds: Phenolic compounds are important for the growth and development of fruits as they are thought to provide protection to plant growth regulators which are required for proper development of fruits. They also provide protection against spoilage micro-organisms as there is a correlation between the degree of resistance to fungal attacks and the level of phenolics present in fruits. Complex phenolics of higher molecular weight and their oxidation products are known as effective inhibitors of the pectic enzymes and hence delay softening. They are also responsible for the astringent taste of many fruits. Phenolic compounds generally decrease during ripening as a result of which the fruits become less astringent and more susceptible to softening and attack by spoilage micro-organisms. Astringency is due to the presence of tannins of low molecular weight (polyphenols) which react with the proteins of the mucous membrane of mouth upon being chewed. During ripening the tannins polymerises into larger molecules and lose their capacity to react with the proteins of the mucus membrane and hence a decrease in astringency is observed.

Volatiles: During ripening fruits produce numerous small molecular weight compounds some of which may be volatile under ambient conditions. The volatile compounds mainly include esters, alcohol, aldehydes and ketones. These compounds may not be important quantitatively, but they are important in producing the characteristic flavour and aroma of fruits. Most fruits may produce more than 100 different volatiles but the characteristic aroma to each fruit is provided by the presence of only one or two compounds. Their relative importance depends upon threshold concentration, which can be as low as 1 ppb, and interaction with other compounds. The characteristic aroma of apple, ripe banana, lemon and orange is due to the production of ethyl 2-methyl butyrate, eugenol, citral and valencene, respectively.

Ripening Indices for Fruits

In different climacteric and non-climacteric fruits ripening indices has been established for the benefits of fruits growers and for timely marketing of fruits.

Fruit	Ripening indices	
Avocado	After 4-5 days of harvesting	
Banana	Colour changes from green to complete yellow	
Citrus	Musambi	30:1 TSS/acid ratio
	Blood Red	14:1 TSS/acid ratio
	Pineapple	14:1 TSS/acid ratio
	Jaffa	14:1 TSS/acid ratio
	Valencia	10:1 TSS/acid ratio
	Kinnow	12:1 to 14: 1 TSS/acid ratio
	Daisy	21:1
	Marsh Seedless	7:1 TSS/acid ratio
	Foster	6:1 TSS/acid ratio
	Duncan	6:1 TSS/acid ratio
	Red Blush	6:1 TSS/acid ratio
	Star Ruby	6:1 TSS/acid ratio
Grapes	TSS 16%, TSS acid ratio	30:1
Jamun	Deep purple or black	
Litchi	Deep golden yellow colour	
Mango	Colour changes from green to yellow or TSS more than 16%	
Papaya	Skin colour changes from green to yellowish green	
Pineapple	Orange yellow colour	
Plum	TSS 13%	
Sapota	Fruits become soft	

LITERATURE CONSULTED

Dhatt, A.S.2002. Physiology of growth, development and ripening of fruits. In:Post-harvest Handling of Fruits and Vegetables.A.S.Sandhu and J.S.Bal (eds)PAU, Ludhiana. Pp11-16.

Dhillon,W.S.,Kamboj, J.S.and Bal, J.S., 2002. Maturity indices and harvesting of fruit crops. In; Post harvest Handling of Fruits and Vegetables. A.S.Sandhu and J.S.Bal (eds) PAU,Ludhiana.Pp:21-28.

Giovannoni, J. 2001. Molecular biology of fruit maturation and ripening. *Annual Rev. Plant Physiol. Plant Mol. Biol.,* 52: 725-49.

Kader, A.A. 1999. Fruit maturity, ripening, and quality relationship. *Acta Hort.,* 485: 203-208.

Kader, A.A, Kasmire, R.F., Mitchell, F.G., Reid, M.S., Sommer, N.F. and Thompson, J.F. 2002. Post Harvest Technology of Horticultural Crops. University of California, Agriculture and Natural Resources Publication No 3529.

Mitra, S. K. 1997. Post Harvest Physiology and Storage of Tropical and Sub-tropical Fruits. CAB International.

Seymour, G.G., Taylor, J.E. and Tucker, G.A. (eds). 1993. Biochemistry of Fruit Ripening. Chapman and Hall, London.

Thompson, A. K. 1996. Post harvest Technology of Fruits and Vegetables. Blackwell Science.

Verma, L.R. and Joshi, V.K. (eds.) 2006. Postharvest Technology of Fruits and Vegetables. Vol 1. Indus Publishing Company, New Delhi.

Watada, A.E, Herner, R.C., Kader, A.A., Romani, R.J. and Staby, G.L. 1984. Terminology for the description for the description of developmental stages of horticultural crops. *HortScience,* 19: 20-21.

Wills, R.B.H., McGlasson, W.B., Graham, D. and Joyce, D.C. 2007. Postharvest- An Introduction to the Physiology and Handling of Fruits, Vegetables and Ornamentals. CAB International.

18

Fruit Grading and Packaging

B.V.C. Mahajan and J.S.Bal

Grading is defined as sorting of fruits into different grades according to the size, shape, colour, and volume to fetch high price in market. Grading of fruits after harvesting is an essential step in post-harvest management. Grading of fruits is done on the basis of physical characteristics like weight, size, colour, shape, specific gravity and freedom from diseases. The known methods of grading of fruits and vegetables are manual grading and size grading. Sizing produce is usually a part of grading and is optional for direct marketers, but may be worthwhile if certain size grades receive a higher price than others. Round produce units can be easily graded by sizing rings. Several types of mechanical sizers are available for small-scale operations. Sorting by maturity can maximize the market value of crops such as stone fruits and mangoes .The grades of different fruits suggested by Directorate of Marketing and Inspection (DMI) are given in Table 1.

Advantages of Grading

1. Losses in the selling price due to presence of substandard products or specimen can be easily avoided.
2. It increased marketing efficiency by facilitating buying and selling a produce without personal selection.
3. Grading enhanced to set good price for graded products.
4. Heavy marketing cost in packing and transportation can be avoided by grading.
5. By grading there is fairness to both Buyers and Sellers.
6. Properly graded fruits are purchased by the consumer easily without inspection.

Grading: Fruits and vegetables have different shape (round, long, oblong elliptical etc) and size . Similarly they vary in colour also such as green, red and yellow etc. The round fruits can be easily graded either manually or through mechanical grading machines. The coloured fruits and vegetables can also be sorted through colour sorter devices attached with mechanical grading system.

1. **Manual grading**: Grading of fruit and vegetable is performed primarily by visual inspection. Manual grading lacks accuracy and has lower efficiency because there is individual difference in visual inspecting which is affected by workers experience.

2. **Mechanical grading:** Mechanical fruit graders are classified by their working principles as size grader, weight grader, fruit colour grader and fruit colour and weight grader. Fruit grading by their size and colour is the advanced technology of the post-harvest fruit handling nowadays in the world. Its working principle is a combination of automatic size grading and colour grading. At first, size grading of fruit is done on conveyer belt with changeable apertures. Fruit down from the belt are exposed to lamps under conveyer belt and yield reflex. Then the reflex signal was passed to computer and different grades of fruit such as the whole green, the half-green and half-red and the red under every conveyer belt were obtained according to different reflex signals. Finally, the graded fruit were transported by different conveyer belts.

Grading and sorting machine vision system consist of feeding unite, a belt conveyor to convey the fruit, a colour CCD camera located in an image acquisition chamber with lighting system for image capturing, control unite for open and close gates according to signals from computer unite and a computer with an image frame grabber to process the captured image.

GRADES OF DIFFERENT FRUITS

The grading is very important aspect of handling and marketing of fruits to get higher premium out of the produce. Most of the fruit growers do not make grading after harvesting of fruits and bring the produce in the market as such. This is not a healthy practice in disposing fruits in the market. It is strongly advocated that the fruits should be graded according to their size and colour development.The detailed grading of fruits in various crops are given for the benefits of growers and contractors.

Grades Suggested by Directorate of Marketing and Inspection – DMI.

i) Kinnow

Size code	Fruit Size (diameter) mm	No. of fruits in 10 kg pack
A	60-64	84
B	65-69	72
C	70-72	60
D	72-74	54
E	75-79	51
F	80-85	45

ii) Mango

Grade	Fruit weight (g)	Max. permissible difference between fruits within packages (gm)
A	100-200	50
B	201-350	75
C	351-550	100
D	551-800	125

iii) Grapes

Grade	Large berries bunch weight (g)	Small berries bunch weight (g)
Extra class	200	150
Class I	150	100
Class II	100	75

iv) Litchi

Grade	Fruit Diameter (mm)
Extra class	33
Class I	28
Class II	23

v) Pomegranate

Grade	Fruit weight (g)	Diameter (mm)
A	400	90
B	350	80
C	300	70
D	250	60
E	200	50

vi) Pineapple

Provisions concerning sizing

Size is determined by the average weight of the fruit with a minimum weight of 700 g. except for small size varieties (such as Victoria and Queen) which can have a minimum weight of 250 g. in accordance with the following table:

Size code	Average weight (+ or -12%) (in grams)	
	With Crown	Without crown
A	2750	2280
B	2300	1910
C	1900	1580
D	1600	1330
E	1400	1160
F	1200	1000
G	1000	830
H	700	560

vii) Guava

Size is determined by the weight or maximum diameter of the equatorial section of the fruit, in accordance with the following table:

Size code	Weight (g)	Diameter (mm)
A	>350	>95
B	251-350	86-95
C	201-250	76-85
D	151-200	66-75
E	101-150	54-65
F	61-100	43-53

viii) Plum

Sizing is determined by the maximum diameter of the equatorial section:

Size code	Large fruit varieties (Min.) (mm)	Other varieties (Min.) (mm)	Mirabelles & Damsons varieties (Min.) (mm)
Extra Class	40	28	20
Class I	35	28	20
Class II	30	25	17

ix) Papaya

Size is determined by the weight of the fruit with a minimum weight of 200g. in accordance with the following table.

Size code	Weight (g)
A	200-300
B	301-400
C	401-500
D	501-600
E	601-700
F	701-800
G	801-1100
H	1101-1500
I	1501-2000
J	>2001

x) Apple

Size code is determined by the diameter in millimetre of the equatorial section or weight of the fruit in grams as follows: -

Size code	Diameter (mm)	Weight (g)
A	More than 80	More than 241
B	76-80	211-240
C	71-75	181-210
D	66-70	151-180
E	61-65	121-150
F	56-60	91-120
G	50-55	60-90

Colouring Criteria For Apples

Following shall be the colouring criteria for different grades of apples of different varieties except for green and yellow apple varieties.

Grade	Percentage characteristic colour of the variety
Extra class	85 or more
Class-I	70 or more
Class-II	Less than 70

xi) Strawberry

Size is determined by the maximum diameter of the equatorial section of the fruit. Strawberries must have following minimum sizes.

Class	Diameter (mm)
Extra class	25 mm
Class I&II	18 mm

xii) Cherry

Size is determined by the maximum diameter of the equatorial section of the fruit. Cherries must have following minimum sizes.

Class	Diameter (mm)
Extra class	20 mm
Class I &II	17 mm

xiii) Pear

Size is determined by maximum diameter of the equatorial section. The minimum size by diameter is required for all grades as follows:

Class	Diameter (mm)
Extra class	60 mm
Class I	55 mm
Class II	50 mm

xiv) Sapota

Size is determined by the weight of the fruit.	Class Weight (g)
Extra class	101 & above
Class I	81 – 100
Class II	60-80

xv) Custard Apples

Size is determined by the weight of the fruit	Class Weight (g)
A	150-200
B	201-300
C	301 and above

Grading and Sizing in Ber Fruits

The following grades in ber fruits (Table 1) are suggested before marketing of fruits(Bal *et al.*, 2002).

Percent distribution and physical characteristics of graded ber fruits cv. Umran

The ber fruit should be graded into four grades namely,A grade-large sized; B grade-medium sized; C grade-small sized and D grade-under/over riped, deformed, misshapened and cankered fruits. Highest distribution of fruits is found in B grade (32.6%), which is closely followed by A grade (27.2%). The extent of C and D grade fruits is 21.2% and 19%, respectively. The fruits of Umran cultivar belonging to A and B grades which accounted for 60% of the total crop have deep golden yellow colour are more acceptable to the consumers and considered the best from marketing point of view.

PACKAGING OF FRUITS

Packaging fresh fruits and vegetables is one of the most important steps in the long and complicated journey from grower to consumer. Bags, crates, baskets, cartons, bulk bins, and palletized containers are convenient containers for handling, transporting and marketing fresh produce.

According to UK institute of packaging, packaging can be defined as a coordinated system of preparing goods for transport, distribution, storage, retailing and end use.It can also be defined as a means of ensuring safe delivery to the ultimate consumer in sound conditions at minimum cost.

The function of a package is primarily to contain and protect the produce. Due to the perishable nature of fruits and vegetables, the harsh environment conditions and the lack of adequate infrastructure 25-30% produce is lost between farm gates to consumer. Therefore, proper packaging of fruits and vegetables is essential to maintain product quality during storage, transportation and marketing. A package must with stand:

1. High humidity during pre-cooling, transit and storage.
2. Rough handling during loading and unloading.

Table 1: Grading and Sizing in Ber Fruits

Grade	Specification of grade (diameter)	Distribution(%)	Fruit size		Fruit weight(g)	Fruitcolour
			Breadth(cm)	Length (cm)		
A	Large-sized (3.30-3.70 cm)	27.2	3.49	4.67	32.0	Deep golden yellow
B	Medium-sized (2.90-3.30cm)	32.6	3.04	4.21	22.9	Deep golden yellow
C	Small-sized (2.50-2.90 cm)	21.2	2.73	3.86	17.0	Golden yellow
D	Under/over riped,deformed misshapend & cankered	19.0	-	-	-	Greenish yellow to golden yellow

3. Compression from overhead weight of other containers.
4. Impact and vibration during transportation.

Package Materials

Most commonly used package materials for horticultural produce are wood bins/boxes, jute sacks, plastic crates and corrugated fibre board (CFB) boxes.

Types of CFB packages: There are many types of packing containers for horticultural produce and they come in a range of sizes. The most commonly used material is corrugated fibre board (CFB). The following types of CFB boxes are in common use for overseas transportation:

- One piece box
- Two piece box with cover
- Full telescopic box
- Bliss type box
- One piece telescopic box
- Self-locking tray
- Interlocking box

Holes are provided in most fibre board boxes to provide ventilation of product heat (respiration) and allow cold air circulation to the product.

A. One piece Box

B. Full Telescope Box

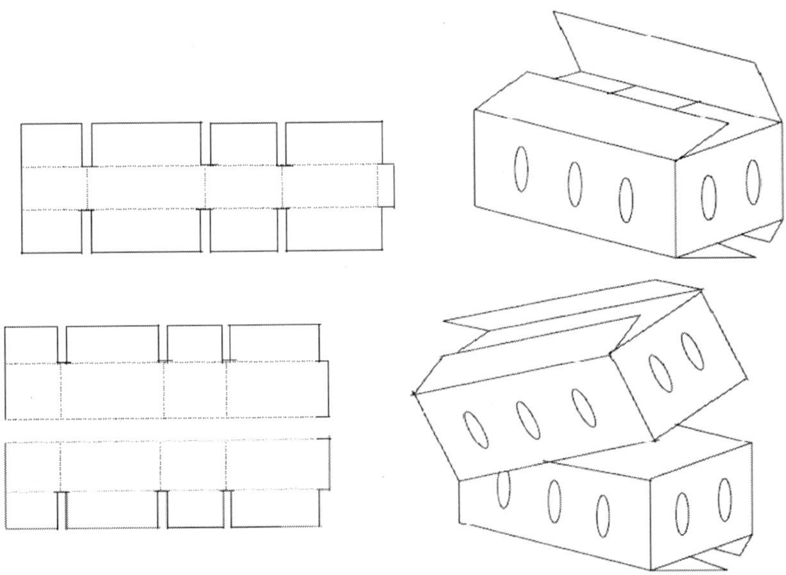

Source: Fibre Box Association

Half of full-telescopic containers have the highest stacking strength and protect against bulging but are more costly. The choice of package style will depend on what commodity is to be packed and how to use the container throughout the post harvest system (during cooling, for long-term storage, as a display).

The container known as a Bliss box has very strong corners, but is not collapsible. Bliss boxes for fresh fruits and vegetables most commonly have either flaps that meet in the middle of the top or that form an open topped tray, which is covered with a separate lid.

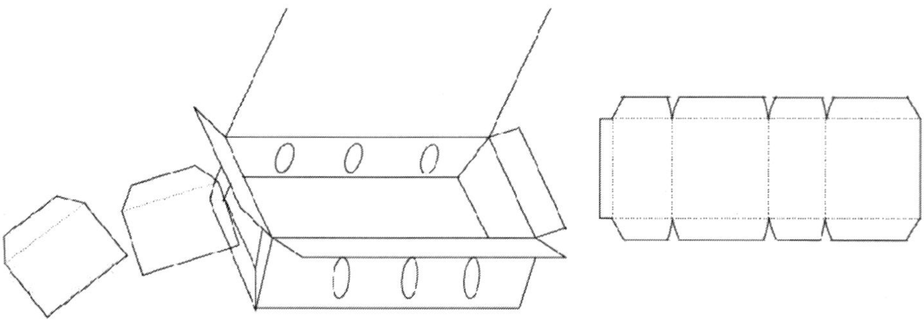

Bliss Box

A simple wooden tray with raised corners is stackable and allow plenty of ventilation for fragile crops such as ripe tomatoes.

Stackable carton reinforced with wooden corners supports

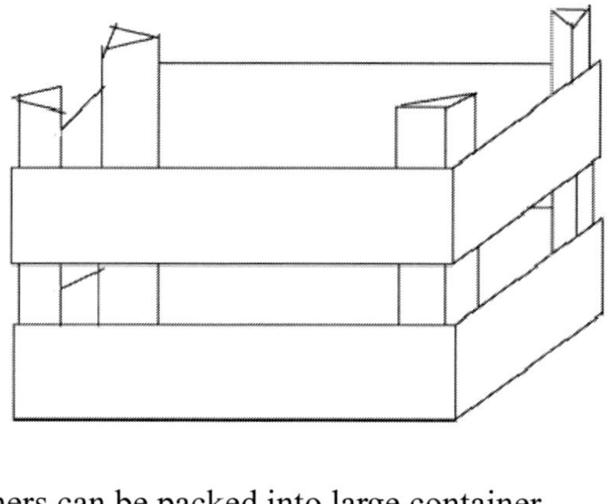

Smaller consumer sized containers can be packed into large container.

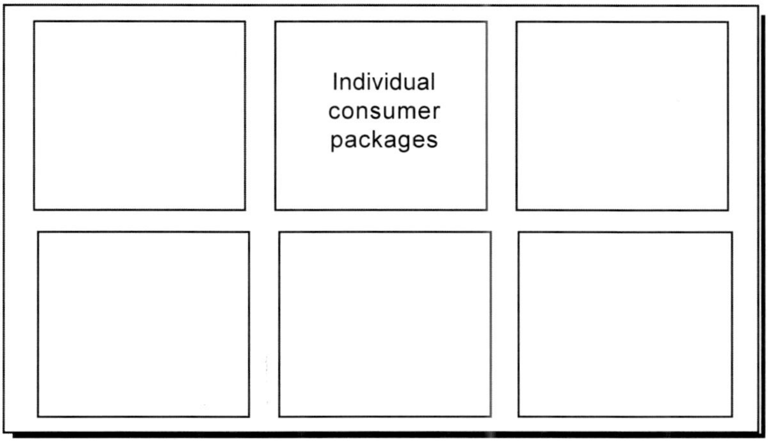

Cartons designed to hold six small containers per layer

Use of Package Inserts, Fillers and Liners

- Adding a fiberboard divider to a carton will increase strength. The use of simple dividers is common with heavy crops such as melons, and prevent melons from vibrating against one another during handling and transport.
- Wooden inserts, or fiberboard folded into tight triangles and placed in all four corners can be especially useful when a carton needs strengthening.
- Using a polyethylene liner in a fiberboard carton can help protect produce and reduce water loss in commodities such as cherries, nectarines, kiwifruits, bananas and herbs.
- Water vapor given off by the product is contained within the liner, increasing the RH around the product and decreasing the rate of water loss.
- The liner reduces abrasion damage that results from fruit rubbing against the inside of the box

Precautions

- It is important to keep produce cool to prevent damage in cartons lined with polyethylene due to gas composition changes related to increased respiration rates.
- Never use unperforated plastic bags to store onions or garlic- the relative humidity (RH) inside the bag is too high for proper handling and storage.

Losses due to rooting, sprouting and postharvest decay will increase dramatically.

- When locally made containers have sharp edges or rough inner surfaces, a simple, inexpensive liner can be used to protect produce from damaging during handling.
- **Labeling**: Labeling packages helps handlers to keep track of the produce as it moves through the postharvest system and assists wholesalers and retailers in using proper practices. Cartons can contain information on maximum stacking heights, recommended temperature range, RH conditions and warning against storage with ethylene producing crops etc.

For produce exporters the knowledge about International regulations is must. Labelling of consumer packages is mandatory under FDA regulations. Labels must contain the name of the product, net weight, and name and address of the producer, packer or distributor.

General dos and don'ts for packaging high quality produce

The packages should be sturdy and capable of standing up to handling, cooling and storage conditions.

- Rough packages such as baskets and wooden crates should be lined with cardboard inserts.
- Packages used for heavy produce should be reinforced with corner supports or folded dividers.
- Avoid using very large packages since the produce suffers more damage during handling of large packages.
- For delicate produce such as berries, grapes, summer squash and ripe stone fruits use shallow packages having single layer or double layers.
- Avoid overfilling or under-filling packages.
- For adequate ventilation of packages about 5% of the surface area per side should be vented.
- For immobilizing the produce in a package, use packaging materials such as trays, cups, wraps, liners and pads.

- To decrease the rate of water loss from produce like cherries perforated plastic film liners should be used.
- Take care while using fillers and liners in packages so that the ventilation holes are not blocked.
- Labelling containers with your logo or farm name help in creating a brand name for quality produce.
- Use large packages to pack 'consumer packages' such as gift packs or display trays for providing better protection to the containers during stacking, transport and marketing.
- For reducing produce damage, packages with handles to carry should be considered.

Modified Atmospheric Packaging (MAP)

Modified atmospheric packaging of commodity refers to the technique of sealing actively respiring produce in polymeric film packages to modify the O_2 and CO_2 levels within the package headspace. Polymeric films have been used to package fresh products for over 35 years. The most application of polymeric films for MAP are in flexible package structures. The most commonly used polymeric films for MAP are Polyolefin, Low-Density Polyethylene (LDPE), Linear Low-Density Polyethylene (LLDPE), High Density Polyethylene (HDPE), Polypropylene (PP), Polyvinyl Chloride (PVC), Polyesters, Polyethylene Terephthalate (PET), Polyvinylidene Chloride (PVDC), Ethylene-Vinyl Alcohol (EVOH), Polyamide (Nylon), Polychlorotrifluoroethylene (PCTFE), Polyvinyl Alcohol (PVOH), Ethylene-Vinyl Acetate (EVA), Ionomers, Polycarbonate Films, Polystyrene, Cellulose-Based Plastics, Biodegradable Polymers. These materials provide a range of permeability to gases and water vapour together with the necessary package integrity needed for MAP.

The shelf-life of perishable fruits and vegetables is limited in the presence of normal air by two principal factors-the chemical effect of atmospheric oxygen and the growth of aerobic spoilage micro organisms. These factors either individually or in association with one another bring about changes in odour, flavour, colour and texture leading to an overall deterioration in quality. Chilled storage will slow down these undesirable changes but will not necessarily extend the shelf-life sufficiently for retail distribution and display purposes.

The modified atmosphere concept for packaged goods consists of modifying the atmosphere surrounding a food product by controlled permeability of the pack thus controlling the biochemical, enzymatic and microbial actions so as to avoid or decrease the main degradations that might occur. This allows the preservation of the fresh state of the fruit and vegetable without the temperature or chemical treatments.

Principal: The normal composition of air is 21% oxygen, 78% nitrogen and less than 0.1% carbon dioxide. Modification of the atmosphere within the package by reducing the oxygen content while increasing the levels of carbon dioxide and/or nitrogen has been shown to significantly extend the shelf-life of perishable foods at chill temperatures.

MAP was first recorded in 1927 as an extension of the shelf-life of apples by storing them in atmospheres with reduced oxygen and increased carbon dioxide concentrations. MAP techniques are now used on a wide range of fresh or chilled foods, including raw and cooked meats and poultry, fish fresh pasta, fruit and vegetables and more recently coffee, tea and bakery products.

Advantages of Modified Atmospheric Packaging

a) Increased shelf-life

b) Reduction in retail waste

c) Improved presentation-clear view of product and all round visibility

d) Little or no need for chemical preservatives

LITERATURE CONSULTED

Bal, J.S., Randhawa, J.S. and Singh, H. 2002. Grade specification and colour variation studies in ber fruits. *Proc. National Workshop on" Post-harvest Management of Horticultura Produce*". NPC, Chandigarh. pp161-167.

DMI, 2009. Fruit and Vegetable grading and marketing rules. Directorate of Marketing and Inspection, Department of Agriculture and Cooperation, Ministry of Agriculture, http://agmarknet.nic.in/dmiwelcome.html.

Kader, A. A. 2002. Post harvest Technology of Horticultural Crops. 3rd Edition, University of California, Agricultural and Natural Resources.

Kitinoja, L. and Gorny, J. 1998. Post-harvest technology for fruits and vegetables Produce Marketers: Economic opportunities,. Quality and Food Safety by, Department of Pomology, University of California, Davis. A joint publication of UC Post harvest Outreach Program and Punjab Horticultural Post-harvest Technology Centre, USAID/ACE.

Mahajan, B.V.C. and Pongener, A. 2015. Pre and Postharvest management. In Managing Postharvest Quality and Losses in Horticultural Crops. (Ed Chadha, K.L and Pal, R.K), Daya Publishing House, New Delhi.

Pongener, A. and Mahajan, B. V. C. 2016. Advances in Packaging of Fresh Fruits and Vegetables. In Postharvest Biology and Technology (Ed Sidiqui, M.W.), Apple Academic Press Inc. 9 Spinnaker Way, Waretown, NJ 08758, USA, pp. 231-264.

Thompson, A.K. 1996. Post harvest Technology of Fruits and Vegetables. 1st Edition, Blackwell Science, Inc. USA.

Wills, R., McGlasson, B., Graham, D. and Joyce, D. 1998. Post Harvest: An Introduction to the Physiology and Handling of Fruit, Vegetables and Ornamentals. 4th Edition, CAB International, UK.

19

Treatments for Enhancing Shelf Life

B.V.C. Mahajan

In post-harvest management of horticultural crops, post-harvest treatments play an important role in maintaining the quality and enhancing the shelf life. There are number of treatments viz non-chemical, chemical and biological treatments which are being used commercially depending on the requirement of trade. Fresh fruits are living tissues subject to continuous change after harvest. Some changes are desirable from consumer point of view but most are not. Post-harvest changes in fresh fruit cannot be stopped, but these can be slowed down within certain limits to enhance the shelf life of fruits

1. Low Temperature Treatment

Pre-cooling: Pre-cooling refers to the rapid removal of field heat from freshly harvested fruits in order to slow ripening and reduce the deterioration prior to storage and transportation. The major effect of pre-cooling consists in reducing respiration rate, retarding the growth of decay organisms and reduces wilting and shrivelling, since transpiration and evaporation occur more slowly at low temperature. The harvesting of many perishable crops is done at evening or early morning to avoid excessive field heat.

To achieve maximum storage for a crop or to reduce losses during its marketable life it is essential to keep it at the most appropriate temperature. This temperature is just above that which cause chilling or freezing injury. To maximize the effect, the crop should be brought to that temperature as quickly as possible after harvest. The most important methods of pre-cooling are: hydro-cooling, forced air-cooling and vacuum cooling.

i) **Hydro-cooling:** This method uses cold water for pre-cooling the produce. It is a quick cooling method and suits a numbers of products. The transmission of heat from solid to a liquid is faster. Therefore, cooling the produce with cooled water can be very quick and result in no loss in weight. The hydro-cooler normally used are of two types i.e. immersion type and shower type. Hydro-cooling helps in cleaning the produce. However, the water can become contaminated with microorganism which can result is increased levels of spoilage during subsequent storage and marketing. To avoid this problem, chlorine can be added to the water.

ii) **Forced Air-cooling:** The crop is placed into the cold room but airflow pattern is arranged in such way so that it directed through the crop. The heat given out from the surface of the crop is then carried away in the steam of cold air, thus setting up a temperature gradient and cooling the crop more quickly. Forced air cooling is energy efficient method and most suitable for small scale operations and for a range of highly perishable commodities.

iii) **Vacuum cooling:** Cooling is achieved by the latent heat of vaporization rather than conduction. It refers to cooling by reducing atmospheric pressure. The speed and effectiveness of cooling is related to the ratio between the mass of the crop and the surface area. This system results into very even and quick cooling throughout the crop. Vacuum coolers are made from heavy duty and are expressive. This system is fast and uniform cooling takes place. It is suitable for leafy crops.

2. Hot Water Treatment: Postharvest heat treatments are used for insect disinfestations, disease and decay control of fruits. In mangoes, hot water treatment is recommended at 50-52 °C for 5 minutes to reduce the fungal infection during ripening or storage. Hot water dips have generally been utilized for fungal pathogen control, since fungal spores and latent infections are either on the surface or in the first few cell layers under the peel of the fruit or vegetables. However, if the temperature and time is not monitored properly at the time of hot water treatment, it may lead to internal injury to fruit such as flesh darkening in avocado, citrus, litchi and nectarine, and poor colour development, abnormal softening, lack of starch breakdown and development of internal cavities in mango and papaya.

Table 1: Hot water treatments

Commodity	Pathogens	Temp. (°C)	Time (min)	Possible injuries
Apple	*Gloeosporium* sp. *Penicillium expansum*	45	10	Reduced storage life
Grapefruit	*Phytophthora citrophthora*	48	3	
Lemon	*Penicillium digitatum* *Phytophthora* sp.	52	5-10	
Mango	*Collectotrichum gloeosporioides*	52	5	No stem rot control
Melon	*Fungi*	57-63	0.5	
Orange	*Diplodia sp.Phomopsis* sp. *Phytophthora* sp.	53	5	Poor degreening
Papaya	*Fungi*	48	20	
Peach	*Monolinia fructicola* *Rhizopus stolonifer*	52	2.5	Motile skin

3.Vapour Heat Treatment (VHT): Vapour heat, as a method of heating fruit with warm air saturated with water vapour at temperatures between 40 and 50 °C to kill insect eggs and larvae, is used as a quarantine treatment before fresh market shipment. Vapour heat treatment (VHT) proves very effective in controlling infection of fruit flies in fruits after harvest. The boxes are stacked in a room, which are heated and humidified by injection of steam. The temperature and exposure time are adjusted to kill all stages of insects (egg, larva, pupa and adult), but fruit should not be damaged. VHT, as a quarantine measure, is mandatory for export of mangoes. VHT can also reduce the incidence of anthracnose during postharvest storage of mango. To achieve complete control of anthracnose, it is necessary to apply a combination of VHT (95% RH and above) such that fruit core temperature of 46.5 °C is maintained for 10 minutes.

4. Irradiation: Ionizing radiation can be applied to fresh fruits and vegetables to control micro-organisms, insects and parasites, and inhibit/or prevent cell reproduction and some chemical changes. Food irradiation is a physical means of food processing that involves exposing the pre-packaged or bulk foodstuffs to gamma rays, X-rays, or electrons. Foodstuffs are generally irradiated with gamma radiation from a radioisotope source, or with electrons or X-rays generated using an electron accelerator. Radiation doses are measured in Grays (Gy). This technology is regarded as a mature and safe procedure, although its approval for use is country and product specific. Internationally, food irradiation has been considered a safe and effective technology by the World Health Organization (WHO), the Food & Agriculture Organization (FAO) and the International Atomic Energy Agency in Vienna . The Codex Alimentarius standard provides that 'the maximum absorbed dose delivered to a food shall not exceed 10 Gy except when necessary to achieve a legitimate technological purpose'. A combination of hot water treatment (55 °C for 5 minutes) followed by 30 Gy irradiation has been found to be the best treatment for shelf-life extension and quality maintenance of mangoes. After this treatment the fruit exhibits a storage life of 38 days (at 15 °C) with no irradiation injury. Irradiation can also be used to control postharvest diseases of other fresh fruit and vegetables.

In Hawaii irradiation at a minimum of 250 Gy was recommended for the commercial control of fruit fly in papaya, preferably combined with hot water treatment. The role of irradiation to reduce postharvest losses, control insect pests and microorganisms, prolong shelf life and improve quality have been tested and found beneficial in a number of horticultural produce, *viz*. mango, papaya, grapefruit, apple, strawberries *etc*.

Table 2: Doses of irradiation for postharvest disease control

Crop	Minimum dose required (Gy)	Maximum dose tolerated (Gy)
For postharvest disease control		
Apple	150	100-150
Apricot, peach, nectarine	200	50-100
Avocado	—	15
Lemon	150-200	25
Orange	200	200
Strawberry	200	200
Grapes	—	25-50

5. Biological Control: Increasing public and scientific concerns about presence of synthetic chemicals in food supply and in the environment have provoked the need to look for safer, effective, and more eco-friendly alternatives for postharvest disease control in horticultural crops. Biological control agents or microbial antagonists have been pronounced in control of various postharvest rot pathogens. The yeast *Candida guillermondi* was shown to be antagonistic towards *Penicillium* spp and could be successfully incorporated into commercial waxes. *Bacillus* spp. on their own or combined with a fungicide could be used to control the postharvest disease of mangoes, avocadoes and litchis. *Bacillus subtilis* and *B. licheniform* isolated from mango leaves and applied to fruit as a warm water dip controlled anthracnose levels. Microbial antagonists exert their biocontrol activity through antibiosis, direct parasitism, competition for nutrients and space or through induction of natural defense mechanisms such as phytoalexins. Some established reports on postharvest biological control are given in Table 3.

Table 3: Biocontrol agents used for postharvest decay control in horticultural crops

Commodity	Disease	Biological Control
Apple	Blue mould	*Pseudomonas syringae, P. cepacia, Cryptococcus* spp., *Pichia guilliermondii*
	Gray mould	*Pichia guilliermondii, Cryptococcus laurentii, C. Flavus*
Citrus	Green mould	*Pichia guilliermondii, Bacillus subtilis,*
	Blue mould	*Pichia guilliermondii, C. laurentii*
	Sour rot	*Bacillus subtilis, Pichia guilliermondii, Trichoderma sp.*
Pear	Blue mould	*Pseudomonas cepacia*
	Gray mould	*Pseudomonas cepacia, P. gladioli*
	Mucor rot	*Cryptococcus laurentii, C. flavus*
Peach/Nectarine	Brown rot	*Bacillus subtilis*
Apricot	Brown rot	*Bacillus subtilis, Pantoea agglomerans*
Plum	Brown rot	*Bacillus subtilis, Rhodotorula* sp.
Grape	Gray mould	*Trichoderma herzianum, Pichia guilliermondii*
	Rhizopus rot	*Candida guilliermondii*
Strawberry	Gray mould	*Trichoderma* sp.

6. Waxing or Coating: An ideal coating is defined as one that extends the storage life of fresh fruits without anaerobic conditions and reduces the decay without affecting the quality of the produce. Edible coatings are thin layers of edible material applied to the product surfaces as a replacement for natural protective waxy coatings and providing a barrier to moisture, oxygen and solute movement from the food. Edible coatings are a simple technology by which fruits and vegetables can by physically protected and have their respiration and in some cases, ripening regulated as with passive modified atmospheric packaging (MAP). This is because, like plastic packaging, coatings create a modified atmosphere within the fruit or vegetable. Use of coatings has been practiced for decades in the United State on apples and oranges as well as on other crops.

Waxing is, therefore, a common post-harvest practice in those fruits which shrivel rapidly and lose consumer appeal during storage and marketing. Food grade waxes are used to replace some of the natural waxes removed during harvesting and sorting operations and can help reduce water loss during handling and marketing. It also helps in sealing tiny injuries and scratches on surface of fruits and vegetables. It improves cosmetic appearance by providing shine and prolongs the storage life of fruits and vegetables.

Materials used for edible coatings include lipids, polysaccharides, proteins, resins, or combinations of these materials or composite coatings. Most fruits have a waxy cuticle that resists moisture loss, but often this is partially removed or disturbed during washing. This necessitates a coating or packaging treatment to prevent water loss and subsequent shriveling and shrinking of fresh produce.

Any 'Generally Recognized as Safe' (GRAS) material that is approved for use in coatings without restriction are considered by the U.S. Food and Drug Administration (FDA) to be "edible".

Edible coatings have been found beneficial on number of fruit crops such as citrus, apple, plum, guava etc. to reduce shrivelling and improve textural and organoleptic qualities. The edible coatings approved by Food Safety Standards Authority of India (FSSAI) are: Shellac coatings, Carnauba coatings and Bees wax coatings. The edible coatings can be applied in fruits either manually using foam pad or mechanically with power operated automatic machines.

7. Chemical Treatments

(i) **Chlorine solution:** Clean water containing an appropriate concentration of sanitizers is required to minimize the potential transmission of pathogens from water to produce, from infected to healthy produce within a single lot and from one lot of produce to another, over time. Chlorine treatment (100-150 ppm available chlorine) can be used in wash water to help control/ reduce microbial inoculum build-up during packing operations. The pH of wash solution should be maintained between 6.5 and 7.5. Now-a-days, Chlorine dioxide (ClO_2) has been increasingly adopted as an alternative to chlorine for sanitization in the fresh and fresh-cut produce industry. ClO_2 used as a disinfectant has several advantages over chlorine, including larger oxidant capacity, lower reactivity with organic matter and high effectiveness at low concentrations.

(ii) **Calcium application:** Postharvest application of calcium plays an important role in enhancing the storage and marketable life of fruits by maintaining their firmness and quality. Calcium application delays aging or ripening, reduces postharvest decay, controls the development of many physiological disorders and increases the calcium content, thus improving their nutritional value. Vacuum infiltration of mangoes with 2-8% $CaCl_2$ solution results in delayed softening of fruit during storage. Calcium develops resistance to fungal attack by stabilizing or strengthening cells walls, thereby, making them more resistant to harmful enzymes produced by fungi. Calcium dips have been used as firming agents to extend postharvest shelf life of a wide range of fruits *viz.* apples, strawberries, pears, blueberries, peaches, oranges, and pineapples.

(iii) **1-Methylcyclopropene:** 1-Methylcyclopropene (1-MCP) is a powerful inhibitor of ethylene action that has been added to the list of options for extending the shelf life and quality of fruit. 1-MCP is a cyclic olefin, non-toxic, odourless and stable at room temperature. This compound has been manufactured in powder form releasing 1-MCP when diluted with water. Presently, this chemical is being traded under the name of Smart fresh™ The magic gas, l-MCP, has been used as a tool to enhance the shelf life of perishables, as it irreversibly binds to the ethylene receptors, thereby bringing about a conformational change in the receptors and prevents ethylene effects in plant tissues for extended period. There are numerous scientific reports which established the role 1-MCP in delaying the process of ripening and senescence in fresh horticultural produce.

Table 4: Effects of 1-Methylcyclopropene on post-harvest storage behaviour and quality of horticultural produce (Watkins, 2008; Schotsmans *et al*., 2009)

Crop	1-MCP	Altered postharvest physiology and biochemistry
Apple	1 $\mu L\ L^{-1}$	Reduced softening, delayed ripening, inhibited superficial scald and decay.
	2 $\mu L\ L^{-1}$	Reduced softening, and reduced superficial scald
Apricot	0.5 $\mu L\ L^{-1}$	Decreased respiration rate and tissue softening.
Banana	0.01-1.0 $\mu L\ L^{-1}$	Delayed peel colour change and fruit softening, extended shelf life.
Ber	1 $\mu L\ L^{-1}$	Delayed senescence, maintained quality, reduced decay.
Grape	0.5-1.0 $\mu L\ L^{-1}$	Increased shelf life, and improved quality.
Guava	0.6 $\mu L\ L^{-1}$	Improved shelf life, ameliorated chilling injury, and reduced decay.
Kiwifruit	0.5 $\mu L\ L^{-1}$	Suppressed ethylene production and fruit softening
Lime	1 $\mu L\ L^{-1}$	Maintenance of green colour of fruit.
Mandarin	5 $\mu L\ L^{-1}$	Inhibited chlorophyll degradation and senescence.
Mango	2 $mg\ kg^{-1}$	Delayed ripening, increased antioxidant enzymes.
Peach	10 $\mu L\ L^{-1}$	Inhibited degradation of starch, prolonged shelf life.
Pear	0.3-0.6 $\mu L\ L^{-1}$	Lower ethylene production and delayed softening.
Pineapple	0.1 $\mu L\ L^{-1}$	Effectively controlled internal browning or black heart.
Plum	0.5 $\mu L\ L^{-1}$	Slowed fruit ripening, reduced fruit softening .
Sweet Cherry	1 $\mu L\ L^{-1}$	Maintained firmness, enhanced shelf life.

(iv) **Diphenylamine (DPA):** Diphenylamine is a colourless, aromatic amine possessing antioxidant properties. Postharvest drenching of fruits with DPA provides good control against scald, a serious physiological disorder of apples and pears during storage manifested as brown discolouration that makes the fruit unmarketable and limits its storage potential. In addition, postharvest treatment of pear with DPA prevents normal rates of respiration and ethylene production during ripening and the fruits remain hard and green during storage.

(v) **Fumigation:** The fumigation with sulfur dioxide (SO_2) is successfully used for controlling postharvest diseases of grapes. This is achieved by placing the boxes of fruit in a gas tight room and introducing the gas to the appropriate concentration from a cylinder. Special sodium metabisulphite pads are available which can be packed into individual boxes of fruit to give a slow release of SO_2. The primary function of treatment is to control *Botrytis cinerea*. The storage life of grapes packed along with sulfur dioxide generating pads can be extended upto 6 weeks at 0-2 °C and 90-95% RH.

Sulphur dioxide treatment in litchi is commercially practiced by exporters to control pericarp browning. Sulphur dioxide is a well-known inhibitor of polyphenol oxidase and it combines with anthocyanin to form a stable complex. Fumigation with 1.2% SO_2 for 10 minutes was shown to be effective in reducing skin discolouration in fresh litchis, especially if it is combined with a 2 minute dip in 1N HCl acid directly afterwards. Paper pads or wraps impregnated with biphenyl fungicides are commonly applied to citrus fruits. The chemical vaporizes slowly protecting the fruit from fungal infection.

Despite the good efficacy of SO_2 treatment against decay pathogens, there has been growing concerns against the use of SO_2 in food due to the health issues related to sulfite residues. Concerted efforts are needed to find safer alternatives to SO_2 fumigation for overcoming decay during postharvest storage of horticultural produce.

(vii) **Gibberellic acid:** The role of GAs in delaying the process of senescence in plants and plant tissues is well established. Postharvest application of gibberellic acid (GA_3) has been popularly used as a tool to delay the process of ripening and senescence in various fruits. The anti-senescence effect of GAs is manifested through their effect on delay of colour changes, alleviation of chilling injury, reduction in microbial decay and increase in storage life of fresh produce. The beneficial use of GA_3 as a postharvest application in various horticultural produce is presented in Table 5.

(viii) **Jasmonates:** 'Jasmonates' are a class of endogenous plant growth regulators that have unique and potentially useful properties that affect plant growth and development in response to environmental stresses. Jasmonate is a broad term covering numerous linolenic acid-derived compounds. Postharvest application of MeJA as a gas can activate defences and can positively modify postharvest physiology in a number of crops. Some research on horticultural uses of MeJA has focussed on pre-harvest and postharvest treatments to protect against microbial development on harvested tissue. Postharvest exogenous application of MeJA has also been reported to suppress fruit decay, improve quality and enhance storage life of a number of horticultural crops such as strawberry, raspberry, grapefruit, peach. Application of MeJA positively reduces the rate of fruit damage by chilling temperature in pomegranates, avocado, grapefruit, guava *etc* through enhanced antioxidant potential and reduced ion leakage .

Table 5: Postharvest application of GA_3 for enhanced shelf life and quality through physiological and biochemical changes in fruits.

Crop	Concentration	Altered postharvest physiology and biochemistry
Banana	1.5 g kg^{-1}	Delay of fruit ripening.
Grapefruit	100 ppm	Maintained peel quality, reduced decay, and increased storage life.
Guava	100 ppm	Delayed onset of senescence and maintained fruit quality.
Mango	200 mg L^{-1}	Retarded ripening of fruits.
Papaya	100 ppm	Delayed ripening and improved shelf life.
Peach	100 mg L^{-1}	Maintained fruit firmness and delayed ripening.
Pear	2000 µg mL^{-1}	Improved the efficacy of biocontrol against blue mold rot.
	50 ppm	Improved palatability and sensory quality of fruits.
Persimmon	200 µg mL^{-1}	Induced resistance to black spot caused by *Alternaria alternata.*
Sapota	150 ppm	Maintained fruit quality and improved shelf life.
Sweet Cherry	10 mg L^{-1}	Higher fruit firmness and lower fungal infection.

(ix) **Ethylene gas:** Ethylene, a gas at physiological temperature, is a natural hormones involved in fruit ripening. The autocatalytic production of endogenous ethylene is considered to immediately trigger off the ripening process in fruits like banana, mango and papaya etc. These fruits are commercially ripened by traders with calcium carbide popularly known as *Masala*. This practice is banned U/S 44-AA of Prevention of Food Adulteration (PFA) Act, 1954 with PFA rules, 1955, due to health reasons. The ripening of fruits with ethylene gas is a safe and internationally acceptable technique. The fruits are exposed to low level of ethylene gas (100 ppm) in an air-tight ripening chamber for 24 hours to induce the fruits to ripen. The most important thing in this technique is temperature and relative humidity control inside the ripening chamber, which should range between 16 -24°C and 90-95% RH, depending upon the fruit type. This technique of fruit ripening is approved by Food Safety and Standard Authority of India (FSSAI), Ministry of Health and Family Welfare. It guarantees uniform ripening with development of good flavour and colour.

LITERATURE CONSULTED

Kader, Adel A .2002. Post harvest Technology of Horticultural Crops. 3rd Edition, University of California, Agricultural and Natural Resources.

Kitinoja, L and Gorny, J. 1998. Post-harvest technology for fruits and vegetables Produce marketers: Economic opportunities. Quality and Food Safety by, Department of Pomology, University of California, Davis. A joint publication of UC Post harvest Outreach Program and Punjab Horticultural Post harvest technology Centre, USAID/ACE.

Schotsmans, W.C., Prange, R.K. and Binder, B.M. 2009. 1-Methylcyclopropene: Mode of action and relevance in postharvest horticulture research. *Horti.Reviews*, 35: 263-313.

Thompson, A.K.1996. Post harvest Technology of Fruits and Vegetables. 1st Edition, Blackwell Science, Inc. USA.

Wills, R., McGlasson, B; Graham, D. and Joyce, D .1998. Post harvest: An Introduction to the Physiology and Handling of Fruit, Vegetables and Ornamentals. 4th Edition, CAB International,UK.

Witkins, C.B. 2008. Overview of 1-Methylcyclopropene trials and uses for edible horticulture crops *HortiScience,* 43 (1): 86-94.

20

Storage Techniques for Fruits

B.V.C. Mahajan

INTRODUCTION

Once a crop is harvested, it is almost impossible to improve its quality. Therefore, proper storage conditions—*temperature and humidity*—are needed to lengthen storage life and maintain quality once the crop has been cooled to the optimum storage temperature. In modern times, the storage of perishables has become more sophisticated because of the demand for a wide range of fresh fruits and vegetables at all the times of the year and because of need to spread peaks of production over longer periods to maximize profit and reduce waste. But in our conditions, storage of fruits and vegetables is practiced to avoid the low prices during the glut season in anticipation that prices will increase during the lean period or off-season. Certain crops like potatoes and apples can be stored for long period whereas for the crops peach and plum, the storage period is very short. However, for both long-term as well as short-term storage, high level of technical know-how about the crop is necessary for successful storage. There are a few important factors, which need to be taken into account before storage of the produce.

- Suitability of crops for storage
- Knowledge of appropriate storage conditions such as temperature and relative humidity
- Compatibility with other crops for storage

Temperature : Fresh fruits and vegetables are living tissues, although they are no longer attached to the plant. They breathe, just as humans do, and their composition and physiology continue to change after harvest. They continue to ripen and finally they begin to die. Cellular breakdown and death (senescence) are inevitable, but can be slowed with optimal storage conditions. Fresh fruits and vegetables need low temperatures (32 to 55°F) and high relative humidities (80 to 95 percent) to lower respiration and to slow metabolic and transpiration rates. By slowing these processes, water loss is reduced and food value, quality and energy reserves are maintained.

Respiration and metabolic rates are directly related to room temperatures within a given range. The higher the rate of respiration, the faster the produce deteriorates. Lower temperatures slow respiration rates and the ripening and senescence processes, which prolongs the storage life of fruits and vegetables. Low temperatures also slow the growth of pathogenic fungi which cause spoilage of fruits and vegetables in storage.

Temperature is one of the main factor that affects the respiration and enzyme activity of fruits and vegetables. In the practice of storage, low temperature storage is a basic method. Currently, the most common preservation method of fruit and vegetable at home and abroad is combining controlled atmosphere storage and low temperature storage methods. Storability refers to the quality of fruit and vegetable that has no significant change over storage period, the quality loss is minimal. Storability is not only determined by a single kind of character, it is the comprehensive reflection of all kinds of physical, chemical, physiological and biochemical natures. To solve these mentioned problems, plants' respiration must be controlled. In order to store them for long term, their living state must be remained and at the same time, their breathing must be weakened. Low temperature can weaken respiration and prolong storage time. Recommended temperature and relative humidity and their storage life of fruits is given in Table 1.

Relative humidity: Transpiration rates (water loss from produce) are determined by the moisture content of the air, which is usually expressed as relative humidity. At high relative humidity, produce maintains salable weight, appearance, nutritional quality and flavor, while wilting and softening are reduced.

Relative humidity needs to be monitored and controlled in storage. A hygrometer or a sling psychrometer, not the appearance of the produce, should be used to monitor humidity. Control can be achieved by a variety of methods:

1. Operating a humidifier in the storage area.
2. Regulating air movement and ventilation in relation to storage room load.
3. Maintaining refrigeration coil temperature within 2°F of the storage room air temperature.

4. Using moisture barriers in the insulation of the storage room or transport vehicle, and in the lining of the packing containers.
5. Wetting the storage room floor.
6. Using crushed ice to pack produce for shipment.
7. Sprinkling leafy vegetables, cool-season root vegetables, and immature fruits and vegetables with water.

Table 1: Recommended temperature, RH and approximate storage life of some important fruits.

Commodity	Temperature (°C)	Relative Humidity (%)	Approx. Storage life
Apples	0-1	90-95	1-6 months
Apricots	0-1	90-95	1-3 weeks
Bananas	13.0-14.0	85-90	1-4 weeks
Ber	7.5	90-95	2 weeks
Cherries, Sweet	0-1	90-95	2-3 weeks
Coconuts	0-1.5	80-85	1-2 months
Figs	-0.5-0	85-90	7-10 days
Grapefruit	10-15	85-90	6-8 weeks
Grapes	0-1	90-95	1-6 months
Guavas	5-10	90-95	2-3 weeks
Kiwifruit	-0.5-0	90-95	3-5 months
Lemons	10-13	85-90	1-6 months
Limes	9-10	85-90	6-8 weeks
Litchi	1.5	90-95	3-5 weeks
Mandarin, Kinnow	5-7	90-95	2 months
Mangoes	12-13	85-90	2-3 weeks
Papayas	7	85-90	1-3 weeks
Peach, plum	0-1	90-95	2-4 weeks
Pears	0-1	90-95	2-7 months
Pineapples	7-13	85-90	2-4 weeks
Pomegranates	5	90-95	2-3 months

Ethylene in the Storage Environment

Ethylene causes a wide range of damage symptoms in sensitive commodities. Avoid handling or storing ethylene sensitive commodities with ethylene producing ones.

Undesirable effects of ethylene

- Accelerated ripening (in climacteric fruits)
- Loss of green colour (snap beans, cucumber, leafy greens)
- Induction of leaf disorders (russet spotting in lettuce, death of leafy greens)

- Isocoumarin formation (bitter taste in carrots)
- Sprouting (potatoes)
- Browning of pulp and/or seeds (eggplant, sweet potato)
- Poor flavour (watermelon, sweet potato)

Overcoming ethylene effects

- Elimination of sources of ethylene (combustion engine exhaust, ripening fruits)
- Ventilation (one air change per hour)
- Chemical removal of ethylene by filtering air in storage (potassium permanganate scrubbers, UV lamps, activated charcoal, catalytic oxidizers).

Status of Fruits and Vegetables Grouping on the Basis of Temperature and Relative Humidity

The various fruits and vegetables are grouped into 1-6 groups on the basis of temperature and relative humidity for enhancing shelf life and better consumers acceptance (Table 2).

Table 2: Grouping of fruits and vegetables stored at different temperature and humidity

Group	Temperature	Crops	Status of commodities
Group 1	0-2°C and 90-95% RH	Apple, Apricot, Peach, Asian, Pear Grapes, Litchi, Plum, Prunes, Pomegranate.	Produce ethylene.
Group 2	0-2°C and 90-95%RH	Asparagus, Leafygreens, Beans, Broccoli, Peas, Spinach, Cabbage, Carrot, Cauliflower, Cherries.	Sensitive to ethylene.
Group 3	0-2°C and 65-70% RH	Garlic, Onions dry.	Moisture will damage these crops.
Group 4	4.5°C and 90-95% RH	Mandarin, Tangerines.	
Group 5	10°C, 85-90% RH	Beans, Potatoes, Cucumber, Brinjal, Okra Pepper.	
Group 6	13-15°C, 85-90% RH	Mangoes, Banana, Grapefruit	

Precautions

- The recommended temperatures and RH values should only serve as general guidelines. However, operators should closely monitor changes in temperature and RH for specific product for its optimum storage as conditions such as production techniques, type of cultivar, harvesting methods, packaging and package materials, pre-storage treatments etc. also effect quality of stored produce.

- Apple, pears being ethylene producer should not be stored with ethylene sensitive products like cabbage, carrot, cauliflower, cucumber, brinjal, okra and peas, otherwise these products are likely to be spoiled.
- Separate cold chambers for produce groups having similar storage conditions are desirable for the maintenance of high quality of produce during storage.

General do's and don'ts for Storage

- Only high quality produce, free of damage, decay and of proper maturity (neither over-ripe nor under-ripe) should be considered for storage.
- Before storing, know the requirements for the commodities you want to put into storage, and follow recommendations with regard to proper temperature and relative humidity.
- Root, tuber and bulb crops should be cured before storage.
- To avoid damage to commodities from freezing or chilling, do not use storage temperatures lower than recommended.
- Storage rooms should not be overloaded and adequate ventilation should be provided in the storage room.
- Keep storage rooms clean. The storage facilities should be protected from rodents by keeping the immediate outdoor area clean and free from trash and weeds.
- Storage rooms should have thermometers at a several locations so that storage temperatures are properly monitored.
- Always store onions or garlic in low humidity environments (less than 70% RH).
- Always store crops in a dark room. Light stimulates solanine production in potatoes, which is a toxic compound and is not destroyed by cooking.
- Always store ethylene sensitive commodities away from those that produce ethylene.
- Produce known for emitting strong odours (apples, garlic, onions, turnips, cabbages, potatoes) should not be stored with odour-absorbing commodities.
- Regularly inspect stored produce for any signs of injury, water loss, damage and disease and remove damaged or diseased produce to prevent the spread of problems.

Chilling Injury

The fruits exihibit different types of chilling injuries if they are stored below their prescribed lowest safe temperature.The common chilling injuries occurred in various fruits during low temperature storage are given below in Table 3.

Table 3: Chilling injury symptoms of different fruits Commodity.

	Lowest safe temperature ^{o}C	Chilling injury symptoms
Apple	2-3	Soft scald, brown core
Banana	12-13	Dull colour when ripened
Guava	4-5	Pulp injury, decay
Lemon	11-13	Pitting, membrane staining
Mango	10-13	Uneven ripening, greyish skin
Ber	7.5	Surface pitting, appearance of brown

Controlled Atmosphere (CA) storage

The term implies the addition or removal of gases resulting in an atmospheric composition different from that of normal air. Thus the levels of carbon dioxide, oxygen, nitrogen, ethylene, and metabolic volatiles in the atmosphere may be manipulated. Controlled atmosphere storage generally refers to keeping produce at decreased oxygen and increased carbon dioxide concentrations and at suitable range of temperature and RH. Systems where atmospheric control is accurately controlled are generally called CA storage and where degree of control "less accurately" monitored are called MA (modified atmosphere) storage. In MAP (modified atmospheric packaging) produce is enclosed in polymeric films and is allowed to generate its own atmsophere (passive MAP) or air of known composition is flushed into the bag (active MAP) and depending upon gas / vapour transmission characteristics of the film on appropriate atmosphere develops in the package to prolong shelf life. MAP is ideally combined with temperature control for maximum benefit.The recommended CA or MA conditions of selected fruits and vegetables is given below in Table 4.

Table 4: Recommended CA or MA conditions for selected fruits and vegetables Commodity.

	Temperature (°C)	% O_2	% CO_2
Apple	0-5	1-2	0-3
Banana*	12-16	2-5	2-5
Cherry, sweet	0-5	3-10	10-15
Mango*	10-15	3-7	5-8
Peach, clingstone	0-5	1-2	3-5
Pear, European	0-5	1-3	0-3
Beans, green	5-10	2-3	4-7
Cantaloupes	2-7	3-5	10-20
Onions (bulb)	0-5	1-2	0-10
Pepper (bell)	5-12	2-5	2-5
Radish (topped)	0-5	1-2	2-3
Tomato (green) ripe	12-20 10-15	3-5 3-5	3-5 3-5

Benefits of CA storage

- Slow down respiration and ethylene production rates, softening and retard senesence of horticultural produce.
- Reduce fruit sensitivity to ethylene action
- Alleviate certain physiological disorders such as chilling injury of various commodities, russet spotting in lettuce, and some storage disorders including, scald of apples.
- Reduction in microbial activity especially molds.

Harmful effects of CA storage

- Initiation or aggravation of certain physiological disorders can occur, such as blackheart in potatoes, brown stain on lettuce, and brown heart in apples and pears.
- Irregular ripening of fruits, such as banana, mango, pear and tomato, can result from exposure to O_2 levels below 2% or CO_2 levels above 5% for more than 2 to 4 weeks.
- Off- flavors and off-odours at very low O_2 or very high CO_2 concentration may develop as a result of anaerobic respiration and fermentative metabolism.

Hypobaric Storage or Low-pressure System

Hypobaric storage is a form of controlled atmosphere storage in which the produce is stored in a partial vacuum. The vacuum chamber is vented continuously with water saturated air to maintain oxygen levels and to minimize water loss. Ripening of fruit is retarded by hypobaric storage, due to the reduction in the partial pressure of oxygen and for some fruits also to the reduction in the ethylene levels. A reduction in pressure of air to 10 kilopascals (0.1 atmosphere) is equivalent to reducing the oxygen concentration to about 2 per cent at normal atmospheric pressure. Hypobaric stores are expensive to construct because of the low internal pressures required, and this high cost of application appears to limit hypobaric storage to high value produce such as cut flowers. Secondly control of gases during the storage cannot be manipulated.

Dynamic Control System

Dynamic control system is based on stress response of the fruits, which can be defined as ethanol production or increased chlorophyll fluorescence status.Important advantages of these systems are disorder prevention and firmness retention. The Dynamic control system aims at storage of fruit and vegetable just above oxygen levels where fermentation occurs (<1%).

Dynamic Controlled Atmosphere (DCA) technology is a healthy, environmentally-friendly way to keep fruit fresh using low oxygen levels to prevent the development of scald. DCA works by monitoring the stress level of the fruit in storage and adjusting the oxygen (O_2) and carbon dioxide (CO_2) levels to maximise fruit storage potential while maintaining sound, marketable fruit.

LITERATURE CONSULTED

Hardenburg, R.E., Watada, A.E. and Wang, C.Y. 1986. *The Commercial Storage of Fruits, Vegetables, and Florist and Nursery Stocks.* USDA-ARS Agriculture Handbook Number 66 (revised) 136p.

Kader, Adel A. 2002. Post harvest Technology of Horticultural Crops. 3rd Edition, University of California, Agricultural and Natural Resources.

Kitinoja, L. and Gorny, J. 1998. Post-harvest technology for fruits and vegetables Produce marketers: Economic opportunities. Quality and Food Safety by, Department of Pomology, University of California, Davis. A joint publication of UC Post harvest Outreach Program and Punjab Horticultural Post harvest technology Centre, USAID/ACE

Thompson, A.K. 1996. Post harvest Technology of Fruits and Vegetables. 1st Edition, Blackwell Science, Inc. USA.

USDA ,2014. The Commercial Storage of Fruits, Vegetables, and Florist and Nursery Stocks. Agriculture Handbook Number 66 (HB-66). Agricultural Research Service, Beltsville Area / BARC. http://www.ba.ars.usda.gov/hb66//

21

Post-harvest Transportation of Fruits

Jagmohan Singh

INTRODUCTION

As per National Horticulture Database published by National Horticulture Board, during 2014-15 India produced 89.15 million metric tonnes of fruits. This value however differs from the consumption value depending upon the post harvest losses, storage losses or transportation losses. Transportation and distribution of fruits being the most important factor, plays a key role in determining the net income from horticulture trade. It is an essential link in the modern food chain and exists as complex matrix where not only fresh goods finished products are transported but there is also transportation of bulk raw materials. Locally grown produce may be transported directly from farms to retail outlet. For local market, the produce is brought by bullock cart, tractor, trollies. Mainly the long distance transportation is by trucks, rails which is costly. In India the transportation by Railway is much less than the other means. The road transport is 3-4 times more expensive than rail, yet it is preferred for shipment of fresh fruits simply because of the faster movement of perishable commodities.

Minimum damage during transportation is important for successful marketing of produce because harvested fresh fruits perform life sustaining physiological and metabolic reactions even when detached from mother plant. Biological changes that occur after harvest and during transport and storage are generally called post harvest behaviour. Various pre harvest factors can affect reserve status and post harvest behaviour of fruits, their shelf life, fruit quality, climate, soil condition and horticultural aspects.

IMPORTANT FEATURES FOR EFFICIENT TRANSPORTATION

The following are the important points for efficient transportation of fruits after harvesting.

- Loading and unloading of horticulture produce should be done carefully.
- Vehicles produce should not be overloaded. Adequate ventilation should be available in vehicle in order to prevent heat gain.
- Transit time should be short.
- It should provide shelter from sun and rains at loading and unloading areas.
- There should be minimum water losses by the produce during transportation.
- A wind catcher may be designed so that free air management is possible into vehicle during transportation. In refrigerated transportation vehicle should be well insulated and doors should be sealed tightly.

FACTORS WHICH AFFECTS PRODUCE QUALITY DURING TRANSPORTATION

Various factors which affect the horticultural produce quality during transport are discussed as:

Initial Quality: The packing of fruits must be well suited to the different handling conditions while transportation to markets. The fruits which are to be sold in a local market are kept fairly mature whereas, the fruits for distant market are less mature. Also, mechanical damage or other prevalent barriers are checked at the time of transporting.

Temperature: Temperature is most important factor for determining the fruit quality. During respiration, the fruits utilize carbohydrates and oxygen in order to produce carbon dioxide, water and energy. High respiration leads to rapid carbohydrates depletion which further reduces the shelf life of fruits. This rate of respiration is primarily regulated by the temperature e.g. strawberries, raspberries and blackberries have a shelf life of 7 days at 0°C and only 1 day at 20°C. In case of plum and peach, physiological disorders may develop if the storage temperature is too low. If the temperature falls below the advocated level, chilling injuries or freeze damage can occur. This is very often observed in case of tropical or sub tropical areas. The distinct symptoms of chilling injury include tissue darkening, surface pitting, abnormalities in ripening, poor flavour or increased susceptibility to decay.

Humidity and Water Loss: The relative humidity required for superlative shelf life in case of fruits varies between 90-95%. Wilting or shrivelling due to low humidity reduces the market value of fruits. In addition to this, water loss weakens the plant cell, making it susceptible to fungal activity. Fungal growth further increases, ethylene production leading to loss of chlorophyll. This causes undesirable yellowing of the fruit. There is lack of relative humidity (RH) control capability in refrigerated highway trailers. This produces a high chance of wilting. However, by using waxed packages or plastics boxes can reduce water loss. On the other hand, marine containers are equipped with spray humidification systems. Such containers are more airtight than highway trailers and consist of an adjustable fresh air exchange vent.

Atmospheric Composition: The carbon dioxide and ethylene levels increased subsequently due to respiration causing simultaneous decrease of oxygen level in the containers. The carbon dioxide level should be kept below 0.3% in a tightly sealed container. Highway trailers usually consist of air leakage to prevent accumulation of carbon dioxide. The marine containers are equipped with devices for a controlled atmosphere. This helps to maintain the carbon dioxide and oxygen concentration depending on the type of fruit. It also increases the storage life by 30%.

Physical Injury: Physical injuries can be due to compression, vibrations or any impact leading to the damage of the fruit. It can be minimized by using appropriate packaging and correct choice of refrigerated vehicle. Compression may occur due to overfilling of the boxes which are not properly palletized. Compression damage can be prevented by properly aligning all boxes and ensuring that the edge of pallet load aligns with the pallet edge. Vibration damage arises due to constant vibration of the vehicle while it is moving. The use of air ride suspension reduces the vibration damage drastically.

SYSTEMS FOR MAINTAINING FRUIT QUALITY DURING TRANSPORTATION

Optimum conditions should be maintained during transportation. For this purpose, various temperature control systems and air circulation systems are installed in the vehicle container.

Cooling Requirements: The transport vehicle must have sufficient cooling capacity. Most of the heat input is due to air infiltration and heat conducted across the walls of the vehicle. The three main sources of heat in case of fruit transport are internal heat , external heat and residual heat loads. Therefore the cooling capacity of the refrigeration system is the sum total of all three heat loads multiplied by a safety factor.

Internal Heat Loads: The residual field heat and the heat of respiration produced by the fruits, account for the internal heat loads. The fruits should be properly pre cooled before loading and a recommended temperature is maintained during transport.

External Heat Loads: External heat loads depends on the external environmental conditions and may enter the vehicle by means of conductions, convection, radiation and infiltration. These are the main sources of heat gain or loss in refrigerated transport. Conduction is the movement of heat through solid objects. It can occur through roof, floor and doors of refrigerated vehicles. Insulating material can be used to prevent conduction.

Residual Heat Loads: Residual heat loads include heat initially contained in the transport vehicle and heat load not included as internal or external loads. When trailer doors are open at an unrefrigerated loading dock, the refrigeration system should be turned off and fruits should be loaded rapidly and doors should be closed immediately.

Refrigeration Systems

Various types of refrigeration systems are available for transportation. The most commonly used cooling systems include mechanical refrigeration, ice and cryogenic cooling. However, heating systems may be used to protect fruits from chilling injuries during cold winters.

Mechanical Refrigeration: Mechanical refrigeration is used most frequently in case of refrigerated transport vehicles. Chlorofluorocarbon refrigerants were used earlier due to their excellent physical characteristics. But in order to promote eco friendly refrigeration, new families of refrigerant fluids with no ozone depleting material are used presently. In case of transport vehicles, the refrigeration system runs on diesel engine. The engine and condenser are installed outside the container while the evaporator and blower are placed inside.Two control modes are mainly available in mechanical refrigeration systems:

1. Continuous: It includes the fan and compressor which are operating continuously. This mode is recommended for easily perishable fruits.
2. Automatic: It includes only the fan.

Ice Cooling: Ice cooling is one of the simplest and most primitive methods of maintaining low temperature for fruits transport. Ice has a cooling capacity of 335 J/kg and one unit weight of ice will drop the temperature of three times the weight of fruits by about 28 degrees celsius. Ice also helps to maintain high humidity and minimizes the water loss from fruits .Finely crushed ice, flaked ice of liquid ice can be blown on top of the fruit boxes, this is called top-icing. Crushed ice or liquid ice can also be applied on top of the pallet loads or can even be injected into individual boxes. Flaked ice or crushed ice is produced on site and stored in an ice bunker. This enables the manufacturing equipment to be operated with inexpensive off peak electricity. However the use of ice for fruit cooling and transport has declined due to number of disadvantages; it is very energy inefficient. Ice melting can cause freeze injury to the fruits. To prevent this, use of commercially available ice gels is promoted as they possess a lower melting temperature than water ice.

Cryogenic Cooling: Cryogenic cooling is done with the help of liquid nitrogen, liquid carbon dioxide or solid carbon dioxide (dry ice) to maintain temperature of fresh fruits. Due to difficulty in regulating the atmospheric conditions and development of efficient refrigeration, this type of cooling is limited to frozen products, local deliveries and air transport only. Dry ice is quite favourable as it does not produce a liquid and rather sublimates to a gas. It is basically solidified carbon dioxide having 92% more refrigeration capacity pre mass (645 kJ/kg) than water ice (335 kJ/kg). However dry ice has a temperature of minus seventy eight degrees celsius which is too cold to be packed with horticultural produce. Therefore it must be kept away from the fruits to prevent chilling injuries. Often dry ice is put on a sheet of foam insulation placed above the fruits in order to allow the carbon dioxide to mix with the air before coming in direct contact with the fruits.

Heating under low temperature condition: As mentioned earlier, during cold weather conditions, transport vehicles must have a functional heating system to avoid fruit chilling. Mechanically refrigerated vehicles (trucks, trailers, railcars or containers) are heated by electric heating or by running the refrigeration unit in reverse cycle. However, in case of reverse cycle, the efficiency of mechanical refrigeration is one third of the normal effective capacity of unit set at zero degrees celsius in a thirty degrees celsius ambient temperature. Therefore a supplementary electric heating is required while transporting in extremely cold conditions.

MODE OF TRANSPORTATION

Following factors should be taken into consideration for selection of mode of transportation

- Produce Perishability
- Destination
- Produce value
- Quality of the produce to be transported
- Recommended transport condition
- Climate conditions and time required to reach the destination.

Horticultural produce can be transported by roads, rails by air and ship.

1. JEEP/CAMPER

This means is used when the destination market is near 30-5-km from the production centre. This means is suitable for short distances and can enter roads where trucks cannot enter. Smaller growers can afford to hire small vehicles rather than truck and rail. This transportation is faster than trucks but it is not suitable for long distance.

(See colour version on page 424)

2. REFRIGERATED TRUCKS

When the transportation is to be carried at low temperature these trucks are most suitable. In India this transport is used over a limited extend. It provides better shelf life to perishable produce such as strawberries etc., the main disadvantages of this truck is costly and unaffordable for a common grower and is not flexible in use.

(See colour version on page 424)

3. NON REFRIGERATED TRUCKS

In India it is common type transportation adopted for most of horticultural crops. Trucks generally prefer to move during night due to less traffic on roads. This type of transportation is used upto about 500-600 kms.They can move long distance and don't have to wait according to the timing of the railway station. They can reach at different places within the cities. Main disadvantage of this transportation is to face the mechanical problem in rural areas where repairing facilities are not available. Damage to produce is caused during summer and raining seasons. Injury can also occur due to jerks and undulated roads during journey.

(See colour version on page 424)

4. RAIL TRANSPORT

Rail transport has steadily declined with the increased use of highway transport. But where the roads or highways are not well developed this mode is still used in those regions. Different vehicles can be used for rail transport such as ice refrigerated rail car, mechanically refrigerated rail car, flat car and inter modal container.

The refrigeration system used in rail cars is similar as in case of trucks and containers. The ice refrigerated rail car is being replaced by mechanically refrigerated system. This type of transportation can be faster than the trucks and is suitable for long distances. There are less jerks. Main disadvantage is it cannot reach at all places within city and don't reaches most of the places.

5. MARINE TRANSPORT

It is most economical mode of transportation over long distance. This type of transport is used to carry produce having larger shelf life (kiwis, citrus fruits and bananas). Majority of the exported fruits are shipped by the sea e.g. 96%

of bananas, 75% of exotic fruits 60% of citrus fruits. This type of transport is carried at in conventional refrigerated ships or in containership. Conventional refrigerated ships are completely insulated having series of holds divided into rooms. These ships carry refrigerated good, frozen foods and non perishable products in different holds or rooms.

In refrigerated ships, two cargo handling techniques are used manual and palletized handling. Ships must have sufficient refrigeration capacity and air circulation and lower the temperature of cargo. In addition to proper air circulation, fresh air must be added to remove Carbon dioxide, ethylene gases, produced during respiration. Container ships have vertical guides to stack containers below deck. Generally container ship accommodate all types of container, insulated port hole, mechanically refrigerated and dry cargo. Container ships are equipped with special cranes and handling equipments which can lift and lower containers for loading and unloading in ports. Perishable foods can be carried in insulated port hole containers or in mechanically refrigerated containers. Mechanically refrigerated containers are equipped with refrigeration unit which operate with a diesel engine during highway travel and then are plugged on shipboard to operate with an electric motor. Mechanically refrigerated containers are nevertheless suitable for marine transport, but because of flexibility of use the most popular type of sea containers are used. The main disadvantage of ship is slow means of transportation and is not suitable for highly perishable produce.

6. AIR TRANSPORT

Generally this transport is used to carry horticultural produce over a long distance having short shelf life. Thus these produce cannot be transported easily by roads, rails or sea. Whenever there is shortage of particular fruit in market air flight can be used. This transport is more expensive than other means such as roads, rail or marine transport. Horticultural produce can be carried either a passenger flight or in cargo flights.

Loading and unloading of aircrafts, hold must be done rapidly because flights arrival and departure is short. The cargo is generally stacked in unit load device (ULDs) to shorten the time requires for loading and unloading. ULDs are light weight, and two main categories of ULDs are used aircraft pallets and dry craft containers.

Generally aircraft containers can always be used for perishable fruits. In order to maintain produce temperature containers can be lined with expanded polystyrene panels. For journeys of 48 hrs or more main source of refrigeration is required such as dry ice. Dry ice can be used to maintain cold temperature in aircraft container. In order to prevent from freezing dry ice must be packaged air spaced above the cargo in the container. Main disadvantage of air transport is very costly and unaffordable for common grower.

LOADING OF TRANSPORT VEHICLE

Important steps must be taken into prior consideration for loading of horticultural produces.

1. Product must be well cooled.
2. After harvest, pre cooling removes produce field heat so various methods such as room cooling, forced air cooling, hydro-cooling and vacuum cooling are used to pre cool horticultural produce before shipment.
3. Proper air circulation around the transport load.
4. Cleanliness which prevents occurrence of bacterial, chemical and odour contamination within vehicle. Floor should be washed properly before loading. Floor ducts should be free of debris to provide clear air circulation. Ground coffee is helpful in absorbing odours.
5. Transport vehicle should be pre cooled before loading to reduce initial cooling.
6. During winter pre heating of vehicle reduces the chance of chilling or freezing injuries of horticulture produce.

Horticultural produce can be handled as palletized or unitized loads. Due to ease of handling and less labour requirement, the palletized load is most commonly used for loading arrangements. The unitized load allows better air circulation throughout the cargo, when boxes are stacked, an air passage is left between the rows of boxes. The basic guideline for loading pattern is that the conduction heat transfer is minimized and air circulation is maximized.

UNLOADING AND HANDLING AT DESTINATION

Stair simple set can be constructed for both loading and unloading produce. Stairs can be folded and pushed back under the back of truck when vehicle is in motion. The steps should be made of wood or steel matting.

Before sale if produce is to be stored then it requires well insulated storage room. If produce is held at destination for short time handler can maintain quality and reduce losses by storing the horticultural produce at suitable temperature. Relative humidity should be maintained at between 85-95 percent and ethylene level is kept below 1 ppm by ventilating or using ethylene absorbers.

CONCLUSION

Mainly four modes of transportation are presented in this chapter. Equipments used for the transportation of horticultural crop varies greatly among the various modes of transport. Only high quality of produce should be selected for transportation. It should be protected against temperature variations, water loss, harmful gases and physical injuries throughout the journey. These considerations should keep in mind throughout the entire distribution of fresh horticultural crops.

LITERATURE CONSULTED

International Institute of Refrigeration. 1995. Guide to refrigerated transport. International Institute of Refrigeration, Paris, France.

Kader, A.A.2002. Post harvest technology of horticultural crops, 3rd ed. Coop. Ext., Div. Agri and Nat. Res. University of California, Davis (CA), Publication no. 331.

LeBlanc, D.I. and Hui, K.P.C. 2005. Land transportation of fresh fruits and vegetables: An update. *Stewart Postharvest Review*, 1.4.1-1.4.13.

McGregor, B.M. 1989. Tropical products transport handbook. Agriculture Handbook No. 668, Revised Edition. Office of Transportation, US Department of Agriculture, Washington D.C.

Raghavan, G.S.V., Vigneault, C., Gariepy, Y., Markarian, N.R. and Alvo, P. 2004. Refrigerated and controlled/modified atmosphere storage. In: Processing Fruits, Science and Technology, D.M. Barrett, L. Somogyi, and H. Ramaswamy (Eds), 2nd ed. CRC Press, Boca Raton (FL), 23.

Ryall, A.L. and Pentzer, W.T. 1982. Handling, Transporation and Storage of Fruits and Vegetables. Vol 2, 2nd ed. West Port, CT: AVI.

Sudhher, K.P. and Indira, V. 2007. Postharvest Technology of Horticultural Crops. *Horticultural Science Series* vol. 7 ed. by. K.V. Peter and K.A.U. Thrissur, Kerala pp: 229-238.

Thompson, J.F, Bishop, C.F.H. and Brecht, P.E. 2004. Air transport of perishable products. Agriculture and Natural Resources, Unversity of California, Davis (CA), Publication no. 21618

Vigneault, C.,Thompsan, J., Stefanine , W. U., Cathanine, K.P. and Lebrane D.I. 2009. Post harvest technology for the horticultural crops vol.2:1-24 ed noureddine benkeblia.

Vigneault, C., Goyette, B. and Raghavan, G.S.V. 1995. *Can. Agric Eng.*, 37(3): 225.

Colour Plates

Chapter 2: Planning and Planting of an Orchard

High density planting in plum at 6 X 1.5 m spacing

High density planting in peach at 6 X 1.5 m spacing

Chapter 5: Fruit Plant Canopy Management

Plate 1: Open Centre (Nectarine)

Plate 2: Vertical Axis training in apple

Plate 3: Tall Spindle training in apple

Chapter 9: Growth and Development of Fruits

Plate 1: Fruit surface colour development in July Elberta peach following the application of 250 ppm paclobutrazol, 6 weeks after full bloom

Chapter 12: Tissue Culture and Biotechnology

Plate 1-4: Micro-propagation in fruit plants at various stages in Laboratory

Chapter 13: Seed Germination and Dormancy

P. 1: Open field stratification

P. 2: Refrigerated Stratification

P. 3: Stratified seed of apple (Plump and swollen)

P. 4: Properly Stratified seed of apple (inside view)

Chapter 21: Seed Germination and Dormancy

About the Editor

Dr. J.S.Bal, born on 8 October, 1951,belongs to Ajnala in district Amritsar did B.Sc (Agri.) in 1973 from Khalsa College, Amritsar and M.Sc (Hort.) in 1975 from Punjab Agricultural University, Ludhiana.He joined PAU, Ludhiana as Res. Assoc. (Hort.) in September 1977 and Asstt. Prof. Horticulture in April, 1980. He completed Ph.D (Hort.) in 1991 from GNDU, Amritsar. Dr. Bal elevated to the position of Professor of Horticulture in 2001 and served as Head, Department of Horticulture, PAU in 2009.

Dr. Bal has written 5 books in horticulture viz. Fruit Growing, (Raising Fruit Nursery), Mulching in Ber Orchards, Ber Varieties and (Post-harvest Handling of Fruits and Vegetables) which are very popular among students, growers, nurserymen, scientists and extension workers. He has also compiled 4 under-graduate and 3 post-graduate Practical Manuals in horticulture for students and contributed 6 Book Chapters. He has taught 528 credit hours to UG/PG students at Punjab Agricultural University which is a record itself in teaching career. Dr. Bal has guided 5 UG Advisory Groups, 18 M.Sc / Ph.D students as Major Advisor and 78 M.Sc / Ph.D students as Minor Advisor. He has evaluated and conducted viva-voce of 24 M.Sc and 23 Ph.D students of different universities as external expert. He has selected 10 Asstt.Prof., 9 Assoc.Prof.,24 Prof. in various universities and 25 HDO in other states as an expert member.He acted as Chairman Academic Affairs Committee (Teaching) in the Department of Horticulture from 2004 to 2009.

Dr. Bal was team leader of *ber* crop at PAU for 19 years He has published 192 Research Papers in National and International Journals and compiled & edited 4 Research Reports. His seventeen recommendations are included in Package of Practices for the benefits of fruit growers. He has earned distinction for evaluating a draft report on genetic resources of *ber* by International Bureau of Plant Genetic Resources, Rome (Italy). He has attended 29 National and International Seminar/ Symposia/Conference in horticulture. Dr. Bal has attended 2nd International Jujube Symposium at Xinzheng, Henan (China) from 3-7 Sept. 2011 and presented Keynote Report on *ber* in Inaugural Session and Chaired Session on "Post-harvest Physiology, Nutrition and Processing". He has attended 3rd International Jujube Symposium at Brisbane(Australia) from 17-22 August 2014 and delivered Keynote Address and Chaired Oral Session. Acted as a member, International Scientific Committee, Jujube 2016 Workshop, Bucharest,

Romania, October 9-12,2016. Recently nominated member, International Scientific Committee, XXX. International Horticultural Congress. S-8 Jujube (4th International Symposium) to be held on 12-16 August 2018, Istanbol-Turkey. He was instrumental in rejuvenation of 440 years old Historic *ber* plants at Golden Temple, Amritsar.

He has edited monthly "The Horticultural Bulletin" in English and Vernacular for 7 years and monthly "Horticulture Newsletter" from January 2008 to October 2011 for the benefits of fruit growers. He has published 109 Extension Articles, 9 Bulletins and 4 Newsletters in horticulture. He was Incharge of PAU orchard for 8½ years. He also participated in the course "Parliament Process and Procedures" at Lok Sabha Secretariat, New Delhi. Dr. Bal has also actively participated in various curricular activities of PAU such as NCC Officer, Programme Officer NSS, President Arts and Literature/Young Writers Association for 10 years in College of Agriculture and Hostel Warden.

Dr. Bal was awarded Shri Hans Raj Pahwa Award 1998 by PAU; honoured by Punjab Minister at his home town Ajnala in September 2001; Distinguished Alumni Award 2002 by Khalsa College Amritsar; Shri G.L.Chadha Memorial Gold Medal 2004 by 'The Horticultural Society of India'; The Best Citizen of India Award 2006 by IPH New Delhi and Best Paper Award by Punjab Academy of Sciences in February 2007 and by PAU in October 2010 in National Seminar on 'Impact of Climate Change on Fruit Crops'.

Dr. Bal has total 337 publications to his credit. He has completed 35 years service in horticulture (34 years in PAU, Ludhiana and 1 year as Prof. of Horticulture in LPU, Phagwara).

Dr. Bal has total 337 publications to his credit and completed 35 years service in horticulture (34 in PAU and 1 in LPU).